W0235156

Lecture Notes in Physics

For information about Vols. 1–99, please contact your bookseller or Springer-Verlag.

Vol. 100: Einstein Symposion Berlin. Proceedings 1979. Edited by H. Nelkowski et al. VIII, 550 pages. 1979.

Vol. 101: A. Martin-Löf, Statistical Mechanics and the Foundations of Thermodynamics. V, 120 pages. 1979.

Vol. 102: H. Hora, Nonlinear Plasma Dynamics at Laser Irradiation. VIII, 242 pages. 1979.

Vol. 103: P. A. Martin, Modèles en Mécanique Statistique des Processus Irréversibles. IV, 134 pages. 1979.

Vol. 104: Dynamical Critical Phenomena and Related Topics. Proceedings, 1979. Edited by Ch. P. Enz. XII, 390 pages. 1979.

Vol. 105: Dynamics and Instability of Fluid Interfaces. Proceedings, 1978. Edited by T. S. Sørensen. V, 315 pages. 1979.

Vol. 106: Feynman Path Integrals, Proceedings, 1978. Edited by S. Albeverio et al. XI, 451 pages. 1979.

Vol. 107: J. Kijowski, W. M. Tulczyjew, A Symplectic Framework for Field Theories. IV, 257 pages. 1979.

Vol. 108: Nuclear Physics with Electromagnetic Interactions. Proceedings, 1979. Edited by H. Arenhövel and D. Drechsel. IX, 509 pages. 1979.

Vol. 109: Physics of the Expanding Universe. Proceedings, 1978. Edited by M. Demiański. V, 210 pages. 1979.

Vol. 110: D. A. Park, Classical Dynamics and Its Quantum Analogues. VIII, 339 pages. 1979.

Vol. 111: H.-J. Schmidt, Axiomatic Characterization of Physical Geometry. V, 163 pages. 1979.

Vol. 112: Imaging Processes and Coherence in Physics. Proceedings, 1979. Edited by M. Schlenker et al. XIX, 577 pages. 1980.

Vol. 113: Recent Advances in the Quantum Theory of Polymers. Proceedings 1979. Edited by J.-M. André et al. V, 306 pages. 1980.

Vol. 114: Stellar Turbulence. Proceedings, 1979. Edited by D. F. Gray and J. L. Linsky. IX, 308 pages. 1980.

Vol. 115: Modern Trends in the Theory of Condensed Matter. Proceedings, 1979. Edited by A. Pekalski and J. A. Przystawa. IX, 597 pages. 1980.

Vol. 116: Mathematical Problems in Theoretical Physics. Proceedings, 1979. Edited by K. Osterwalder. VIII, 412 pages. 1980.

Vol. 117: Deep-Inelastic and Fusion Reactions with Heavy Ions. Proceedings, 1979. Edited by W. von Oertzen. XIII, 394 pages. 1980.

Vol. 118: Quantum Chromodynamics. Proceedings, 1979. Edited by J. L. Alonso and R. Tarrach. IX, 424 pages. 1980.

Vol. 119: Nuclear Spectroscopy. Proceedings, 1979. Edited by G. F. Bertsch and D. Kurath. VII, 250 pages. 1980.

Vol. 120: Nonlinear Evolution Equations and Dynamical Systems. Proceedings, 1979. Edited by M. Boiti, F. Pempinelli and G. Soliani. VI, 368 pages. 1980.

Vol. 121: F. W. Wiegel, Fluid Flow Through Porous Macromolecular Systems. V, 102 pages. 1980.

Vol. 122: New Developments in Semiconductor Physics. Proceedings, 1979. Edited by F. Beleznay et al. V, 276 pages. 1980.

Vol. 123: D. H. Mayer, The Ruelle-Araki Transfer Operator in Classical Statistical Mechanics. VIII, 154 pages. 1980.

Vol. 124: Gravitational Radiation, Collapsed Objects and Exact Solutions. Proceedings, 1979. Edited by C. Edwards. VI, 487 pages. 1980.

Vol. 125: Nonradial and Nonlinear Stellar Pulsation. Proceedings, 1980. Edited by H. A. Hill and W. A. Dziembowski. VIII, 497 pages. 1980.

Vol. 126: Complex Analysis, Microlocal Calculus and Relativistic Quantum Theory. Proceedings, 1979. Edited by D. Iagolnitzer. VIII, 502 pages. 1980.

Vol. 127: E. Sanchez-Palencia, Non-Homogeneous Media and Vibration Theory. IX, 398 pages. 1980.

Vol. 128: Neutron Spin Echo. Proceedings, 1979. Edited by F. Mezei. VI, 253 pages. 1980.

Vol. 129: Geometrical and Topological Methods in Gauge Theories. Proceedings, 1979. Edited by J. Harnad and S. Shnider. VIII, 155 pages. 1980.

Vol. 130: Mathematical Methods and Applications of Scattering Theory. Proceedings, 1979. Edited by J. A. DeSanto, A. W. Sáenz and W. W. Zachary. XIII, 331 pages. 1980.

Vol. 131: H. C. Fogedby, Theoretical Aspects of Mainly Low Dimensional Magnetic Systems. XI, 163 pages. 1980.

Vol. 132: Systems Far from Equilibrium. Proceedings, 1980. Edited by L. Garrido. XV, 403 pages. 1980.

Vol. 133: Narrow Gap Semiconductors Physics and Applications. Proceedings, 1979. Edited by W. Zawadzki. X, 572 pages. 1980.

Vol. 134: $\gamma\gamma$ Collisions. Proceedings, 1980. Edited by G. Cochard and P. Kessler. XIII, 400 pages. 1980.

Vol. 135: Group Theoretical Methods in Physics. Proceedings, 1980. Edited by K. B. Wolf. XXVI, 629 pages. 1980.

Vol. 136: The Role of Coherent Structures in Modelling Turbulence and Mixing. Proceedings 1980. Edited by J. Jimenez. XIII, 393 pages. 1981.

Vol. 137: From Collective States to Quarks in Nuclei. Edited by H. Arenhövel and A. M. Saruis. VII, 414 pages. 1981.

Vol. 138: The Many-Body Problem. Proceedings 1980. Edited by R. Guardiola and J. Ros. V, 374 pages. 1981.

Vol. 139: H. D. Doebner, Differential Geometric Methods in Mathematical Physics. Proceedings 1981. VII, 329 pages. 1981.

Vol. 140: P. Kramer, M. Saraceno, Geometry of the Time-Dependent Variational Principle in Quantum Mechanics. IV, 98 pages. 1981.

Vol. 141: Seventh International Conference on Numerical Methods in Fluid Dynamics. Proceedings. Edited by W. C. Reynolds and R. W. MacCormack. VIII, 485 pages. 1981.

Vol. 142: Recent Progress in Many-Body Theories. Proceedings. Edited by J. G. Zabolitzky, M. de Llano, M. Fortes and J. W. Clark. VIII, 479 pages. 1981.

Vol. 143: Present Status and Aims of Quantum Electrodynamics. Proceedings, 1980. Edited by G. Gräff, E. Klempt and G. Werth. VI, 302 pages. 1981.

Lecture Notes in Physics

Edited by H. Araki, Kyoto, J. Ehlers, München, K. Hepp, Zürich
R. Kippenhahn, München, H. A. Weidenmüller, Heidelberg
and J. Zittartz, Köln

182

Laser Physics

Proceedings of the Third New Zealand Symposium on Laser Physics held at the University of Waikato Hamilton, New Zealand, January 17–23, 1983

Edited by J. D. Harvey and D. F. Walls

Springer-Verlag
Berlin Heidelberg GmbH 1983

Editors

J. D. Harvey
Physics Department, University of Auckland
Auckland, New Zealand

D. F. Walls
Physics Department, University of Waikato
Hamilton, New Zealand

ISBN 978-3-540-12305-7 ISBN 978-3-540-39891-2 (eBook)
DOI 10.1007/978-3-540-39891-2

This work is subject to copyright. All rights are reserved, whether the whole or part of the material is concerned, specifically those of translation, reprinting, re-use of illustrations, broadcasting, reproduction by photocopying machine or similar means, and storage in data banks. Under § 54 of the German Copyright Law where copies are made for other than private use, a fee is payable to "Verwertungsgesellschaft Wort", Munich.

© by Springer-Verlag Berlin Heidelberg 1983

Originally published by Springer-Verlag Berlin Heidelberg New York Tokyo in 1983

2153/3140-543210

PREFACE

This volume contains the lecture notes delivered at the third New Zealand Symposium on Laser Physics, held at the University of Waikato, Hamilton, from January 17th to 23rd 1983. This meeting like the previous ones held in 1977 and 1980 brought together a group of about 60 physicists working in both experimental and theoretical laser physics from many countries. One of the strengths of these meetings lies in the new interactions and collaborative efforts generated by lectures and discussions between physicists working in what sometimes seem to be disparate areas.

Currently one of the most interesting and fast developing areas in laser physics concerns optical bistability and the transition to chaotic behaviour of optical systems, and this theme runs through many of the papers presented here.

The editors would like to express their gratitude to the lecturers for providing detailed notes for publication shortly after the meeting, and to the various organisations who provided financial support. These include the New Zealand Institute of Physics (Inc.), the British Council, the Royal Society of New Zealand, the University of Waikato and the following Companies:- Spectra Physics Inc., Quentron Optics Pty, Radiation Research, Exciton, Spex, Oriel, Lumonics, Burleigh Instruments and Scientec.

The success of this meeting and the relaxed atmosphere of the University Campus in mid-summer have established the Symposium as a triennial event. We look forward to another stimulating and productive meeting in 1986.

Hamilton, New Zealand — J.D. Harvey

March 1983 — D.F. Walls

TABLE OF CONTENTS

OPTICAL BISTABILITY IN SEMICONDUCTORS

S.D. Smith and B.S. Wherrett

Physics Department

Heriot-Watt University

Edinburgh EH14 5AS Scotland

1. INTRODUCTION

Since the first observations of optical bistability in semiconductor materials[1][2] in 1979 considerable progress has been made in extending to a variety of wavelengths, different materials and more practical temperatures compared to the original observations. Present observational parameters are listed below with holding intensities estimated very approximately.

Material	Wavelength	Temperature	Intensity and Comment
InSb	5.4μm	77K	W/cm^2 cw
InSb	9-11μm	300K	$100KW/cm^2$ pulsed
GaAs	0.8μm	120K	$50KW/cm^2$
GaAs(MQW)	0.8μm	300K	$50KW/cm^2$ Multiple Quantum Well Structure
Te	10.6μm	300K	MW/cm^2
Si	1.06μm	300K	MW/cm^2
ZnS, ZnSe interference filters	0.7μm	300K	W/cm^2 Thermal?
$GeSe_2$	0.7μm	300K	Photostructural Slow?

The majority of these observations have been reviewed[3] or reported at the Munich International Quantum Electronics Conference;[4] we therefore report here on new experimental results in InSb, which material continues to act as the "hydrogen atom" of solid state bistable devices because of its capability to be operated cw or addressed by external pulses.

On the theoretical side, the microscopic understanding of the third order nonlinearities involved in optically bistable devices has developed within the envelope of dynamical nonlinear effects introduced by promotion of free carriers into otherwise unoccupied states;[5] there remains much confusion as to the origin of these effects in the literature. We attempt to review and derive these various processes from a 4-stage virtual transition approach which indicates the predominant importance of the effects of blocking interband transitions.

2. EXPERIMENTAL

i) The Optical AND Gate

We report the novel operation of a natural reflectivity InSb bistable resonator at 77K, pumped with a cw CO laser at $1819cm^{-1}$, as a single pulse detector with definite threshold energy and as an optical AND gate. The two switching pulses for the AND gate logic operation are 30 psec single, switched-out pulses from a mode-locked Nd:YAG laser. Introduction of a variable time delay between the logic pulses provides a unique technique for measurement of the photogenerated carrier lifetime and yields a recombination time of $\sim$90nsec for a cw CO holding intensity of $\sim 80W/cm^2$.

A Fabry-Perot cavity was constructed from an InSb crystal (n-type $\sim 4 \times 10^{14} cm^{-3}$) of $\sim$ 210µm thickness and with natural reflectivity ($R \simeq 0.36$) polished faces. Radiation from an Edinburgh Instruments PL3 cw CO laser operating at $1819cm^{-1}$ incident over a $1/e^2$ intensity beam diameter of $\sim$200µm produced the bistable transmission characteristic shown in Figure 1(a).

This device was externally switched[6] from OFF to ON resonance using a single, 30 psec switched-out pulse from a mode-locked Nd:YAG laser. The bistable element was held off-resonance, between the two switching thresholds, close to switch-up with a constant CO holding power of $\sim$26mW. Switch-on was then achieved whenever a pulse from the Nd:YAG laser above a measurable threshold energy was incident on the effective area of the sample defined by the CO beam as shown in Figure 1(b). Once this condition had been established the device remained on-resonance until the CO beam was interrupted to return the transmission to the lower branch. Thus the device was acting as a memory element which registered and stored the incidence of a single 30 psec pulse above a given threshold energy.

The change in refractive index per absorbed carrier per unit volume has previously been deduced from experiment[5] and the characteristic response time for this non-linearity is determined by the rate at which carriers are excited into states where they can be effective; which is controlled by the incident intensity. Carriers are introduced by the 1.17eV Nd:YAG photons by two means. They will be raised from the split-off valence band ($\sim$ 1.13eV below the conduction band at 77K) and also excited from the heavy hole valence band to higher levels in the conduction band and subsequently scattered to the band minimum.

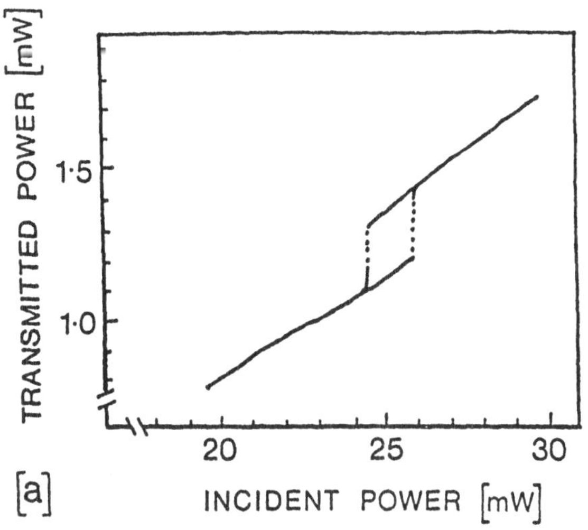

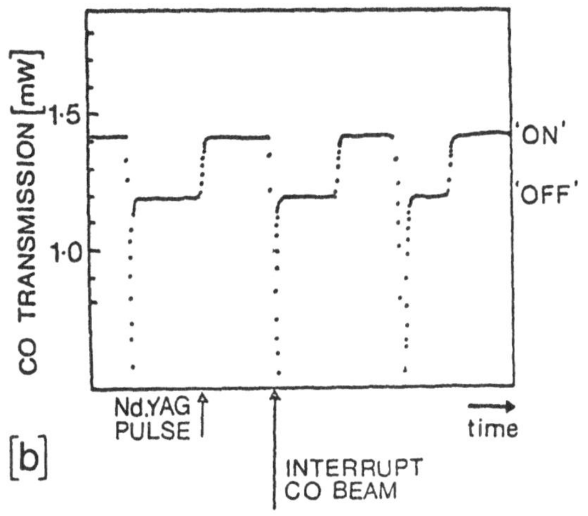

Figure 1 (a) : Transmitted power plotted against incident power for a cw CO laser beam (wavenumber ∿1819cm^{-1} and spot size ∿200μm) passing through a natural reflectivity InSb cavity (thickness ∿210μm, carrier concentration ∿4 x 10^{14}cm^{-3} (n-type)) at 77K.

(b) : Transmission of incident cw CO laser beam at ∿26mW showing 'on-switching' caused by single switched-out, 30 psec pulses from a Nd:YAG laser. Switch-off is caused by interrupting the CO holding beam.

Experimentally, consistent switch-up was achieved with $\sim$5nJ of Nd:YAG laser energy over the effective area of the focussed CO beam. The large absorption coefficient of 1.06μm radiation ($\sim 10^4 cm^{-1}$) results in the carriers altering the effective optical thickness (nl) of the cavity in the immediate vicinity of the absorbed pulse. This is however as effective as an even distribution of carriers throughout the entire length (since $\Delta(nl) \simeq \int_0^L \Delta n(l)dl$). Thus calculation of the induced change in refractive index caused by the Nd:YAG generated carriers is shown to be sufficient to cause switching within the duration of the switching pulse. The effect of saturation of the nonlinearity, carrier diffusion and recombination will all have a bearing on the details of the switching process but leave the main conclusion unchanged: that the lower limit on the switching-on speed should be the optical field build-up time within the cavity (τ_c) given by $\tau_c \simeq 2nL/c(1-R)$ which is $\sim$8 psec for the 210μm thick cavity operating here.

Whilst the switch-on time may be controlled as discussed above, the switch-off time is controlled by the carrier relaxation rate from the excited state when the cavity field is reduced below the ON-resonance condition. The lifetime of these excited carriers was investigated by splitting the Nd:YAG switching pulse into two component pulses and recombining them at the crystal. No switching was observed when either one of the component pulses was incident individually on the sample. When both pulses were incident simultaneously switching to the upper branch and holding was observed. Thus the device acts as an optical AND gate only providing a high transmission of the CO beam on simultaneous arrival of two Nd:YAG pulses of sufficient combined energy.

The excited lifetime of these externally introduced carriers must be sufficiently long to enable the build-up of CO laser field inside the cavity to maintain the on-resonance condition after completion of the Nd:YAG pulse otherwise the device would switch-off again.

By temporally separating the two component pulses by a variable delay, the excited state lifetime of the carriers introduced by the initial pulse was determined. The cumulative effect of the two pulses is dependent on the rate at which the carrier population introduced by the initial pulse has decayed on introduction of the delayed pulse. A novel procedure was adopted which avoided any absolute measurement of the pulse intensities by employing the threshold nature of the switching. The delayed pulse alone was attenuated with standard calibrated filters to a known percentage below the threshold switching energy. Using a delay line up to 250 feet long, a carrier lifetime of $\sim$90ns was determined for the experimental sample.

ii) Optical Bistability in InSb at Room Temperature

We report the observation of optical bistability in an InSb resonator at room temperature. This effect and fringe shifts were caused by nonlinear refraction induced by two photon absorption of radiation from a single longitudinal mode injection-locked pulsed CO_2 laser operating at 9.6 to 10.6μm. Intensities as low as $100kW/cm^2$ were found to be sufficient to tune the 250μm thick cavity through a fringe maximum. From our results we deduce a value of $\chi^{(3)}$ of the order of 10^{-4} e.s.u. over the range of intensities investigated.

Figure 2(i) shows the raw data obtained for the sample at different resonator tunings. These clearly show the self-tuning of the InSb crystal due to laser intensity where each peak in the transmitted pulse corresponds to tuning through a fringe maximum.

For each set of data we have plotted (Fig. 2(ii)) the corresponding relationship between the input and transmitted instantaneous intensities. These display hysteresis loops associated with optical bistability. These plots have also been obtained dynamically by connecting the incident and transmitted signals to the x and y plates of a Tektronix 7104 with identical results. We also deduce that the resulting change in peak position indicates that refractive index is decreasing with intensity whereas if the effect is thermal we would expect an increase. This result, combined with the observed speed of switching confirms an electronic effect; further experiments up to $20MW/cm^2$ also indicate negligible absorptive effects up to $1MW/cm^2$.

We can see from these plots that there is a sharp switch up from off-resonance level to on-resonance level consistent with a nonlinear cavity with optical feedback. The switch down is slower due to the long lifetime of the carriers. Transient loops are also obvious, consistent with the lifetime of the carriers being of the order of the laser pulse length. The transient effects are smaller at high intensities where the carrier lifetime decreases with the increased population.

In conclusion we have demonstrated that with relatively modest intensities ($\sim 100kW/cm^2$) we can induce at room temperature optical tuning in n-InSb and also construct optically bistable devices. From our measurements we obtain values of dn/dI as $0.2cm^2/MW$ at 100 kW/cm^2 and $0.12cm^2/MW$ at $500kW/cm^2$. These correspond to a value of $\chi^{(3)}$ of the order of 10^{-4} e.s.u. over the range of intensities investigated.[7]

3. MICROSCOPIC THEORY OF THIRD ORDER OPTICAL NONLINEARITY IN SEMICONDUCTORS

There is a confusion present in much of the recent literature on semiconductor nonlinear refraction. This confusion originates from the development of several apparently independent theories. Here we consider five theoretical approaches to the calculation of the nonlinear susceptibility ($\chi^{(3)}$) and show that each is associated with the effect of interband transitions, even though some are conventionally referred to as free-carrier theories. By comparing groups of terms in $\chi^{(3)}$ the physical equivalence of the theories is established. New near-resonance calculations are presented, and the circumstances in which each of the various theories can be applied is discussed.

Nonlinear refraction has recently been employed to produce optical bistability in the semiconductor InSb.[1-3] In similar materials degenerate four-wave mixing has been

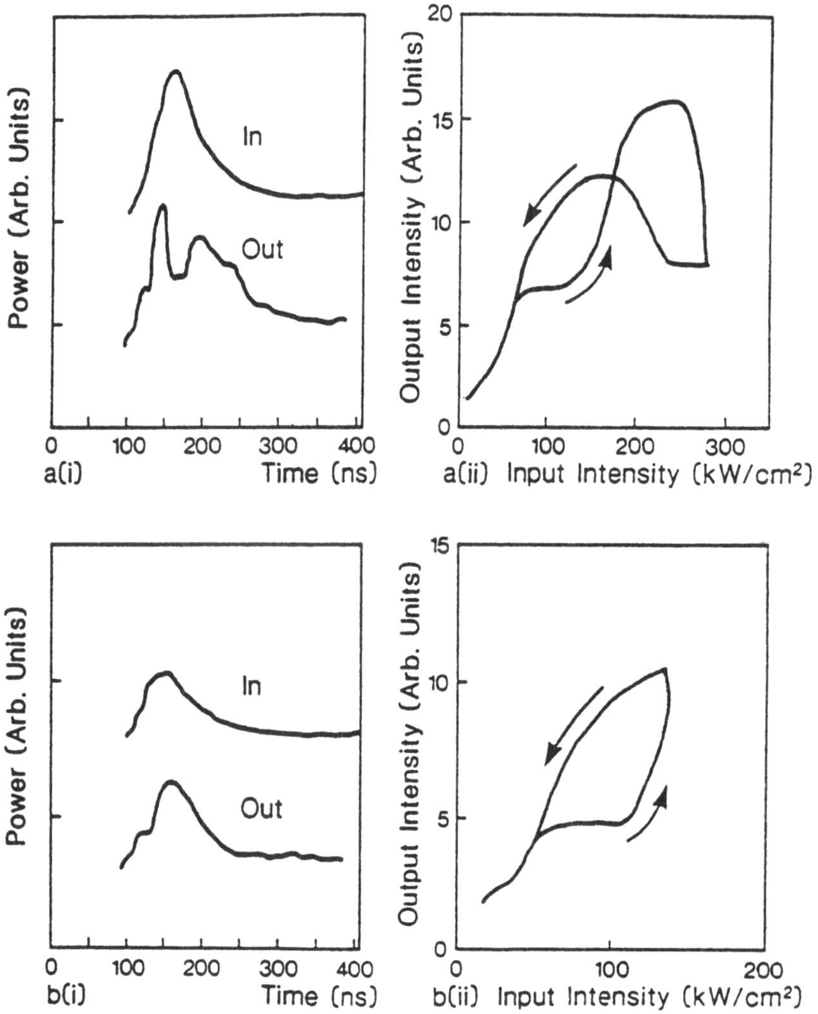

Figure 2 i) : Variation of transmitted pulse shape with cavity tuning (a) 7°, (b) 13° for peak incident intensity ∿ 600KW/cm^2.

ii) : Relation between incident and transmitted instantaneous intensities.

achieved and phase-conjugation observed. In all the above experiments the radiation frequency lies just below the fundamental band edge. These phenomena are described theoretically through the nonlinear susceptibility $\chi^{(3)}$ $(\omega,-\omega,\omega)$. It is interesting though that in their interpretations of the dominant contributions to $\chi^{(3)}$ various authors select apparently different mechanisms. It is the purpose of this paper to show in detail how the various models can be compared and to discuss the regions in which each may be valid.

The models which will be discussed are: (i) the saturation and (ii) interband blocking models, both used by Wherrett and Higgins (WH),[8] which predict the enormous $\chi^{(3)}$ values observed in InSb,[5] (iii) the intraband model introduced by Jain and Klein[9] with respect to their four-wave mixing experiments; (iv) the nonparabolicity calculation of Wolff and Pearson (WP)[10] for $(2\omega_1 - \omega_2)$ mixing processes, which has been adapted in order to interpret phase-conjugation experiments in HgCdTe; and (v) the density-matrix calculations of Jha and Bloembergen (JB)[11] and Wynne[12] for $(2\omega_1 - \omega_2)$ mixing. These latter calculations are at first sight the most closely related of the above theories to the original $\chi^{(3)}$ density-matrix theory discussed by Butcher and MacLean (BM)[13], Armstrong et.al.[14], and later by many authors. It will be to this original work and to conventional perturbation theory expressions (cf. Ward[15]) that we must return in order to obtain a unified picture from which all the above models can be drawn.

The range of experimental $\chi^{(3)}$ values is particularly intriguing. In $(2\omega_1 - \omega_2)$ - mixing experiments in Si and Ge Wynne and Boyd[16] obtained $\chi^{(3)}$ of order 10^{-12} to 10^{-11} e.s.u. Similarly small magnitudes were obtained in the perturbation calculations of Jha and Bolembergen.[11,17] Also, in mixing experiments on n-type III-IV materials, Patel et.al.[18] observed $\chi^{(3)}$ values from 10^{-11} to 10^{-9} e.s.u., in agreement with the WP free-carrier susceptibility calculations.

In contrast it is found from conjugation experiments in HgCdTe values of 10^{-8} to 3×10^{-7} e.s.u. and Jain et.al.[5,6] quote 10^{-7} e.s.u. for 1.06 m degenerate four-wave mixing in Si and 5×10^{-6} e.s.u. in HgCdTe. (More recently they report 10^{-2} e.s.u. in the latter material).

Finally Miller et.al.[5,19] observe nonlinear indices compatable with $\chi^{(3)}$ values from 10^{-2} to 1 e.s.u., in InSb.

The sources of these remarkable differences must be found in the material and pump laser characteristics: in the bandgap and effective masses, the quality of the material, the laser power levels and most notably in the proximity of the radiation frequency to the band edge.

4. SURVEY OF MODELS

i) The optical-bistability experiments[1] and the beam-profile experiments of Weaire et.al.[20] have been interpreted on the basis of the effect of interband absorption. That is, whilst the pump photon frequency (ω) lies just beneath the nominal band edge (ω_g) there is still assumed to be a measure of excitation across the gap. (This may be described phenomenologically by introducing a T_2 - broadening of the electron states). The excited carriers will in turn affect the absorption coefficient itself and, by causality, must also lead to a nonlinearity in the refractive index. If intraband nonradiative scattering is ignored then one is concerned with the direct saturation of essentially independent two-level systems each consisting of one specific k-state in the valence band and one in the conduction band. The resulting individual contributions to $\chi^{(3)}$ are highly resonant and indeed diverge unless a recombination process is included. Denoting the conduction to valence recombination time as T_1 then one finds for example,[8,19]

$$\mathrm{Re}\chi^{(3)}_{(i)}\ (\omega,-\omega,\omega) \quad - \frac{1}{4\pi\hbar^3}\left|\frac{ep_{cv}}{m\omega}\right|^4\left(\frac{2m_r}{\hbar}\right)^{3/2}\frac{T_1}{T_2}\,(\omega_g-\omega)^{-3/2} \qquad (1)$$

p_{cv} is the interband, momentum matrix element.

ii) The interband blocking model is very similar to the above except in that intraband non-radiative scattering is accounted for. Such scattering is taken to lead to independent thermal distributions of the excited carriers within the valence and conduction bands. Thus only the precise distribution of excited carriers need differ in these first two models - the $\chi^{(3)}$ values so calculated differ only by a numerical factor. The blocking model can however, be taken one stage further towards an empirical model. As the distribution of carriers is not determined by the manner in which they were created, only by their total number, then the interband absorption can be treated entirely empirically by assuming an absorption coefficient α_{int}[8,21];

$$\mathrm{Re}\chi^{(3)}_{(ii)}\ (\omega,-\omega,\omega) \simeq -\frac{nc}{2\pi\hbar}\left|\frac{ep_{cv}}{m\omega}\right|^2\frac{\alpha_{int}}{\hbar\omega}\,T_1(\omega_g-\omega)^{-1} \qquad (2)$$

An alternative description of the blocking model is that the radiation, of intensity I, creates a steady-state change in the free carrier population $\Delta N = \alpha_{int}\, IT_1/h\omega$ and one observes the contribution to the linear refraction due to these electrons. The nonlinearity appears only because of the manner in which the ΔN are created.

In both of the above models we would note that it is the effect of the excited carriers in preventing (in part or wholly) further interband transitions that causes the nonlinearity. In both cases, because the pump frequency lies below the transition frequencies one is reducing the positive refractive index contribution that would be present had the transitions remained. Thus the expected nonlinear index (n_2) or suceptibility ($\mathrm{Re}\chi^{(3)}$) is negative. The resonance behaviours as ω approaches the band edge are $(\omega_g-\omega)^{-3/2}$ and $(\omega_g-\omega)^{-1}\alpha_{int}(\omega)$ respectively. Finally one does not require the existence of free carriers prior to the application of the electromagnetic radiation. If they are present then their effect is merely to modify the above results in a small way.[8]

iii) Jain and Klein[9] invoke a different result for the effect of the interband-excited free-carriers on the linear dispersion.

$$\mathrm{Re}\chi^{(3)}_{(iii)} = -\frac{nce^2}{2\pi\omega}\,\frac{\alpha_{int}T_1}{\hbar\omega}\,\frac{1}{m_r} \qquad (3)$$

This comes straight from standard free-carrier linear dispersion theory, with ΔN given as above. In order to compare this with result (2) we need only think of the origin of the reduced mass m_r in (3). In the simple, two-band model used to obtain (2) the effective masses of the conduction (c) and valence (v) bands are given by

$$\frac{m}{m_{c,v}} = 1 \pm 2\,\frac{|p_{cv}|^2}{m\hbar\omega_g}$$

Now only the most resonant contribution to $\chi^{(3)}$ was included in (2). There are in principle two forms of non-resonant terms.[8] One originates from the vector potential ($\tilde{A}$) term in the current density (cf. Butcher and MacLean[13]). The second nonresonant term is proportional to $(\omega_g + \omega)^{-1}$. Strictly speaking then the resonance behaviour is $[2\omega_g/(\omega_g^2 - \omega^2)]$. Including these terms, and taking the limit $\omega \ll \omega_g$ then result (2) reduces exactly to (3). That is, the presence of m_r, rather than the free electron mass, in (3) is again a consequence of the removal of specific inter-band transitions from the absorption spectrum. Equation (3) is only valid in the limiting case of small frequencies.

iv) Turning to the 'non-parabolicity' model for $\chi^{(3)}$ $(\omega_1,-\omega_2,\omega_1)$; this should equate to the nonlinear refraction susceptibility when ω_2 is set equal to ω_1. The non-parabolicity calculation of Wolff and Pearson[10] is made on the basis that the group velocity of the free carriers is nonlinear in the momentum $\tilde{p}$ and hence contains E^3 terms when an electric field E is applied. Thus if we specify the E direction as x, the p_x-dependence of the electron velocity is,

$$v(p_x) = v(p_{xo}) + \left.\frac{\partial v}{\partial p_x}\right|_o \Delta p + \left.\frac{\partial^2 v}{\partial p_x^2}\right|_o \frac{\Delta p^2}{2} + \left.\frac{\partial^3 v}{\partial p_x^3}\right|_o \frac{\Delta p^3}{6} + \ldots$$

As Δp, the field-induced change in p, is proportional E the cubic term will generate an E^3 term in v providing $\partial^3 v/\partial p_x^3$ is non-zero, and hence a nonlinear current density conductivity, and finally electric susceptibility, $\chi^{(3)}$. Considering an initial population of N_0 conduction band carriers[10]

$$\mathrm{Re}\chi^{(3)}_{(iv)} = \frac{N_0 e^4}{4m_c^2 E_g \omega^4}\,\frac{(1 + 8E_F/5E_g)}{(1 + 4E_F/E_g)^{5/2}} \qquad (4)$$

The inference of this calculation is that the mechanism for the nonlinearity is purely a free-carrier one. It is not immediately clear whether interband transitions play any role in the derivation of $\chi^{(3)}_{(iv)}$. Neither is it clear whether one should expect any resonance behaviour at frequencies close to ω_g. At first sight however, one is tempted to say that because the nonparabolicity has nothing to do with the radiation frequency, nor with interband radiative effects, we are dealing with purely intraband processes, which would not display any resonance enhancement. However, the fermi energy E_F is almost always small compared to E_g, and therefore to a good approximation $\chi^{(3)}_{(iv)}$ is directly proportional to the carrier concentration; electrons near the band minimum contribute equally to those of higher energies. This seems to contradict the impression that nonparabolicity, which is significant usually only for the high $\underset{\sim}{k}$ states, is the cause of the nonlinearity. The answer to these apparent contradictions is that rather than talk of the nonparabolicity as being the cause of the nonlinearity one should have in mind that the source of nonparabolicity - a fourth order $\underset{\sim}{k} \cdot \underset{\sim}{p}$ perturbation - is directly equivalent to the source of this $\chi^{(3)}$ - a fourth order $\underset{\sim}{A} \cdot \underset{\sim}{p}$ interaction - and therefore leads to similar expressions. As only the latter involves the radiation frequency we need return to $\underset{\sim}{A} \cdot \underset{\sim}{p}$ perturbation calculations to bring in any resonance behaviour. Our first task then is to relate the nonparabolicity calculation to more conventional susceptibility calculations, broached in terms of virtual transitions.

We note here that Khan et.al.(22) do quote a resonant enhancement, of form $\omega_g^2/(\omega_g - \omega)^2$, but do not provide its derivation. They find that such a factor is fairly consistent with their experimental results for reflected power for phase-conjugation in HgCdTe.

Even if one accepts this enhancement it is significant that the resulting $\chi^{(3)}_{(iv)}$ expression contains no mention of T_2, of T_1, or of the interband absorption coefficient. Therefore one is still left with the question, 'under what conditions should one use $\chi^{(3)}_{(iv)}$ and under what conditions is a form such as $\chi^{(3)}_{(ii)}$ more valid?'

It is significant also that $\chi^{(3)}_{(iv)}$ relies on the initial presence of free carriers; also there is no long-term interband excitation process involved, and the sign of this nonlinearity is positive. These facts are all in contrast with those of the previous models. As we shall see however virtual interband transitions still provide the mechanism for this nonlinearity.

At this stage it is useful to generalise upon the remark made in connection with the Jain quote of $\chi^{(3)}_{(iii)}$. Just as the presence of effective masses (i.e. of $\partial v/\partial p$) in the expression for the linear susceptibility reflects the blocking of interband absorption so do we anticipate that the presence of $\partial v^3/\partial p^3$ terms in the nonlinear susceptibility also reflects such blocking. The logic of this statement becomes more apparent when we note, with Wolff and Pearson, that $\underset{\sim}{v}$ is itself given by $\partial E/\partial \underset{\sim}{p}$ so that one is considering $\partial^4 E/\partial k^2$ - type terms in the nonparabolicity calculations and $\partial^2 E/\partial k^2$ (effective mass) terms in the linear case. The resulting nonlinearity ends up the same order in both cases because the free-carriers are considered to be present initially in the former calculation but radiatively induced in the latter.

The significance of these energy differentials was first discussed by Butcher and MacLean.[13] Providing one can work in the limits that ω is small compared to all interband transition frequencies and providing all lifetime effects can be ignored, they show that the nth-order nonlinear susceptibility is given by:

$$\chi^{(n)}(\omega_1 \ldots \omega_n) = \frac{1}{V[\omega_1\omega_2 \ldots \omega_n(\omega_1 + \ldots + \omega_n)]}\left(\frac{e}{i\hbar}\right)^{n+1} \sum_{\ell,\underset{\sim}{k}} f_{\ell \underset{\sim}{k}} \frac{\partial^{n+1}E}{\partial k^{n+1}} \qquad (5)$$

The proof of this expression originates from the density-matrix treatment of conductivities/susceptibilities. As lifetime effects must be ignored it is hardly surprising that one cannot use this method to generate the saturation results. What is required is to consider individual microscopic mechanisms that contribute predictable terms in $\chi^{(n)}$, i.e., the virtual transition schemes. These terms tend to be completely concealed in general expressions such as (5) and indeed in the other formal expressions presented earlier in the Butcher and MacLean treatment. However, we shall need to consider them very carefully when comparing resonant nonlinear refraction expressions.

(v) The equivalence of the Wolff and Pearson result (4) to expression (5) was noted by Jha and Bloembergen[11] and by Wynne.[12] Pointing out that the expression (4) cannot be valid at optical frequencies, the former authors go on to calculate $\chi^{(3)}$ numerically by summing over interband virtual transitions. As certain combinations of the frequencies they consider exceed the bandgap they require to include dephasing in order to avoid divergences due to exact resonance. Their numerical results though differ from those of WP by somewhat less than a factor of two. In passing we also note that the dephasing time $\tau_c = 10^{-13}$sec, which they introduce, would in linear absorption, produce band tails that are dramatically larger than observed in practice. Thus any band-edge resonance would be considerably reduced.

The expressions that must be computed in the JB perturbation model are of the form:

$$\chi^{(3)}_{(v)}(\omega_1,\omega_2,\omega_3) = \frac{e^4}{V\hbar^3 m^4 \omega_1\omega_2\omega_3(\omega_1+\omega_2+\omega_3)} \sum_{c,u,t,s,\underset{\sim}{k}} f_{c\underset{\sim}{k}} \times$$

$$\left[\frac{P^{\eta}_{cu} P^{\gamma}_{ut} P^{\beta}_{ts} P^{\alpha}_{sc}}{(\omega_{uc}-\omega_1-\omega_2-\omega_3)(\omega_{tc}-\omega_1-\omega_2)(\omega_{sc}-\omega_1)} + 23 \text{ terms}\right] \qquad (6)$$

The additional terms can be thought of as corresponding to different time-orderings of the radiative interactions. $f_{c\underset{\sim}{k}}$ is the fermi population factor of the state $c\underset{\sim}{k}$. In order to avoid exact resonances JB set $\omega_{sc} = (E_s - E_c)/\hbar - i/\tau_c$, etc. As we shall see, even if a τ_c is introduced various problems arise if the frequencies $\omega_1,\omega_2,\omega_3$ are equal in magnitude, as required for nonlinear refraction.

5. TRANSITION SCHEME TREATMENTS

In order to compare the models considered in the previous section we refer to the individual virtual transition schemes that are inherent in them. If one works with many-electron, Slater - determinant wave functions and with a many-electron radiation interaction that includes c-factors, $\underset{\sim}{p} \rightarrow \underset{\sim}{\pi}$, then $\chi^{(3)}$ can be written formally, using conventional perturbation theory as opposed to density-matrix theory:

$$\chi^{(3)}_{\eta\alpha\beta\gamma} = \frac{e^4}{Vh^3 m^4 \omega_1\omega_2\omega_3(\omega_1+\omega_2+\omega_3)} \sum_{b,c,d}\Big[\frac{\pi^{\eta}_{ad}\,\pi^{\gamma}_{dc}\,\pi^{\beta}_{cb}\,\pi^{\alpha}_{ba}}{(\omega_{da}-\omega_1-\omega_2-\omega_3)(\omega_{ca}-\omega_1-\omega_2)(\omega_{ba}-\omega_1)}$$

$$+ \ 23 \text{ terms}\Big] \qquad (7)$$

Whilst this is an extremely unwieldy formula it has the merit that there is a one-to-one correspondence between each term and each allowed physical event. For example for the term shown we have a four-stage virtual transition scheme in which radiation of frequency ω_1 and polarisation α causes an excitation from the initial (when no radiation is present) many-electron state $|a\rangle$ to an excited state $|b\rangle$, one photon being removed from the radiation field at frequency ω_1; and so on until one photon at frequency $(\omega_1+\omega_2+\omega_3)$, of polarisation η, is emitted. Energy is only conserved between the initial and final states of the entire system. The frequency denominators manifest the energy mismatch at the intermediate states. The 23 unspecified terms correspond to the different time-orderings of the four radiation interactions.

Most importantly every transition must be in accord with Pauli exclusion and indeed every Pauli - allowed four-stage process appears just once in the sum.

Using this approach, and comparing with the density matrix approach, we summarise the results as follows: the same physical processes, namely interband transitions are in fact represented indirectly in all the calculations referred to earlier. The source of $\chi^{(3)}$ is the prevention of interband transitions. Under resonant conditions, effects of damping (T_2) and recombination (T_1) must be included. One is then led naturally from the virtual transition scheme, via the saturation model to the blocking or "dynamic Burstein-Moss" model. This latter gives good order of magnitude agreement with experiment with only the lifetime T_1 as a fitted parameter and now determinable independently from experiments such as described in 2(i).

CONCLUSION

The framework of a fully quantitative theory of $\chi^{(3)}$ in semiconductors which reconciles the large variation in magnitude and apparently differing theories has been laid. Considerable scope for further work both experimentally and theoretically[23] exists as progress is made towards practical all-optical circuit elements.

REFERENCES

1. D.A.B. Miller and S.D. Smith, Opt.Commun., 31, 101, (1979); D.A.B. Miller, S.D. Smith and A. Johnston, Appl.Phys.Lett., 35, 658, (1979).

2. H.M. Gibbs, S.L. McCall, T.N.C. Venkatesan, A.C. Gossard, A. Passner and W. Weigman, Appl.Phys.Lett., 35, 451, (1979).

3. E. Abraham and S.D. Smith, Rep. on Prog. in Phys., 45, 815-887, (1982).

4. Various authors, Appl.Phys.B., 28, 132-140, (1982).

5. D.A.B. Miller, C.T. Seaton, M.E. Prise and S.D. Smith, Phys.Rev.Lett., 47, 197, (1981).

6. C.T. Seaton, S.D. Smith, F.A.P. Tooley, M.E. Prise and M.R. Taghizadeh, Appl. Phys.Lett., Jan. 1983, (to be published).

7. A.K. Kar, J.G.H. Mathew, S.D. Smith, B. Davis and W. Prettl, Appl.Phys.Lett., Feb. 1983, (to be published).

8. B.S. Wherrett and N.A. Higgins, Pro. Roy. Soc. (London) A379, 69, (1982).

9. R.K. Jain and D.G. Steel, Appl.Phys.Lett., 35, 454, (1979); R.K. Jain and M.B. Klein, Appl.Phys.Lett., 37, 1, (1980).

10. P.A. Wolff and G.A. Pearson, Phys.Rev.Lett., 17, 1015, (1966).

11. S.S. Jha and N. Bloembergen, Phys.Rev., 171, 891, (1968).

12. J.J. Wynne, Phys.Rev., 178, 1295, (1969).

13. P.N. Butcher and T.P. MacLean, Proc. Phys. Soc., 81, 219, (1963).

14. J.A. Armstrong, N. Bloembergen, J. Ducuing and P.S. Pershan, Phys.Rev., 127, 1918, (1962).

15. J.F. Ward, Rev.Mod.Phys., 37, 1, (1965).

16. J.J. Wynne and G.D. Boyd, Appl.Phys.Lett., 12, 191, (1968).

17. S.S. Jha and N. Bloembergen, IEEE J. Quantum Electron., QE-4, 670, (1968).

18. C.K.N. Patel, R.E. Slusher and P.A. Fleury, Phys.Rev.Lett., 17, 1011, (1966).

19. D.A.B. Miller, S.D. Smith, B.S. Wherrett, Opt.Commun., 35, 221, (1980).

20. D.L. Weaire, B.S. Wherrett, D.A.B. Miller and S.D. Smith, Opt.Lett., 4, 831, (1979).

21. T.S. Moss, Phys.Status Solidi (b), 101, 555, (1980).

22. M.A. Khan, P.W. Kruse and J.F. Ready, Opt.Lett., 5, 261, (1980).

23. A. Miller, D.A.B. Miller and S.D. Smith, Adv. in Phys., 30, 697-800, (1981).

OPTICAL BISTABILITY WITH TWO-LEVEL ATOMS

H.J. Kimble, D.E. Grant and A.T. Rosenberger,
Department of Physics and Electronics Research Center,
The University of Texas,
Austin, Texas 78712, U.S.A.

and

P.D. Drummond,
Department of Physics and Astronomy,
University of Rochester,
Rochester, New York 14627, U.S.A.

1. INTRODUCTION

One of the attractions of the study of optical interactions is that tractable and to a large extent solvable theoretical models do in fact provide realistic descriptions of many experiments. In both theory and experiment many phenomena in quantum optics are rather stark in their conceptual simplicity and as such serve as ideal proving grounds for ideas from numerous areas of physics. Some of the most striking tests of quantum electrodynamics and indeed of the conceptual basis of quantum mechanics have been provided by experiments in quantum optics (1-4). Given the capability both to realize experimentally and to describe theoretically a diversity of phenomena and to do this in quantitative detail, it is then an exciting prospect to search for the limits of the validity of our understanding.

Perhaps the most well studied of all systems in quantum optics are those that produce laser action (5-7). The basic ideas of the laser have given rise to a number of laser-like offspring, including the laser with saturable absorber and the laser with injected signal. In this article our attention is to be directed to another member of this family, namely optical bistability. The general configuration that we wish to consider is as shown in Figure 1(a); we will restrict our attention to bistability in passive systems. In the case of the laser above a certain threshold inversion density, the characteristics of the output field change dramatically as laser action commences. In the case of bistability a similar critical density exists for the intracavity (noninverted) medium. Above a certain threshold density, the input-output characteristics develop a hysteresis cycle, as shown in Figure 1(b).

In analogy with the development of the theory of the laser, much of the early theoretical work in optical bistability dealt with an intracavity medium composed of "two-level" atoms (8-12). However the first (13) and subsequent observations of optical bistability were made in systems of considerably greater complexity

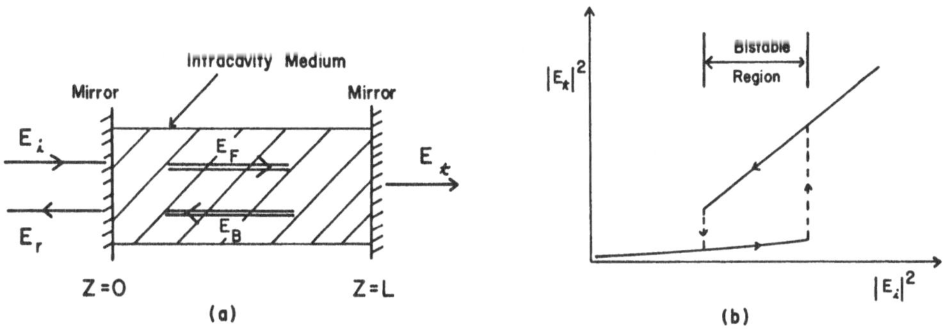

Figure 1 (a) Fabry-Perot interferometer containing a nonlinear intracavity medium. (b) Possible input-output characteristic for such an interferometer illustrating hysteresis and bistability (two values of $|E_t|$ possible for a given input $|E_i|$).

than the simple two-level system, although as has been stressed repeatedly by Gibbs and McCall there is a unifying formalism for the description of the general features of optical bistability that relies only on phenomenological properties of the nonlinear medium (13,14). A great deal of progress has been made in the study of optical bistability, and it is an active area of research with interest derived from its potential application to optical signal processing systems (15,16) and from its relationship to nonequilibrium statistical mechanics (17,18) and to cooperative interactions in atomic physics (19). In spite of this activity there has remained somewhat of a gulf between an extensive theoretical literature on bistability with two-level atoms and actual experimental systems. Several groups including our own have initiated research to try to bridge this gap with experiments that to a good approximation involve two-level atoms (20-23). In such experiments both the phenomenological descriptions and the fully quantized theories of optical bistability can be put to searching tests.

Our presentation is divided into five sections; in Section 2 we develop a simple theory that will serve to establish a basic vocabulary for our subsequent discussions. In Section 3 we describe our apparatus and present observations of the evolution of the steady-state hysteresis cycle in absorptive bistability. Section 4 is devoted to the formulation of a more complete theory of our experiments, including the effects of the standing-wave Gaussian mode structure of the interferometer and the Doppler-broadened absorption of the atomic beams. In Section 5 we conclude with a quantitative comparison of theory and experiment.

2. PLANE WAVE THEORY FOR A FABRY - PEROT RESONATOR

In order to obtain a simple working model for our investigation of optical bistability, we follow the treatment first presented by Szöke et al. (8) and later developed by a number of workers, most notably by McCall (9) and by Bonifacio and

Lugiato (10). Consider a Fabry-Perot resonator such as the one sketched in Figure 1(b) that contains an intracavity medium described by the complex susceptibility $\chi = \chi' - i\chi''$. For the representation of the various fields we choose that of the complex analytic signal (24), with the definitions

$$\vec{E}_i(z,t) = \varepsilon_i e^{-i(\omega t - kz)} \vec{x} + c.c.$$

$$\vec{E}_F(z,t) = \varepsilon_F(z) e^{-i(\omega t - kz)} \vec{x} + c.c. \qquad (2.1)$$

$$\vec{E}_t(z,t) = \varepsilon_t e^{-i(\omega t - kz)} \vec{x} + c.c.$$

and

$$\vec{E}_B(z,t) = \varepsilon_B(z) e^{-i(\omega t + kz)} \vec{x} + c.c. \qquad (2.2)$$

$$\vec{E}_r(z,t) = \varepsilon_r e^{-i(\omega t + kz)} \vec{x} + c.c.$$

The incident field is assumed to be polarized with complex polarization vector $\vec{x}$, and the mirrors of the cavity to be identical, with complex reflection and transmission coefficients for the field amplitudes of $re^{i\phi_r}$ and $te^{i\phi_t}$ for waves travelling into the cavity, and of $r'e^{i\phi_{r'}}$ and $t'e^{i\phi_{t'}}$ for waves travelling out of the cavity. We take the cavity to be filled with a dilute gas composed of "two-level" atoms (25) and assume that the excitation frequency ω is tuned to the atomic resonance. Since $\chi' = 0$ in this case (26) we may write the wave vector k for propagation inside the cavity as

$$k = k_o + i\,\frac{\alpha}{2} \qquad (2.3)$$

where $k_o = \omega/c$ and $\alpha = k_o\chi''$ is an absorption coefficient that includes both the possibility of a nonlinear loss α_a for the atomic medium as well as a distributed, nonsaturable background loss $\overline{\alpha}$,

$$\alpha = \alpha_a + \overline{\alpha}. \qquad (2.4)$$

In the steady state we can solve for the field amplitude ε_t by summing over the successive contributions that result from multiple round trips within the cavity. Following the standard treatment of Born and Wolf (24) we find

$$\varepsilon_t = \varepsilon_o tt' e^{i(\phi_t + \phi_{t'})} e^{-\frac{\alpha L}{2}} [1 + r'^2 e^{2i(\phi_{r'} + kL)} + r'^4 e^{4i(\phi_{r'} + kL)} + \ldots]$$

$$\varepsilon_t = \frac{\varepsilon_o tt' e^{i(\phi_t + \phi_{t'})} e^{-\frac{\alpha L}{2}}}{1 - r'^2 e^{2i\phi r'} e^{2ikL}}. \qquad (2.5)$$

With the definitions $T \equiv tt'$ and $R \equiv r'^2$, the ratio of transmitted to incident intensity is thus found to be

$$\mathcal{T}(\delta) \equiv \frac{|\varepsilon_t|^2}{|\varepsilon_i|^2} = \frac{T^2 e^{-\alpha L}}{(1 - Re^{-\alpha L})^2 + 4Re^{-\alpha L} \sin^2(\frac{\delta}{2})}. \qquad (2.6)$$

The quantity $\delta \equiv 2(\phi_r' + k_o L)$ is the phase compounded in a round trip through the cavity. Since we wish to consider only absorptive bistability in a tuned cavity, we take $\delta = 2\pi p$, p = integer, so that Eqn. (2.6) reduces to

$$\frac{|\varepsilon_t|^2}{|\varepsilon_i|^2} = \frac{T^2 e^{-\alpha L}}{(1 - Re^{-\alpha L})^2}. \qquad (2.7)$$

Note that since the absorption coefficient α_a in general depends upon the local strength of the intracavity field, which is in turn related to ε_t via the boundary condition at the output mirror, Eqn. (2.7) represents a rather complex implicit specification of the transmission function of the Fabry-Perot cavity.

Our reference point for describing the behavior of the cavity will of course be that of the cavity in the absence of the nonlinear medium, for which $\alpha_a = 0$ and $\alpha = \overline{\alpha}$. We will refer to this case as the "empty-cavity" case, and from Eqn. (2.6) calculate the finesse $\mathcal{F}$ as the ratio of the separation in phase $\delta(=2\pi)$ between successive spectral orders to the full width ϵ of the transmission function at half intensity. With $\mathcal{T}(\delta)$ equal to one half of its maximum value at $\delta = 2\pi p \pm \frac{\epsilon}{2}$, Eqn. (2.6) leads to

$$\mathcal{C} = 2\frac{(1-Re^{-\bar{\alpha}L})}{\sqrt{R}\, e^{-\bar{\alpha}L/2}},$$

so that

$$F = \frac{\pi\sqrt{R}\, e^{-\bar{\alpha}L/2}}{(1 - Re^{-\bar{\alpha}L})} \simeq \frac{\pi}{(1 - R) + \bar{\alpha}L}, \tag{2.8}$$

where these results follow for $(1 - R)$, $\bar{\alpha}L << 1$. The peak transmission of the empty cavity occurs at $\delta = 2\pi p$, which from Eqn. (2.6) is

$$T(\delta = 2\pi p) \equiv T_o = \frac{T^2\, e^{-\bar{\alpha}L}}{(1 - Re^{-\bar{\alpha}L})^2} \simeq \frac{T^2}{(1- R + \bar{\alpha}L)^2}, \tag{2.9}$$

again in the limit of small cavity losses. Note that while we have assumed cavity losses of the form $[(1 - R) + \bar{\alpha}L]$, in general the denominators in Eqns. (2.8) and (2.9) represent the total single pass loss in intensity and would include, for example, diffraction losses or losses at intracavity optical elements.

Another feature of the empty cavity that will be important in our discussion of optical bistability is the ratio of intracavity intensity to incident intensity. Taking the intracavity field to be of the form (Eqns. (2.1) and (2.2))

$$\begin{aligned} E(z,t) &= \varepsilon(z)e^{-i\omega t}\,\vec{x} + \text{c.c.} \\ &= [\varepsilon_F(z)e^{ikz} + \varepsilon_B(z)e^{-ikz}]e^{-i\omega t}\,\vec{x} + \text{c.c.} \end{aligned} \tag{2.10}$$

we obtain expressions for ε_F and ε_B by summing over the multiple reflections within the cavity at some point $0 < z < L$. For exact resonance $\delta = 2\pi p$, and such a procedure combined with the definitions of Eqn. (2.10) leads to

$$\frac{|\varepsilon(z)|^2}{|\varepsilon_i|^2} \simeq \frac{2T}{(1 - R + \bar{\alpha}L)^2}[1 + \cos(2k_o z + \phi_{r'})], \tag{2.11}$$

where once again $(1 - R)$, $\overline{\alpha}L \ll 1$. Eqn. (2.11) clearly displays the standing-wave nature of the intracavity field. In the simple analysis of this section we wish to avoid the complexity associated with this spatial variation, and so we will average $|\varepsilon(z)|^2$ over many wavelengths to produce an average intensity $|\varepsilon|^2$, with

$$\mathcal{R} \equiv \frac{|\varepsilon|^2}{|\varepsilon_i|^2} = \frac{2T}{(1-R+\overline{\alpha}L)^2}, \tag{2.12a}$$

or by combining Eqns. (2.8) and (2.9) with (2.12),

$$\mathcal{R} = \frac{2\surd\mathcal{T}_o}{\pi}\,\mathcal{F}. \tag{2.12b}$$

This expression is certainly no surprise; for an ideal cavity with $\overline{\alpha}L = 0$ and $T + R = 1$ (no absorption or scatter loss at the mirrors) Eqn. (2.12) reduces to

$$\frac{|\varepsilon|^2}{|\varepsilon_i|^2} = \frac{2}{(1 - R)},$$

which expresses an enhancement in intracavity intensity due to the "quality factor" of the cavity and to the incoherent sum of two oppositely directed travelling waves.

With the characteristics of the empty cavity as given by Eqns. (2.8), (2.9), and (2.12) in mind, we now return to Eqn. (2.7) in the case $\alpha_a \neq 0$. The particular form for the saturable absorber and hence for α_a that we choose (although this choice has to be modified in the quantitative theory) is that appropriate for a homogeneously broadened intracavity medium driven on resonance, namely (25,26)

$$\alpha_a = \frac{\alpha_o}{1 + I/I_s}, \tag{2.13}$$

with α_o as the absorption coefficient at intensities I much less than the saturation intensity I_s. For a system of "two-level" atoms in the absence of inhomogeneous broadening, with purely radiative relaxation and with aligned dipoles relative to the driving field, α_o and I_s are given by the simple expressions (25-28)

$$\alpha_o = \sigma\rho = \frac{3}{2\pi}\lambda^2\rho \tag{2.14}$$

and

$$I_s = \frac{\hbar\omega_a}{\sigma}\gamma_\perp, \tag{2.15}$$

with $\sigma \equiv$ resonant absorption cross section at the atomic transition frequency ω_a, $\gamma_\perp \equiv$ transverse relaxation rate, and $\rho \equiv$ atomic number density.

For our consideration of an <u>intracavity</u> medium, the intensity I and hence α_a that appear in Eqn. (2.13) are functions of z. In addition as we shall see in Section 4, for moving atoms our whole treatment must be modified to account for atomic motion through the standing-wave field. Nonetheless for our simple theory, we continue to consider the case of a cavity with small losses ($\overline{\alpha}L$, $\alpha_o L$, $(1 - R) \ll 1$) and seek to relate the intracavity intensity I appearing in Eqn. (2.13) to the transmitted field ε_t, in as simple a way as possible. We have

$$I(z) = 2\epsilon_o c|\varepsilon(z)|^2 \tag{2.16}$$

so that

$$I \equiv 2\epsilon_o c|\varepsilon|^2 \simeq 2\epsilon_o c\{2|\varepsilon_F|^2\} \tag{2.17}$$

or

$$I = 4\epsilon_o c\ |\varepsilon_t|^2/T, \tag{2.18}$$

where once again we have averaged over the standing-wave variation of I(z) in going from Eqn. (2.16) to (2.17) and have made use of the boundary condition $|\varepsilon_t| = |\varepsilon_F|t'$ at the output mirror of the cavity. The transmission function for the cavity is obtained by combining Eqn. (2.18) with Eqns. (2.13) and (2.4); the resulting expression for α is inserted into Eqn. (2.7). Expanding this equation and keeping only the lowest order terms in $\overline{\alpha}L$, $\alpha_o L$, and (1-R) leads to the familiar result

$$X_o \left[1 + \frac{2C}{1 + X_o}\right]^2 = Y_o, \tag{2.19}$$

where

$$Y_o \equiv \frac{\mathcal{R} I_i}{I_s}, \tag{2.20}$$

$$X_o \equiv \frac{2I_t}{TI_s}, \tag{2.21}$$

$$C \equiv \frac{\alpha_o L}{2\pi} \mathcal{F}, \tag{2.22}$$

and the intensities (I_i, I_t) are related to the field amplitudes (ε_i, ε_t) by $I = 2\epsilon_o c|\varepsilon|^2$. Here $\mathcal{F}$ and $\mathcal{R}$ are given by Eqns. (2.8) and (2.12) for the empty cavity.

Eqn. (2.19) is one of the well-known state equations of optical bistability. This result was first obtained by Szöke et al. (8) in 1969 and later by Bonifacio and Lugiato (10). The parameter C is the so-called atomic cooperativity parameter (8, 9, 10) and expresses the ratio of unsaturated single pass absorption $\alpha_o L$ due to the atomic medium to single-pass cavity loss given by $\frac{\pi}{\mathcal{F}}$ (Eqn. (2.8)). C is analogous to the pump parameter "a" that occurs in single-mode laser theory (5-7).

At this point we will not discuss in any detail the properties of Eqn. (2.19), but instead refer the reader to an extensive theoretical literature (8-16, 30). Suffice it to say that Eqn. (2.19) produces a cubic equation for X_o as a function of Y_o. For $C > 4$, 3 real roots exist, giving rise to two possible stable output intensities X_o for a given input Y_o. This circumstance is of course the origin of the term "optical bistability".

3. EXPERIMENT

We have stressed in the introduction that the intent of our work is to realize absorptive bistability with an atomic system that as closely as possible approximates a collection of two-level atoms. As pointed out by Ballagh et al. (31) even within the restricted context of the requirement of the induced macroscopic dipole, the possibilities for achieving effective two-state behavior are quite limited. Our strategy has been to follow the well documented work in single-atom resonance fluorescence demonstrating that optical pumping of alkali atoms and in particular of sodium can produce an effective two-level atom (32-36).

A general schematic of the apparatus that we have constructed for our study is presented in Figure 2 below. The essential elements shown in the Figure are the multiple atomic beams of sodium and the high finesse interferometer through which they pass. The atomic beams are optically prepumped before entering the mode volume of the interferometer and in this way are prepared as two-level atoms. Preliminary to a discussion of the bistability that has been observed with this system, we will consider separately the characteristics of the intracavity medium and of the optical resonator.

The source for the sodium atomic beams is a stainless steel oven heated to temperatures in the range 400-600°C and containing an initial charge of 10-15 g of metallic sodium. A group of five square apertures each .5 x .5 mm provides outlets from the oven into the vacuum chamber, and two subsequent sets of .5 x .5 mm apertures act as collimators to form the atomic beams. As previously described (22) this arrangement produces five parallel (primary) atomic beams and two sets of secondary beams. The secondary sets each contain four beams and are offset from the primary beams by an angle $\phi = \pm 8$ mrad. The atomic beams pass through a region of uniform magnetic field of .5 Gauss parallel to the axis of the interferometer which is created by three sets of orthogonal current carrying coils.

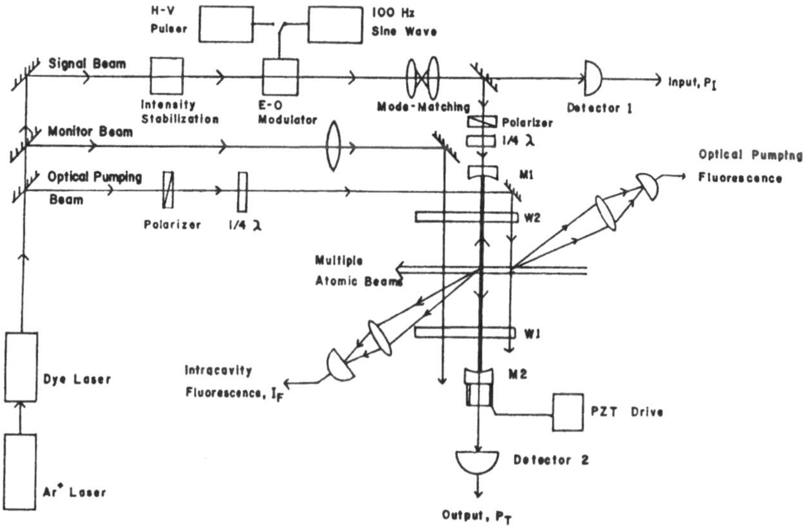

Figure 2 Schematic of apparatus for the investigation of optical bistability. Multiple beams of atomic sodium pass through a high finesse optical resonator.

The particular optical transition that we excite in the sodium beams is one of the lines in the hyperfine structure of the D_2 line of atomic sodium, namely the $3^2S_{1/2}$, $F = 2 \rightarrow 3^2P_{3/2}$, $F = 3$ transition. As is well known (32-36) illumination of this transition with circularly polarized light causes an efficient migration of population into the $m_F = +2$ Zeeman state of the $F = 2$ hyperfine component of the ground state, where σ^+ transitions are assumed. From the $m_F = +2$ state the only dipole-allowed transition in absorption of radiation of the same polarization as the pumping beam is to the $3^2P_{3/2}$, $F = 3$, $m_F = +3$ state, which in turn can decay only to the original Zeeman ground state. To illustrate the effectiveness of the optical pumping process we show in Figure 3 an absorption scan without and with optical prepumping. The experimental setup is as in Figure 2 but without the mirrors of the interferometer. The signal laser beam is focused through the atomic beams such that the waist size (W_0=150μm) is small compared to the cross section of the atomic beams. The peak intensity of the signal laser beam was 1.4 mW/cm^2 for the scans. In Figure 3(a) the signal laser is scanned without the optical pumping laser beam present, and we see the absorption spectrum for the $3^2S_{1/2}$, $F = 2 \rightarrow 3^2P_{3/2}$ $F = 2, 3$ transitions with the multitude of Zeeman levels participating. In Figure 3(b) an identical scan is made but with the optical pumping beam present with a peak intensity of 25 mW/cm^2 and a FWHM of 1.6 mm. Only a single absorption feature of FWHM = 26 MHz appears now at the position

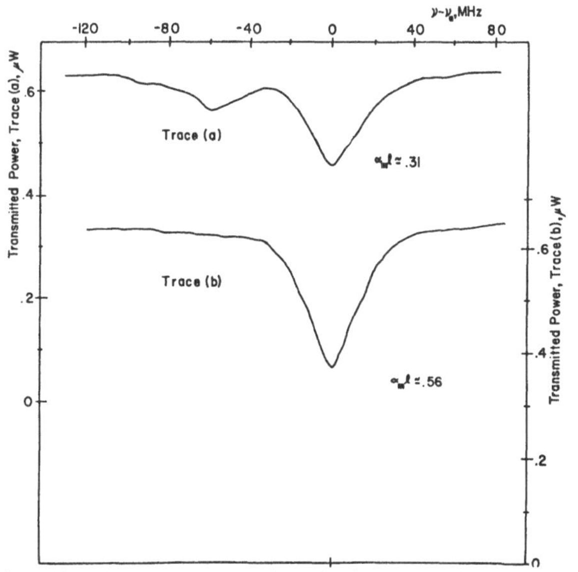

Figure 3 Record of laser power transmitted through the atomic beams as a function of laser frequency. (a) Without optical prepumping, two transitions within the hyperfine structure of Na are seen ($3^2S_{1/2}$, $F = 2 \rightarrow 3^2P_{3/2}$, $F = 2,3$). (b) With optical prepumping, one observes only a single transition corresponding to $3^2S_{1/2}$, $F = 2$, $m_F = 2 \rightarrow 3^2P_{3/2}$, $F = 3$, $m_F = 3$. The scale on the left is for sweep (a); the one on the right is for sweep (b).

of the F = 2 → F = 3 transition. Although the atomic number density is unchanged in going from Figure 3(a) to 3(b), the measured resonant absorption $\alpha_m \ell$, where ℓ is the propagation distance in the beams, has increased by a factor of 1.8. This increase is to be compared to the theoretical ratio of 2.14 obtained by comparing the unpumped to fully pumped absorption cross sections in the absence of inhomogeneous broadening. With changes in the temperature of the source oven, $\alpha_m \ell$ for the atomic beams can be varied in the range $0 < \alpha_m \ell < 1.5$.

We next turn our attention to a characterization of the optical cavity, which was set up as shown in Figure 2. The cavity is operated near the confocal spacing with .25 m radius of curvature mirrors located external to the vacuum chamber and two intracavity antireflection-coated windows. With reference to our discussion of Section 2, the properties of the empty cavity are listed below.

$$\begin{aligned} &\text{Cavity length } L \simeq .25 \text{ m} \\ &\text{Beam waist } W_o = 150\ \mu\text{m} \\ &\text{Finesse } \mathcal{F} = 210 \pm 15 \\ &\text{Transmission } \mathcal{T}_o = .018 \pm .003 \\ &\text{Intracavity enhancement } \mathcal{R} = 18 \pm 2 \\ &\text{Fluctuation in cavity length and hence phase } \delta \simeq 1.6 \text{ mrad } (.15 \text{ MHz}). \end{aligned} \tag{3.1}$$

As indicated by the lenses in Figure 2, an attempt is made to mode match to the cavity (37). It is however difficult to estimate the efficiency with which the fundamental TEM_{oo} mode of the cavity is excited since at the confocal spacing some higher order transverse modes are degenerate with the longitudinal modes of the cavity and are not detected directly in a sweep of transmission versus phase δ(26,37). The excitation source for all our studies is a commercial cw dye laser operating with Rhodamine 6G and pumped by the 5145 Å light from an Ar^+ laser. The dye beam is stabilized in frequency to approximately .25 MHz rms relative to an external reference cavity, with drifts of several megahertz occuring over intervals of a few minutes. The laser intensity fluctuates by roughly ± 3% principally at acoustic frequencies.

Given this discussion of the characteristics of the individual components of our apparatus, we can now imagine assembling the system shown in Figure 2 to undertake a study of optical bistability. The actual procedure followed is rather tedious since all measurements must be made in a single experimental run due to the high rate at which sodium is depleted from the oven. With the interferometer disassembled, adjustments of the collimating apertures are made to maximize the atomic flux at the position of the signal laser beam. The signal and monitor laser beams are brought into alignment perpendicular to the primary atomic beams to within $\pm\ 6 \times 10^{-4}$ rad (±1 MHz). The optical pumping beam is introduced, and the angle of intersection with the atomic beams adjusted until the peak fluorescent signal from the primary atomic beams occurs at the same laser frequency as the peak of the signal beam absorption. We next perform several calibration measurements at various operating temperatures of the source oven (.3

$< \alpha_m \ell < 1.0$). As a function of laser frequency, the transmitted power of the signal laser beam, the optical pumping fluorescence, and the monitor beam absorption are simultaneously recorded. These data allow us to calculate the atomic absorption coefficient $\alpha_m \ell$ for the signal laser beam during the actual measurements of optical bistability from a knowledge of the optical pumping fluorescent signal or from the monitor beam absorption coefficient, both of which are recorded continuously over the course of the experimental run.

The final step before a search for bistability is made is to assemble the interferometer coaxial to the signal laser beam with the oven operated at low temperature and with an intense incident laser. With a manually tuned dc voltage applied to the piezoelectric transducer on which one of the cavity mirrors is mounted and with the dye laser tuned to resonance as indicated by either the optical pumping fluorescence or the monitor beam absorption signal, the temperature of the oven is slowly increased. The incident laser power to the cavity is sinusoidally modulated at 100 Hz (Figure 2), and the transmission characteristics of the cavity are displayed on and recorded from an x-y oscilloscope. The x-input is derived from a diode detecting the input power, and the y-inputs are derived from a photodiode monitoring the transmitted laser power and a photomultiplier tube onto which the intracavity fluorescence from one of the atomic beams is focused.

As the intracavity density of the atomic beams increases, the transmission characteristics evolve qualitatively as predicted from the plane-wave theory of Section 2, Eqn. (2.19). Examples from our data are shown in Figure 4, with Figure 4(a) illustrating the general format of the photographs. Below some critical value of atomic cooperativity, bistability is not observed but rather a region of

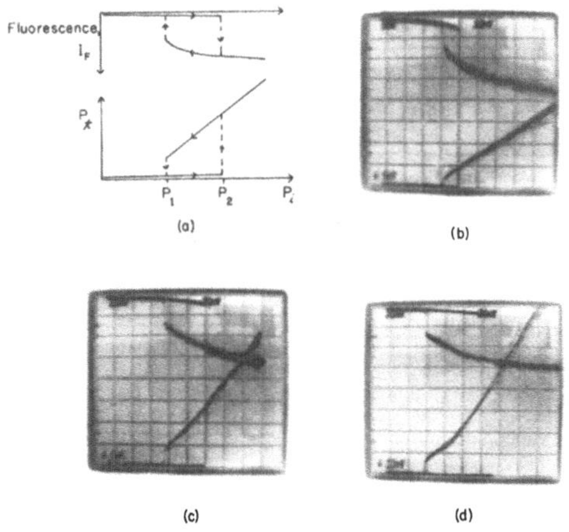

Figure 4 Photographs of the output power P_t and intracavity fluorescence I_F as functions of incident power P_i. The general format for the photographs is given in (a). (b)-(d) are records made for increasing effective cooperativity parameter C_e equal to 22, 32, and 49, respectively.

large differential slope separates the transmission characteristics into low and high transmission regions. At higher densities a definite hysteresis emerges in both the transmitted power and the intracavity fluorescence as functions of incident laser power. The growth of the hysteresis with increasing $\alpha_m \ell$ is illustrated in Figure 4(b)-(d). In Figure 4 and for the other measurements that we report the conditions appropriate to absorptive bistability were approximately met in the following fashion. First the laser was tuned to the atomic resonance as previously described. Next the cavity detuning was set to zero by varying the length of the cavity until the transmission of the cavity in the limit of large input intensities (several times the switching intensities) was a maximum.

From a large number of photographs such as shown in Figure 4 we have determined the dependence of the switching powers P_1 (upper to lower branch) and P_2 (lower to upper branch) on the resonant atomic absorption $\alpha_m \ell$. Figures 5 and 6 display our data and show the actual incident switching powers (P_1, P_2) and the ratio of these powers $S \equiv P_2/P_1$ as functions of atomic cooperativity. For each set (P_2, P_1) a value of the cooperativity parameter has been calculated from Eqn. (2.22) using, in place of $\alpha_o \ell$ the value of $\alpha_m \ell$ determined from the record of monitor beam absorption and optical pumping fluorescence, and the cavity finesse F. We label this C_e, denoting an effective cooperativity parameter. As we shall see in the next section, the theory is generally expressed in terms of the number density and not the actual on resonance absorption $\alpha_m \ell$. In the absence of any broadening mechanism other than radiative relaxation, there is no distinction

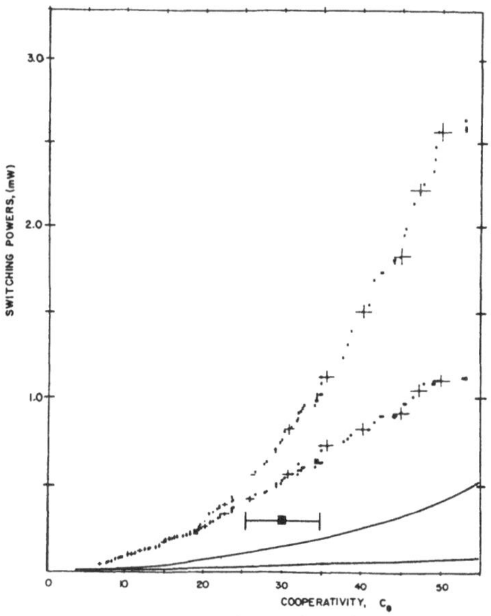

Figure 5 Switching powers (P_1,P_2) as shown in Figure 4 as a function of cooperativity parameter C_e. Relative uncertainties are indicated by error bars at several points. Overall the scale for C_e is uncertain by ±15%. The full curve is the prediction of the plane wave theory of Eqn. (2.19) for the turning points as discussed in the text.

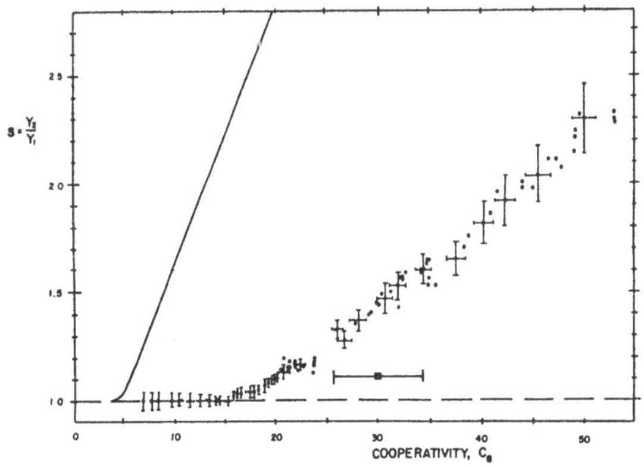

Figure 6 *Ratio of switching powers $S \equiv P_2/P_1 = Y_2/Y_1$ as a function of cooperativity C_e. Relative and absolute uncertainties are discussed in the text. The full curve is the theoretical result obtained from Eqn. (2.19).*

between the two points of view. However in our case the atoms are spread across an inhomogenous line profile.

As indicated by the horizontal error bar at $C_e = 30$ in Figure 5, the overall uncertainty for the scale of the data set as a whole is ±15%, and arises from the lack of precision to which F and $\alpha_m \ell$ are known. Apart from the question of scaling, the relative uncertainty of points within the data set is shown by the smaller horizontal error bars. Likewise relative uncertainties of the incident switching powers are indicated by the vertical error bars. The absolute switching powers are obtained from a calibration of the photodiode input relative to two different and independently NBS traceable power meters and are accurate to ±5%. In Figure 5 a single point at a given value of C_e corresponds to the power at which the maximum differential slope is obtained in the input-output characteristic.

We present the data of Figure 5 in a different format in Figure 6 to more clearly illustrate the dependence of the onset and growth of the hysteresis cycle as a function of C_e. The ratio $S \equiv P_2/P_1$ is set to one for operating conditions in which no hysteresis can be detected; $S > 1$ corresponds to a clearly defined hysteresis cycle. The error bar at $C_e = 30$ again represents the overall uncertainty in the determination of C_e.

It should be stressed that our data represent a good approximation to the steady-state regime in optical bistability. The time taken to sweep up and down in intensity (10 msec) is large compared to either the cavity decay time $\kappa'^{-1} \equiv \frac{2\,L\,F}{\pi c}$ (= 1.1×10^{-7}s) or to the atomic lifetime (16 nsec). As we shall see in the next section the simple theory of Section 2 makes only a qualitative accounting of the various effects in the observed steady state characteristics. A quantitative understanding requires a considerably more advanced analysis. To illustrate this point we have included in Figures 5 and 6 a full curve derived from the state equation (2.19). To scale Y_o of that equation to an actual incident power P_i, we rather arbitrarily associate the incident intensity I_i with .75 times the peak intensity of the Gaussian beam at the center of the interferometer. That is,

$$I_i = \frac{3\ P_i}{2\pi W_o^2} \quad \text{or} \quad Y_o = \frac{3\ P_i\ R}{2\pi W_o^2\ I_s}. \tag{3.1}$$

We also for the moment ignore the distinction between C and C_e. Note that C_e as defined above effectively accounts for the fact that the atoms are distributed over an inhomogeneous profile (52), whereas C does not.

4. THEORY

As is evident from Figures 5 and 6 for our experiments on steady-state optical bistability, there are quantitative departures from plane-wave theory for a homogeneously broadened medium. For example, from Fig. 6 one sees that the critical cooperativity for optical bistability is not at $C = 4$ as in a homogeneous plane-wave theory, but at $C \simeq 14$. Likewise the calculated plane-wave intensities for the switching points are much less than those observed, as shown in Fig. 5.

These quantitative departures are due to several factors which are omitted in the simple homogeneous plane-wave theory of optical bistability, relative to our experiment. One obvious difference is that the experiments do not involve a homogeneously broadened atomic transition, and inhomogeneous broadening tends to reduce the extent of bistability. In addition, we shall see that the departures show the quantitative importance of the transverse mode structure for the interferometer, which in these experiments is the set of Gaussian radially varying modes of a Fabry-Perot interferometer.

Motivated by these comments we will in this section work out a theory for the case of a Gaussian mode, which extends previous work in this area of bistability. In general such a calculation would require the solution of cylindrically symmetric Maxwell-Bloch equations to allow for the large changes possible during propagation (29,40,41). However except at the highest C-values of Figures 5 and 6 the complexity of the full theory can be safely reduced. For this reason propagation effects will be neglected, and a single Gaussian-mode theory will be used (9,38,39).

As for the question of the validity of a steady-state description, we note that there are theoretical predictions relating to plane-wave instabilities in which coupling to other longitudinal modes occurs (42). While these effects are not included in the current theory, they cannot be entirely ruled out at the highest intensities and C values reached, where the ac Stark effect causes a frequency splitting comparable to the longitudinal mode spacing of the 25-cm standing-wave interferometer.

In addition to these instabilities, Ikeda has predicted the existence of instabilities in the theory of nonlinear plane-wave interferometers in the dispersive limit (43,44). One of these occurs for an interferometer round-trip time (τ_R) relatively long compared to the atomic relaxation time T_1 (43). This is not relevant in the current expetiments where τ_R = 1.7 nsec, $T_1 \simeq$ 16 nsec, and where the nonlinearity is purely absorptive. Ikeda has also predicted (44) instabilities for small τ_R, and with T_1 comparable to the interferometer relaxation time (κ'^{-1}). While the dispersive limit used by Ikeda is not applicable here, it is true that the coupled atom-field equations have complex stability properties when $\kappa' T_1 \sim 1$. The simplest case of plane-wave absorptive bistability is analyzed in detail in Ref. (46) where it is shown that for all values of the relaxation rates (κ', $\gamma_\parallel$, $\gamma_\perp$) the positive slope braches of the state equation are stable, and the negative slope branches are unstable, provided the interferometer mode spacing is large enough. To proceed further, we suppose that these stability properties hold even in the general case of inhomogeneous broadening and Gaussian mode structure in absorptive bistability. This is known to be true rigorously (47) in the limit $\kappa' \ll \gamma_\perp$, $\gamma_\parallel$, for single-mode Hamiltonians, although the exact structure of the final state equation becomes radically modified.

The question of transverse mode coupling will not be addressed here in detail, since the higher order transverse modes are implicitly treated as a reservoir in our theory. That is, the effective polarization of the nonlinear medium will be integrated over the zeroth order (TEM_{oo}) mode. It is reasonably clear that due to nonlinearities there must be an excitation of higher order transverse modes, since the induced polarization is not identical in radial structure to the initial mode. The approximation that we adopt is that higher order modes which become excited experience strong relaxation via absorption by atoms outside the waist of the principal TEM_{oo} mode (40).

The theory that we present attempts to include those factors that have a dominant effect on the calculation of switching powers. Specifically, we include: (1) Gaussian-mode radial intensity variations, (2) corrections for standing waves in a Fabry-Perot, (3) inhomogeneous broadening, (4) atomic longitudinal velocities (along the interferometer axis), (5) atomic transverse velocities (perpendicular to the interferometer axis). The Hamiltonian that we employ is the standard one used in studies of optical bistability, except that a time-dependent position and a nonplanar mode structure have to be included:

$$H = H_{Field} + H_{Atom} + H_{Interaction}$$

$$= \hbar\omega_o \hat{a}^+\hat{a} + \sum_j \hbar[\omega_a \hat{\sigma}_j^z/2 - ig_j(\hat{\sigma}_j^+\hat{a} - \hat{\sigma}_j^-\hat{a}^+)]$$

$$+ i\hbar[\varepsilon(t)\hat{a}^+ e^{-i\omega t} - \varepsilon^*(t)\hat{a}\, e^{i\omega t}] + \{\text{RESERVOIRS}\}. \tag{4.1}$$

The reservoirs describe radiative and collisional relaxation, leading to relaxation rates of κ', $\gamma_\perp$, $\gamma_\parallel$ for the interferometer mode, atomic polarization and atomic inversion, respectively. The interferometer mode creation and annihilation operators are $\hat{a}^+$, $\hat{a}$, while the atomic raising, lowering and inversion operators are the Pauli operators $\hat{\sigma}_j^+$, $\hat{\sigma}_j^-$, $\hat{\sigma}_j^z$ (for the j-th atom).

The frequencies (in radian/sec) of the interferometer mode and the individual atomic transitions are ω_o, ω_a, respectively. The input field is proportional to $\varepsilon(t)e^{-i\omega t}$; that is, it has a frequency of ω, and $\varepsilon(t)$ has units of $\sec^{-1}$. The dipole coupling is g_j, which is defined in terms of the Einstein A-coefficient for the relevant transition, and the electromagnetic field mode. Each atom itself has a certain velocity $\vec{v}_j$ and a position $\vec{r}_j^{\,o}$ at $t = 0$, distributed so that there are always η atoms per unit area orthogonal to z. The coupling coefficient is therefore given by:

$$g_j(t) = [\frac{3\pi\gamma_\parallel c^3}{2\omega^2}]^{1/2} |u(\vec{r}_j(t))| \equiv \bar{g}|u(\vec{r}_j(t)| \tag{4.2}$$

with

$$\vec{r}_j(t) = \vec{r}^{\,o}{}_j + \vec{v}_j t = (x^o{}_j + y^o{}_j + z^o{}_j) + (v^x{}_j + v^y{}_j + v_j)t,$$

and where

$$|u(\vec{r})| = (\frac{4}{\pi L W_o^2})^{1/2} \cos(\frac{2\pi z}{\lambda}) \exp(-(x^2 + y^2)/W_o^2).$$

Here $u(\vec{r})$ is simply the mode function of the electric field $\vec{E}$ near the center of a Fabry-Perot interferometer, neglecting the longitudinal variation in the mode radius. The interferometer length is L, with a mode waist of W_o.

In order to include quantum fluctuations, it would be necessary to treat Eqn. (4.1) using a Fokker-Planck approach (47). However, this is not required at present. Instead, the approximate semiclassical equations of motion will be derived in which the quantum fluctuations are regarded as negligible. This approach can be justified as an expansion of the Fokker-Planck equation in powers of $(\frac{1}{N})$ for N atoms, in which the higher order derivatives are dropped, as they involve higher order terms in $(\frac{1}{N})$. The resulting characteristics that solve the reduced Fokker-Planck equation are identical to the semiclassical equations including relaxation. By defining

$$\alpha(t) = \langle \hat{a} e^{i\omega t} \rangle, \qquad \sigma_j^{\pm}(t) = \langle \hat{\sigma}_j^{\pm} e^{\mp i\omega t} \rangle, \qquad \sigma_j^z(t) = \langle \hat{\sigma}_j^z \rangle$$

we obtain for the case $\omega = \omega_a$ (laser tuned to line center),

$$\frac{\partial \alpha}{\partial t} = \varepsilon(t) - \kappa\alpha + \sum_j g_j(t)\sigma_j^-$$

$$\frac{\partial \sigma_j^-}{\partial t} = g_j(t)\, \alpha\sigma_j^z - \gamma_\perp \sigma_j^-, \tag{4.3}$$

$$\frac{\partial \sigma_j^z}{\partial t} = -2g_j(t)(\sigma_j^+\alpha + \sigma_j^-\alpha^*) - \gamma_\parallel(\sigma_j^z + 1),$$

where $\kappa = \kappa' + i(\omega_o - \omega) = \kappa'\,(1 + i\phi)$.

While Eqn. (4.3) can include stationary inhomogeneous broadening, this type of broadening (which does occur in non-Doppler-broadened media) is not relevant to the atomic-beam Fabry-Perot interferometer, as the atomic Doppler shifts do not define a unique frequency. The reason for this is simple: an atom with a velocity component in the z-direction experiences both blue and red frequency shifts relative to the two counter-propagating directions of laser propagation in the interferometer. Hence it is not possible, except in the case of a ring interferometer, to treat the problem of inhomogeneous broadening as originating from a range of effective resonance frequencies.

For long enough transit times of the atoms through the mode volume, only the longitudinal or z-velocity component is relevant, with z along the interferometer axis. Each atom can then be identified by its longitudinal velocity v_j and by its radial coordinate $r_j = (x_j^2 + y_j^2)^{1/2}$. Of course, the radial coordinates change slowly; however provided $\gamma_\perp^{-1}$, $\gamma_\parallel^{-1}$ are very much less than the beam transit time, t_p, this effect can be ignored in a first approximation. The driving field experienced by atoms with velocity v_j is modulated at a frequency of $\Omega(v_j)$ =

$2\pi v_j/\lambda$. If we define $\Delta \equiv \Omega(v_j)/\gamma_\perp$, then the velocity distribution corresponds to a frequency distribution $P(\Delta)$ defined so that $\int P(\Delta)d(\Delta) = 1$.

The modulation of the field seen by the atoms implies that steady-state behavior does not occur for $\sigma^\pm$ and σ^z, and instead we look for periodic solutions to Eqn. (4.3) of the form:

$$\sigma_j^- \equiv \sigma^-(v_j,r_j) = \sum_n J_n(v_j,r_j)e^{in\Omega(v_j)\tau} \qquad \text{(n is odd)}$$

$$\sigma_j^z \equiv \sigma^z(v_j,r_j) = \sum_n D_n(v_j,r_j)e^{in\Omega(v_j)\tau} \qquad \text{(n is even)} \qquad (4.4)$$

with $\tau = t + z_j^o/v_j$ and with $g_j(t)$, $\varepsilon(t)$ defined in Eqns. (4.2), (4.3) by

$$g_j(t) \equiv g(r_j)\cos(\Omega(v_j)\tau) \text{ and } \varepsilon(t) \equiv \varepsilon = \text{ constant.} \qquad (4.5)$$

Solutions of this type were first studied in the theory of the well-known Lamb dip effect, often used in laser frequency stabilization (48-51). This theory was also used recently in optical bistability for the case of a plane wave mode (52). In the current problem it is necessary to combine both Gaussian mode theory, without a perturbative treatment, and the periodic atomic solutions of the type in Eq. (4.4).

Combining Eqns. (4.3) and (4.4), and omitting the (v_j,r_j) arguments for ease of notation, we obtain for any one atom in the beam,

$$\frac{\partial}{\partial\tau}\sum_n J_n\, e^{in\Omega\tau} = \sum_n \left[g\alpha\cos(\Omega\tau)\, D_n - \gamma_\perp J_n\right]e^{in\Omega\tau}$$

$$\frac{\partial}{\partial\tau}\sum_n D_n e^{in\Omega\tau} = -\sum_n \left[2g\cos(\Omega\tau)\left[J_{-n}^*\alpha + J_n\alpha^*\right] + \gamma_\parallel\left[D_n + \delta_{n,o}\right]\right]e^{in\Omega\tau}, \quad (4.6)$$

where $\delta_{n,o}$ is the Kronecker delta function. Hence

$$J_n = g\alpha\left[D_{n-1} + D_{n+1}\right]/\left[2(\gamma_\perp + in\Omega)\right],$$

$$D_n = \{-g\left[\alpha(J_{-n+1}^* + J_{-n-1}^*) + \alpha^*(J_{n+1} + J_{n-1})\right] - \gamma_\parallel\delta_{n,o}\}/(\gamma_\parallel + in\Omega),$$

$$D_n A_n + X_r\left[D_{n+2}\, a_{n+1} + D_{n-2}\, a_{n-1}\right] + \delta_{n,o} = 0. \qquad (4.7)$$

In Eqn. (4.7) the following standard notation (52) is used in order to simplify the recursion relations that define the Fourier component solutions:

$$X_r \equiv g(r_j)^2\ |\alpha|^2\ /\ [\gamma_\perp\gamma_\parallel],\ A_n \equiv 1 + in\Omega/\gamma_\parallel + X[a_{n+1} + a_{n-1}],$$

$$L_n \equiv [1 + in\Delta]^{-1},\ a_n \equiv [L_n + L_{-n}^*]/2. \qquad (4.8)$$

Clearly D_n has a straightforward three-term recursion relation. The solution is very well known from laser theory, and will be reproduced here without further analysis (48-52). We note that only J_1, J_{-1} will be necessary to the final results, and these quantities can in turn be written down using only the values of D_o and D_2 (since $D_{-n} \equiv D_n^*$, as the total inversion has zero imaginary part). The solutions for these equations are:

$$J_{\pm 1}(v_j,r_j) = g\alpha[D_o(v_j,r_j) + D_{\pm 2}(v_j,r_j)]/[2(\gamma_\perp \pm i\Omega)]$$

$$D_o(v_j,r_j) = -1/[1 + S(v_j,r_j)]$$

$$D_2(v_j,r_j) = -X_r F D_o(v_j,r_j) \qquad (4.9)$$

where:

$$S(v_j,r_j) = (|L_1|^2 + |L_{-1}|^2)\ X_r - X_r^2\ \mathrm{Re}\ [F(L_1 + L_{-1}^*)],$$

$$F = \{\frac{a_1}{A_2}\ /\ [1 - X_r^2(\frac{a_3^2}{A_2A_4})\ /\ [1 - X_r^2\ (\frac{a_5^2}{A_4A_6})/\ldots.\}.$$

The arguments of the functions Ω, X_r, F, a_n, A_n are all omitted for ease of notation, although these are functions of (v_j,r_j). Given Eqn. (4.9), the result of calculating the lowest order Fourier components $J_{\pm 1}(v_j,r_j)$ depends only on the continued fraction F, which must in general be evaluated numerically.

With the above results we can now compute the source term in equation (4.3) for the field. This can be calculated in the steady state on taking into account only the non-oscillating components of the source term $\sum g_j(t)\sigma_j^-(t)$. Oscillating terms in this sum will be averaged to zero for sufficiently many interacting atoms, since every atom has a random relative arrival time (z_j^o/v_j). This implies that at any time the atoms are distributed randomly and relatively uniformly along the length of the interaction volume. The resulting state equation for on-resonance excitation is (noting that ε has units of $\sec^{-1}$)

$$\varepsilon = \kappa\alpha - \sum_j \frac{g(r_j)}{2} [J_1(v_j,r_j) + J_{-1}(v_j,r_j)]. \tag{4.10}$$

For a continuous velocity distribution and a large number of atoms, we replace the summation in Eqn. (4.10) by an integral over velocity and radius, given the distribution $P(\Delta)$ defined earlier and the atomic density for a medium of infinite transverse extent. The resulting equation is best expressed in terms of the cooperativity parameter C, and reduced variables X, Y, similar to those defined in Section 2. While C can be unambiguously defined, the definitions of X, Y in the case of Gaussian modes are dependent on the normalization. Here we choose a normalization or scaling that has the property of producing a unified dispersive limit for the state equation (38). Hence we define

$$C \equiv \frac{\eta \bar{g}^2}{2L\kappa'\gamma_\perp},$$

$$X \equiv |\alpha|^2 / n_o \tag{4.11}$$

$$Y \equiv |\varepsilon|^2 / [\kappa'^2 n_o],$$

$$\text{and } n_o = (\tfrac{2}{3}\pi L W_o^2)(\gamma_\perp\gamma_\parallel/(4\bar{g}^2)),$$

with η as the number of atoms per unit cross-sectional area interacting with the mode. These definitions together with Eqn. (4.10) give

$$Y = X|1 + i\phi + 2C \int P(\Delta)\chi(\Delta,X)d\Delta|^2 \tag{4.12}$$

where, for symmetric distributions ($P(\Delta) = P(-\Delta)$),

$$\chi(\Delta,X) = \int \{\frac{1 - X_r F(X_r,\Delta)}{1 + S(X_r,\Delta)}\} \frac{e^{-R}dR}{1 + i\Delta},$$

$$X_r = (\tfrac{2}{3})\, Xe^{-R}, \tag{4.13}$$

$$R \equiv [2r^2/W_o^2].$$

Equation (4.12) is the central result of this theoretical section. We note that in general the procedure of evaluating the state equation is a numerical one, requiring first evaluation of the continued fraction F, and then integration over Δ and R. This then has to be repeated for a large number of X values to obtain the state equation at a given C. Finally, the value of C is varied in order to model the variation in the atomic absorption that occurs over the course of the experiments described in Section 3.

In addition to the above calculation, we have performed several others to determine the validity of the approximations used in deriving Eqn. (4.12). These include checks on the effect of beam transit time, which requires a numerical integration of Eqn. (4.3) during typical atomic trajectories through the interferometer mode, together with integrals over the offset radius of the trajectory from mode center and over the standing waves. For typical velocities of $\simeq$ 1000 ms^{-1}, with a mode waist of 150 μm, the corrections to the switching intensities are of order ±5%. Corrections of similar order of magnitude are found by including the finite atomic beam cross-section (of 500 μm), and by approximately including the effect of absorption at large C-values. Since these corrections are less than the measured uncertainties in C, X, Y, we will not include them in our final calculation of the state equation.

5. COMPARISON OF THEORY AND EXPERIMENT

In order to compare theory and experiment, the first step is to identify the relation between the theoretical quantities C, X, Y and the observed quantities C_e, P_t, P_i, where C_e is the effective cooperativity defined as the resonant atomic absorption times the cavity finesse divided by 2π, and P_i, P_t are the incident and transmitted powers, respectively. We note that C_e would be identical to C as defined in Eqn. (4.11) provided there was no inhomogeneous broadening. However for a finite inhomogeneous width, one has the relation

$$C_e = C \int \frac{P(\Delta)d\Delta}{1 + \Delta^2}. \tag{5.1}$$

In the experiment, $P(\Delta)$ is in fact a rather complicated distribution involving a central peak with broadened non-resonant additional peaks due to imperfect beam collimation. This is simply modelled as a Gaussian,

$$P(\Delta) = \frac{1}{\sigma\sqrt{2\pi}} \exp\left[-\Delta^2/2\sigma^2\right]. \tag{5.2}$$

In Eqn. (5.2) the Gaussian standard deviation (σ) is obtained by fitting the calculated absorption profile derived from Eqn. (5.2) to the actual measured absorption profile as shown in Figure 3(b). Some modifications to a Gaussian distribution were also tried, but only minor changes were found in the characteristics of the state equation.

In order to relate the variables (X,Y) of Eqn. (4.11) to the measured output and input powers (P_t, P_i), we note that κ' is the decay constant for the empty cavity tuned to resonance (Eqn. (4.3)), and hence can be conveniently expressed using the notation of Section 2,

$$\kappa' = \frac{\pi}{F}\frac{c}{2L}.$$

This equation together with the boundary conditions at the mirrors of the cavity allow us to relate X to P_t. Likewise, noting that X = Y for the empty tuned cavity (Eqn. (4.12)), and that P_i and P_t are related in this same case through Eqn. (2.12), we can express Y in terms of P_i. Summarizing these results, we have

$$C = C_e / \int \frac{P(\Delta)d\Delta}{1+\Delta^2} = \frac{\alpha_o \ell}{2\pi} F, \qquad \text{(a)}$$

$$Y = \frac{3\, P_i R}{2\pi W_o^2 I_s}, \qquad \text{(5.3)(b)}$$

$$X = \frac{3P_t}{\pi W_o^2 I_s T}, \qquad \text{(c)}$$

with F, R, T, and I_s as defined in Section 2, with W_o as the beam waist at the center of the cavity, and with ℓ as the length of the absorber. As well we have assumed purely radiative relaxation ($2\gamma_\perp = \gamma_\parallel$) for our two-level system. The effective cooperativity parameter C_e is given in terms of the measured resonant absorption coefficient α_m and the cavity finesse F as

$$C_e = \frac{\alpha_m \ell}{2\pi} F. \qquad (5.4)$$

Cast in this form the theory is directly comparable to our experiment since all quantities that appear in Eqn. (5.3) are measured directly. Using Eqn. (5.3) we can rescale our data of Figure 5 to obtain dimensionless switching intensities (Y_1, Y_2) as a function of C_e. Figure 7 shows our data plotted in this way. Note that because the quantities that relate P_i to Y (Eqn. 5.3(b)) are not known to high precision (Eqn. (3.1)) and since we do not know the absolute efficiency of the excitation of the TEM_{00} mode of the cavity, the transformation from the set (P_1,P_2) to the set (Y_1, Y_2) involves an overall uncertainty in scale. We estimate this uncertainty to be ±30% and indicate this together with the uncertainty in the scale of C_e (±15%).

The dashed curve shown in Figure 7 is the theoretical result obtained from Eqn. (4.12) for an atomic absorption profile of 30 MHz FWHM (Eqn. 5.2) and for ϕ = 0. No adjustment of either the data or the theory has been made. The figure thus represents an absolute comparison between theory and experiment. We see rather good agreement in the region below and around the threshold for bistability for the absolute values of the switching intensities. However at higher C_e values substantial discrepancies appear. We can attempt to remedy this by rescaling the experimental intensities, since there are overall uncertainties in the experiment. The full curve in Figure 7 is the result of such a procedure, showing a greatly improved fit at higher values of C_e, but at the expense of agreement at low C_e. (For convenience, we have scaled the theory up rather than replot the data scaled down by 30%).

A similar analysis has been followed in Figure 8 where we reproduce Figure 6.

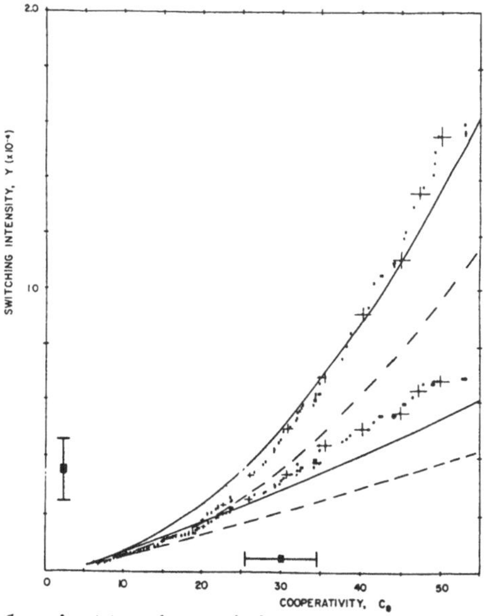

Figure 7 Experimental switching intensities (Y_1, Y_2) as determined from (P_1, P_2) of Figure 5 and from Eqn. 5.3 (b) plotted as functions of the effective cooperativity parameter C_e. The dashed curve is the result of Eqn. (4.12) without adjustment. The full curve represents an adjustment within the experimental uncertainties.

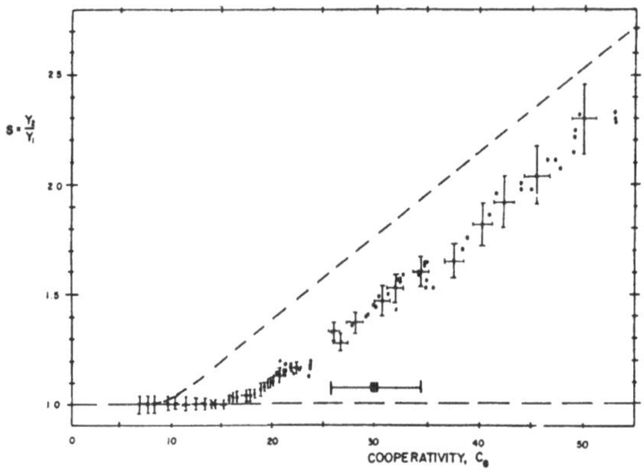

Figure 8 *As in Figure 6 with the dashed curve derived from Eqn. (4.12).*

Since the ratio $S = P_2/P_1 = Y_2/Y_1$ is independent of absolute calibration, we are able to compare more readily the experimental and theoretical results for the evolution of the hysteresis as a function of C_e. The dashed curve corresponds to the dashed curve of Figure 7, without adjustment of theory or experiment. There is quite clearly a discrepancy between theory and experiment, with the theory predicting a lower value of C_e for the critical onset of bistability and larger ratios S for a given C_e than was observed. A possible explanation for this difference is the presence of intensity and frequency fluctuations in the driving laser and fluctuations in the phase ϕ of the cavity. These "noise" sources would tend to reduce the size of the hysteresis cycle and shift upward the critical onset of bistability. As well, systematic errors may exist in our calibration procedures.

In our concluding remarks it is perhaps worthwhile to return to the point of view expressed in the introduction. Optical bistability is quite a general phenomenon occurring in diverse physical systems. A qualitative picture of either absorptive or dispersive bistability can be obtained from a simple theory such as that presented in Section 2. However, a detailed quantitative understanding requires a considerably more vigorous theoretical and experimental effort. While the results that we have presented here represent one of the most careful analyses of optical bistability yet performed, we view them as only a preliminary assessment of both the experimental procedures and the theoretical model. Such a study of the steady state is a necessary first step toward a microscopic understanding of even more complex dynamical processes in optical bistability. We view it as encouraging that a microscopic model can represent our results reasonably well without requiring ad-hoc fitting parameters of any kind.

This work was supported in part by the Joint Services Electronics Program and by the Robert A. Welch Foundation. The research of one of us (P.D.D.) was supported in part by the U.S. Office of Naval Research.

6. REFERENCES

1. J. F. Clauser and A. Shimony, Rep. Prog. Phys. 41, 1881 (1978).
2. L. Mandel in Progress in Optics, Vol. XIII, ed. E. Wolf (North Holland Publishing Co., Amsterdam), p. 29.
3. P.W. Milonni, Phys.Reports 25, 1 (1976).
4. R. Loudon, Rep. Prog. Phys. 43, 913 (1980); H. Paul, Rev. Mod. Phys. 54, 1061 (1982).
5. H. Haken in Handbuch der Physik, XXVc, (Springer, Berlin-Heidelberg, 1970).
6. H. Risken, in Progress in Optics, Vol. VIII, ed. E. Wolf (North Holland Publishing Co., Amsterdam), p. 241.
7. M. Sargent III, M. O. Scully, W. E. Lamb, Jr., Laser Physics (Addison- Wesley Publishing Co., Reading, Mass.), 1974.
8. A. Szöke, V. Daneu, J. Goldhar, and N. A. Kurnit, Appl. Phys. Lett. 15, 376 (1969).
9. S. L. McCall, Phys. Rev. A9, 1515 (1974).
10. R. Bonifacio and L. A. Lugiato, Opt. Commun. 19, 172 (1976); Phys. Rev. A18, 1129 (1978); Lett. Nuovo Cimento 21, 505 (1978).
11. Optical Bistability, edited by C. M. Bowden, M. Ciftan, and H. R. Robl (Plenum Press, N. Y., 1981).
12. IEEE J. Quantum Electron., Special Issue on Optical Bistability, edited by P. W. Smith, QE-17 (1981).
13. H. M. Gibbs, S. L. McCall, and T. N. C. Venkatesan, Phys. Rev. Lett. 36, 1135 (1976).
14. S. L. McCall and H. M. Gibbs, in reference 11 above, p. 1; Opt. Comm. 33, 335 (1980).
15. P. W. Smith and W. J. Tomlinson, IEEE Spectrum 18, 26 (1981).
16. E. Abraham and S. D. Smith, Rep. Prog. Phys. 45, 815 (1982).
17. F. T. Arecchi, in Order and Fluctuations in Equilibrium and Nonequilibrium Statistical Mechanics, XVIIth International Solvay Conference on Physics, ed. G. Nicolis, Guy Dewel, and John W. Turner (Wiley, New York, 1981), 107.
18. R. Bonifacio, M. Gronchi, and L. A. Lugiato, Phys. Rev. A 18, 2266 (1978).
19. R. Bonifacio, ed., Dissipative Systems in Quantum Optics, Topics in Current Physics, Vol. 27 (Springer, Berlin, Heidelberg, New York, 1982).
20. W. J. Sandle and Alan Gallagher, Phys. Rev. A24, 2017 (1981).
21. K. G. Weyer, H. Widenmann, M. Rateike, W. R. MacGillivray, P. Meystre, and H. Walther, Opt. Commun. 37, 426 (1981).
22. D. E. Grant and H. J. Kimble, Optics Lett. 7, 353 (1982).
23. D. E. Grant and H. J. Kimble, Opt. Commun., to be published (1983).
24. M. Born and E. Wolf, Principles of Optics, (Pergamon Press, Oxford, 1980), Section 7.6.

25. L. Allen and J. H. Eberly, Optical Resonance and Two Level Atoms, (John Wiley and Sons, New York, 1975).
26. Amnon Yariv, Quantum Electronics, (John Wiley and Sons, New York, 1975).
27. J. D. Jackson, Classical Electrodynamics, (John Wiley and Sons, New York, 1962), Chapter 17.
28. H. J. Kimble and L. Mandel, Phys. Rev. A13, 2123 (1976).
29. J. H. Marburger and F. S. Felber, Phys. Rev. A17, 335 (1978).
30. G. P. Agrawal and H. J. Carmichael, Phys. Rev. A19, 2074 (1979).
31. R. J. Ballagh, J. Cooper, and W. J. Sandle, J. Phys. B14, 3881 (1981).
32. J. A. Abate, Opt. Commun.10, 269 (1974).
33. M. L. Citron, H. R. Gray, C. W. Gabel, and C. R. Stroud, Jr., Phys. Rev. A16, 1507 (1977).
34. W. Hartig, W. Rasmussen, R. Schieder, and H. Walther, Z. Physik A278, 205 (1976).
35. R. E. Grove, F. Y. Wu, S. Ezekiel, Phys. Rev. A15, 227 (1977).
36. M. Dagenais and L. Mandel, Phys. Rev. A18, 2217 (1978).
37. Michael Hercher, Appl. Optics 7, 951 (1968).
38. P. D. Drummond, IEEE J. Quantum Electron. QE-17, 301 (1981).
39. R. J. Ballagh, J. Cooper, M. W. Hamilton, W. J. Sandle, and D. M. Warrington, Opt. Commun. 37, 143 (1981).
40. W. J. Firth and E. M. Wright, Opt. Commun. 40, 233 (1982).
41. J. V. Moloney and H. M. Gibbs, Phys. Rev. Lett. 48, 1607 (1982).
42. V. Benza and L. Lugiato, in ref. 11 above, p. 9; R. Bonifacio, M. Gronchi, and L. A. Lugiato in ref. 11 above, p. 31.
43. K. Ikeda, Opt. Commun. 30, 257 (1979); K. Ikeda, H. Daido and O. Akimoto, Phys. Rev. Lett. 45, 709 (1980).
44. K. Ikeda and O. Akimoto, Phys. Rev. Lett. 48, 617 (1982).
45. H. M. Gibbs, F. A. Hopf, D. L. Kaplan, and R. L. Shoemaker, Phys. Rev. Lett. 46, 474 (1981).
46. L. A. Lugiato, Z. Physik B41, 85 (1981).
47. P. D. Drummond, in ref. 11 above, p. 481.
48. S. Stenholm and W. E. Lamb, Jr., Phys. Rev. 181, 618 (1969).
49. B. J. Feldman and M. S. Feld, Phys. Rev. A1, 1375 (1970).
50. J. H. Shirley, Phys. Rev. A8, 347 (1973).
51. V. S. Letokhov and V. P. Chebotayev, Nonlinear Laser Spectroscopy (Springer, Berlin, 1977).
52. H.J. Carmichael and G.P. Agrawal, in ref. 11 above, p.237.

OPTICAL BISTABILITY AND NON ABSORPTION RESONANCE

IN ATOMIC SODIUM

W.R. MacGillivray
School of Science
Griffith University
Queensland 4111 Australia

1.0 Introduction

While the performing of theoretical calculations on models of intrinsic optical bistability has been very popular and highly productive over recent years, the number of experiments in the same field has been surprisingly few. This article is concerned with the experiments carried out in our laboratory in which an atomic sodium vapour inside an optical cavity serves as the system under investigation. Firstly, the apparatus is described followed by the results from what were originally envisaged as preliminary experiments but which became a major study. Sections 4 and 5 deal with attempts to model the system and compute the resultant theory.

2.0 The Experiment

For the detection of bistability, the technique of sweeping the laser frequency and searching for distortions in the optical cavity transmission peaks as employed by Sandle and Gallagher [1] was used. However, the experiment conditions were different from theirs. To generate a psuedo two-level system in the D_1 transition, they employed confocal mirrors of high reflectivity in a Fabry-Perot etalon to obtain a high cavity field for large power broadening. As well, they added Argon as a buffer to make the homogeneous width the same order of magnitude as the inhomogeneous width. We have employed neither of these techniques. Our system has incorporated flat mirrors in a low finesse optical cavity and pure sodium vapour. Thus we have essentially the same conditions as used by Gibbs et al [2] who studied the bistability by cyclically varying the injected field intensity when tuned to near resonance with either of the D lines.

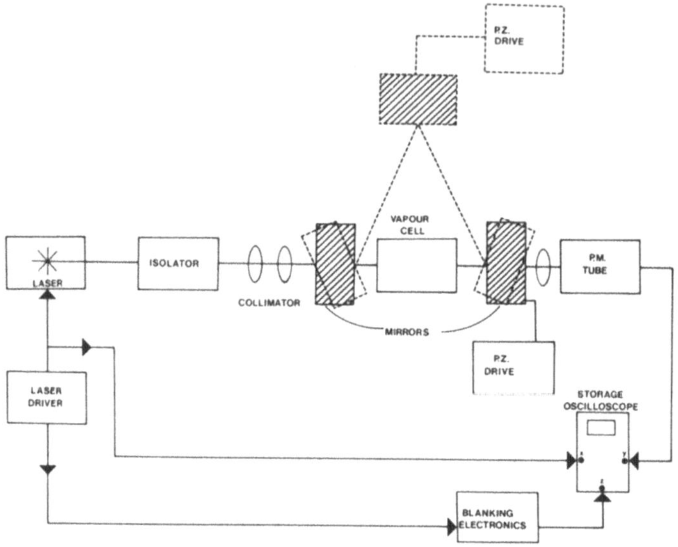

Fig. 1 Experiment Configuration.

The layout of the apparatus is shown in Fig. 1. The Fabry-Perot mirrors were 93% reflecting and flat to λ/100. One mirror was set in a PZT aligner/translator that was driven by a high voltage supply. This allowed both the cavity length to be varied and fine adjustments to the mirror alignment to be made. As well, both mirror assemblies were mounted in star gymbol mounts which allowed coarse adjustments about the horizontal and vertical axes.

The glass vapour cell was made with re-entrant windows. The distance between the windows was 2.5 cm. The re-entrant design enabled a more uniform temperature of the vapour to be obtained in the interaction region. The cell was heated via lagged nicrome wire. Thermo-couples attached to the cell monitored the temperature which, for the data reported here, lay in the range of 140°C to 200°C.

Under operating conditions with the injected light detuned from resonance, the vapour cell/Fabry-Perot system had a free spectral range of 970 MHz and a finesse of 9. A ring cavity was formed by adding a third mirror which was 100% reflecting. A small range cavity scan could be achieved by translating this mirror with a piezo drive. The modifications for the ring cavity are illustrated by the dashed lines in Fig. 1. This ring cavity had a finesse of 9 and a free spectral range of 700 MHz.

The light source was a Spectra-physics 380A ring dye laser pumped all lines by a 164 Ar^+ laser. The jitter bandwidth of this system was less the 10 MHz. The stability of the single-mode output was monitored by a high-finess optical spectrum analyser. 200 mW was the maximum light intensity that could be injected into the optical cavities. The direction of polarisation for the field injected into the ring cavity had to be carefully set perpendicular to the plane of the ring so that more intra-cavity modes were not generated by repolarisation due to reflection at the mirrors. For the Fabry-Perot etalon, either linearly or circularly polarised light was used and the cavity was optically isolated from the laser to prevent retro-reflections entering the laser. For circularly polarised light a combination of a linear polariser and a quarter-wave plate was used as the isolator while for linearly polarised light, a Faraday rotator replaced the quarter-wave plate.

After the isolator, a collimated beam of 1/e width variable between 0.5 and 2.0 mm was formed using a simple telescope arrangement. The beam electric field profile was approximately Gaussian and so the field of view of the detector was restricted to the centre of the transmitted beam so that the data was obtained from an essentially uniform field region. This was accomplished by using a short-focus lens to project a magnified image of the transmitted beam onto a 0.5 mm diameter pinhole in front of the photomultiplier tube. The output of the tube was fed to the Y-amplifier of a storage oxcilloscope.

The symmetric ramp voltage from a sweep generator was used to cyclically change the frequency of the dye laser. The voltage was imposed on the various cavity and etalon translators of the laser via the Spectra-physics 481B Scanner. The same ramp was also fed to the oscilloscope X-amplifier so that a picture of system transmission versus laser frequency could be obtained for the entire sweep. When only a sweep in one direction was required, the Z-input blanking facility of the oscilloscope could be used. The synchronous output of the sweep generator was fed to a gate generator where it was shaped and delayed before being fed to the Z-input.

Data recording consisted of photographing the stored trace.

3.0 Data

Recorded optical bistability in the radiation transmitted through the cavities is shown in Fig. 2.

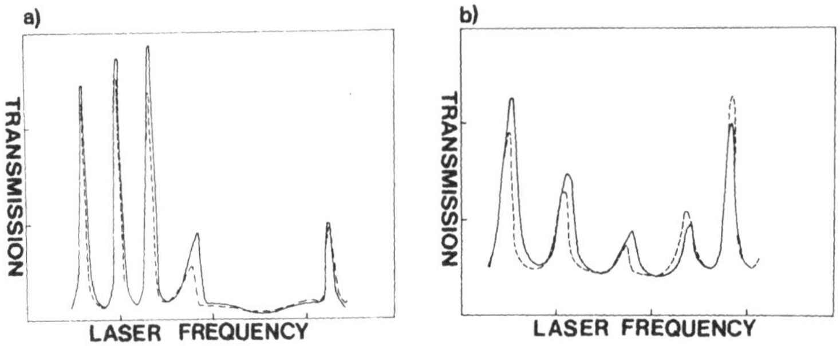

Fig. 2 Bistable transmission profiles for a) the Fabry-Perot etalon and b) the ring optical cavity.

Fig 2 (a) contains data from the Fabry-Perot with the injected radiation frequency swept about the D_2 transition. The solid line is for a sweep to lower frequency while the dashed is to higher frequency. The radiation was linearly polarised and the intra-cavity field intensity is estimated at 50 mW/mm^2 at the centre of the profile. Bistability is indicated both by the distortion of the peaks and the hysteresis in the peak height and position. The frequency at which switching between the transmission branches occurs depends on the direction of the sweep i.e. whether it is towards line centre or away from it. This effect is most clearly seen in the peak near the centre. Instrumental hysteresis has been removed.

Fig 2 (b) illustrates the bistable behaviour of the ring cavity for the D_1 transition. The intensity density at the centre of the beam in the cavity was estimated at 100 mW/mm^2. The peak distortions and the hysteresis are again apparent. The shape of the peaks and the asymmetric nature of the data also indicate that the effect is dispersive. The dependence of the switching threshold on the direction of the sweep with respect to line centre is even more clearly illustrated here. Again, the instrumental hysteresis has been removed. To our knowledge, this is the first observation of bistability in a ring cavity.

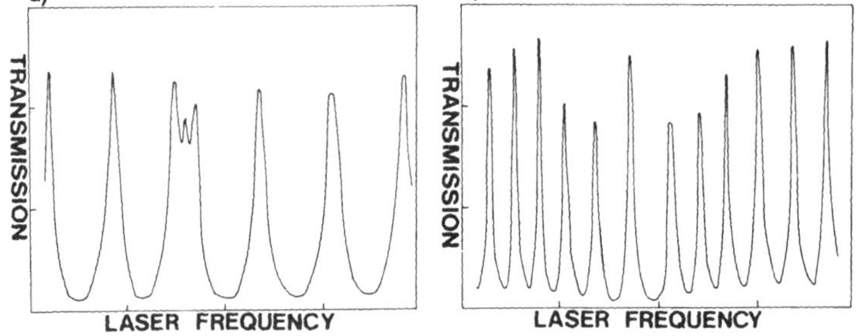

Fig. 3 Transmission of the cavities at lower temperatures a) Fabry-Perot b) ring.

Fig. 3 (a) shows the transmission of the Fabry-Perot at lower temperature. As the intensity of the injected field was increased from a low value, an enhanced absorption dip would appear in the transmission peak situated at a fixed frequency in the Doppler profile. At higher intensities the narrow enhanced transmission feature became apparent. This feature broadened with further increase in intensity. This data is for the D_2 transition and the intensity density of the intracavity beam centre is 100 mW/mm^2. The injected field was linearly polarised.

The transmission of the ring cavity was observed under similar conditions and the results are given in Fig. 3 (b). The intensity density was 85 mW/mm^2. No evidence of enhanced absorption or transmission features was observed although there was saturation at the centre of the absorption profile.

The data presented in this section forms the basis of a manuscript recently submitted [3].

4.0 The Model

Our interpretation of the narrow, enhanced transmission peak in the Fabry-Perot is that it is a nonabsorption resonance phenomenon of the type observed by Alzetta and co-workers [4] and analysed by Oriols [5]. In their case, they had a co-propagating beam consisting of different spatial modes from a free-running laser. Here, we have counter-propagating beams of the same mode. For the typical intracavity fields used in this experiment, the sodium transitions can be modelled as three-level systems of the "lambda" type, that is, one excited level and two ground levels [6]. The two transition frequencies differ by 1772 MHz. The dipole moments of the optical transitions are taken to be equal, i.e.

$$\mu_{13} = \mu_{23} = \mu,$$

and the two ground states are not coupled,

$$\mu_{12} = 0.$$

The accuracy of the three-level model can be illustrated by performing a saturated absorption experiment. A small fraction (~ 5%) of the laser beam is split off and sent through the vapour cell in the opposite direction to the main beam.

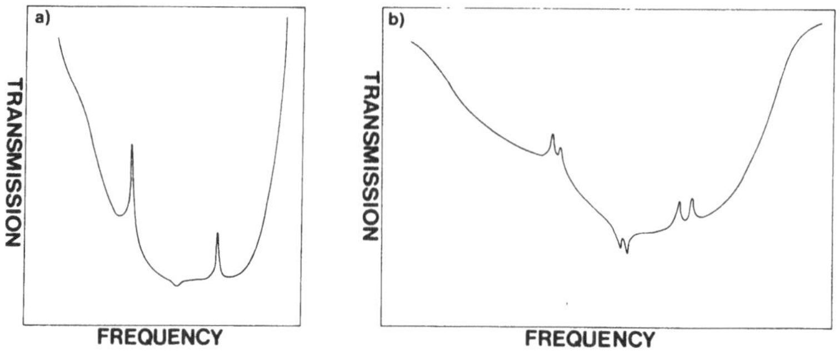

Fig. 4 Absorption of the probe beam a) D_2 b) D_1.

Fig. 4 (a) shows the absorption versus frequency of the probe beam for the D_2 line. When the laser is tuned to either of the transitions, there is an enhanced transmission Lamb dip due to each beam interacting with the zero velocity group. In the centre is an enhanced absorption of the probe beam called a "cross-over resonance" [7]. This results from the velocity group that Doppler shifts the pump beam into resonance with one transition and the probe beam into resonance with the other. The transverse optical pumping effect of the strong beam creates a greater than equilibrium population in the ground state of the transition associated with the probe beam, and so it is more heavily absorbed. Note that there is no evidence of the excited state hyperfine structure.

Fig 4 (b) is of data for the D_1 transition. Notice now that the excited state hyperfine structure is evident at this pump power, so that a three-level model is not as good an approximation for the D_1 transition. However, essentially the same arguments apply.

Using a three-level model, the enhanced transmission feature in the Fabry-Perot data can be explained. Unlike in the saturated absorption experiment, the counter-propagating beams in the cavity are both strong (to a first approximation they are equal). For that velocity group that has one beam in resonance with one transition and the other in resonance with the second transition, the transverse optical pumping at strong fields will create superposition states of the ground states i.e. $\frac{1}{\sqrt{2}}(|1\rangle + |2\rangle)$ and $\frac{1}{\sqrt{2}}(|1\rangle - |2\rangle)$. The excited state has equal probability of relaxing to either state but the second state has a zero excitation amplitude to the excited state. Hence all of the atoms of this velocity group will be trapped in the state $\frac{1}{\sqrt{2}}(|1\rangle - |2\rangle)$ and there will be no absorption of the fields. We claim that this non-absorption resonance is responsible for the narrow enhanced transmission feature in the Fabry-Perot data. No such effect is observed in the ring because there are no counter-propagating beams.

5.0 Theory

a) Fabry-Perot etalon.

The development of even a simple model to describe an ensemble of inhomogeneously broadened three-level atoms in the steady state inside a Fabry-Perot etalon has yet to be attained. The atomic behaviour is described by the density matrix equation

$$i\hbar\dot{\rho} = [H,\rho] + \text{relaxation terms}$$

where ρ is the density matrix which has dimension 3 and the square brackets are commutator brackets.

$$H = H_0 + H_{int}$$

where H_0 is the unperturbed atom Hamiltonian i.e.

$$H_0 = \hbar\begin{pmatrix} \omega_1 & 0 & 0 \\ 0 & \omega_2 & 0 \\ 0 & 0 & \omega_3 \end{pmatrix}$$

and H_{int} represents the interaction between the atom and the classical field

$$H_{int} = \hbar\begin{pmatrix} 0 & 0 & -\alpha^*(t) \\ 0 & 0 & -\alpha^*(t) \\ -\alpha(t) & -\alpha(t) & 0 \end{pmatrix}$$

where $\alpha(t) = \frac{\underset{\sim}{\mu}}{\hbar}\cdot(\underset{\sim}{E}_f(t) + \underset{\sim}{E}_b(t))$;

$\underset{\sim}{\mu}$ being the transition dipole. The subscripts are for forward and backward waves. The time dependence can be extracted as

$$\alpha(t) = \alpha[e^{i(\omega t-kz)} + \text{c.c.} + e^{i(\omega t+kz)} + \text{c.c.}]$$

where $\alpha = \frac{\underset{\sim}{\mu}.\underset{\sim}{E}\hat{e}}{\hbar}$

is the Rabi frequency which measures the strength of interaction between the induced dipole moment of the transitions and the injected field. If linear polarisation is used, the Rabi frequency is a real scalar,

$$\alpha = \frac{\mu E}{\hbar}.$$

The procedure is to solve the equations for the density matrix elements by choosing an appropriate output field. The relevant density matrix elements integrated over the Doppler profile plus the field are then substituted into a Maxwell's field equation to solve for the input field. (There is no analytic solution for output field as a function of input field.) However, for the Fabry-Perot case, field equations are required for both forward and backward travelling waves and the whole situation is complicated by the standing waves due to the interaction of these two fields. So the fields must be solved progressively back through the cavity from the output taking into account the motion of the atoms through the standing waves.

Even to solve a simple model where the mean field approximation is made ($E_f = E_b$) and standing waves are finally neglected requires transforming the density matrix elements to frames of reference where the time dependence is eliminated and an appropriate average over the standing waves can be performed. Our preliminary attempts at finding such a transformation have been unsuccessful.

Some insight into the phenomenon can be gained by solving for the situation where each wave interacts with one transition only; say $\alpha(t)$ is the interaction of the forward travelling field with the $|1\rangle-|3\rangle$ transition and $\beta(t)$ the interaction of the backward travelling wave with the $|2\rangle-|3\rangle$ transition i.e.

$$\alpha(t) = \alpha[e^{i(\omega t-kz)} + c.c.]$$

$$\beta(t) = \alpha[e^{i(\omega t+kz)} + c.c.].$$

Neglecting standing waves, the density matrix component equations in the Rotating Wave Approximation are given by

$$\frac{1}{T_1}\rho_{33} + 2\,\mathrm{Im}\,(\alpha\tilde{\rho}_{13} + \alpha\tilde{\rho}_{23}) = 0$$

$$\frac{1}{2T_1}\rho_{33} - \frac{1}{\tau_1}(\rho_{22} - \rho_{11}) + 2\,\mathrm{Im}\,(\alpha\tilde{\rho}_{23}) = 0$$

$$\frac{1}{2T_1}\rho_{33} - \frac{1}{\tau_1}(\rho_{11} - \rho_{22}) + 2\,\mathrm{Im}\,(\alpha\tilde{\rho}_{13}) = 0$$

$$(\Delta - \frac{i}{T_2})\,\tilde{\rho}_{13} + i\alpha^*(\rho_{11} - \rho_{33}) + i\alpha^*\,\tilde{\rho}_{12} = 0$$

$$(\Delta^1 - \frac{i}{T_2})\,\tilde{\rho}_{23} + i\alpha^*(\rho_{22} - \rho_{33}) + i\alpha^*\,\tilde{\rho}_{21} = 0$$

$$(\omega_{21} + \frac{1}{\tau_2})\,\rho_{12} + i\alpha^*\,\tilde{\rho}_{32} - i\alpha\tilde{\rho}_{13} = 0$$

where $\tilde{\rho}_{13}$ and $\tilde{\rho}_{23}$ are defined as,

$$\rho_{13} = \tilde{\rho}_{13}\, e^{i(\omega t-kz)}$$

$$\rho_{23} = \tilde{\rho}_{23}\, e^{i(\omega t+kz)}.$$

$$\Delta = \omega-\omega_{31}-kv$$

$$\Delta^1 = \omega-\omega_{32}+kv.$$

T_1 describes the relaxations from ρ_{33} to ρ_{11} or ρ_{22} which occur with equal probability. T_2 describes the relaxation of the dipoles associated with the $|1\rangle-|3\rangle$ and $|2\rangle-|3\rangle$ transitions. For dominant spontaneous emission as we have here

$$T_2 = 2T_1 = 32 \text{ nsec.}$$

Similarly, τ_1 and τ_2 relate to relaxation processes between $|1\rangle$ and $|2\rangle$. However, since this is not an allowed transition, a collision mechanism would be dominant and so

$$\tau_1 = \tau_2$$

Furthermore, with the vapour pressures used in this experiment, i.e. 10^{-4} to 10^{-5} torr, τ should be much greater than T_2.

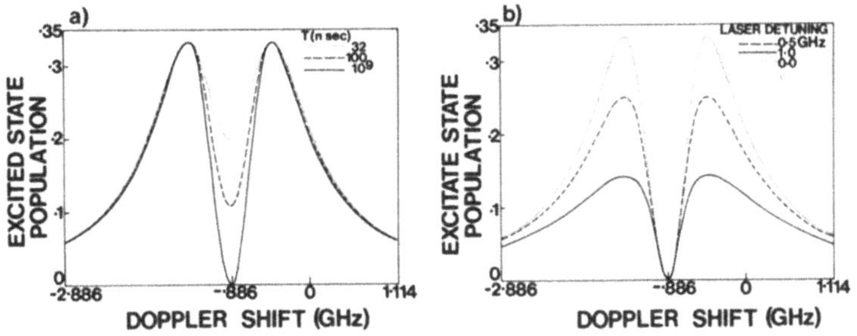

Fig. 5 Excited state population versus Doppler shift for a) fixed laser detuning and various τ and b) fixed τ and various laser detuning.

Fig. 5 (a) is a computer calculated plot of ρ_{33} versus Doppler shift of velocity groups for the injected field tuned to midway between the two transitions. At large τ, the excited state population goes to zero for $kv = -\omega_{21}/2$ indicating complete population trapping. However, as τ diminishes, ρ_{33} becomes finite due to leakage between the ground states. The Rabi frequency for these plots is 500 MHz. This value is an order of magnitude estimate for the intra-cavity interaction under our experimental conditions.

Fig. 5 (b) illustrates the efficiency of the process as a function of laser detuning defined with respect to the frequency mid-way between the two transitions. The population still goes to zero at $kv = -886$ MHz due to energy conservation but the overall effect diminishes.

b) Ring Cavity

The ring optical cavity is modelled as one travelling wave interacting with both transitions, i.e.

$$\alpha(t) = \alpha[e^{i(\omega t - kz)} + c.c.]$$

for both the $|1\rangle - |3\rangle$ and $|2\rangle - |3\rangle$ transitions. The density matrix component equations are the same as given earlier for the Fabry-Perot, except for the definitions of $\tilde{\rho}_{23}$ and Δ^1; and of course there are no standing waves to be neglected. We have

$$\rho_{23} = \tilde{\rho}_{23}\, e^{i(\omega t - kz)}$$

$$\Delta^1 = \omega - \omega_{32} - kv.$$

The equations are solved for a given α.

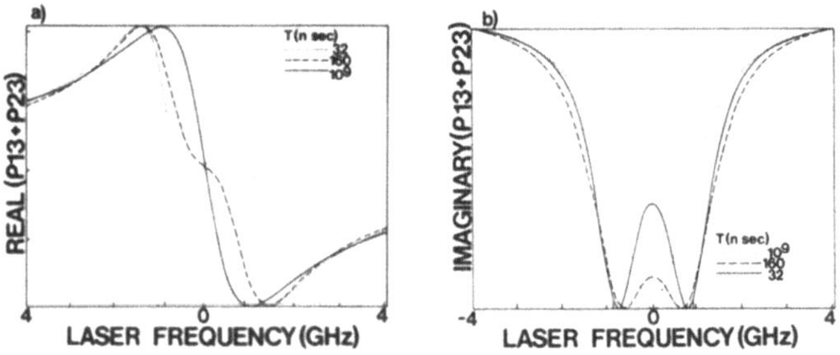

Fig. 6 a) Dispersion and b) absorption versus laser frequency for various values of τ. The longer marks on the frequency axis indicate the positions of the transitions.

Fig. 6 show the dispersion and absorption for strong, weak and very weak coupling between the ground states. Note that at very weak coupling, the system behaves as a broad two-level scheme. The Rabi frequency of the interaction is 400 MHz.

The Rabi frequency associated with the input field, Ω_I, is calculated by solving the steady-state field equation,

$$\Omega_I - \alpha + \frac{i2C}{T_2}\left(\tilde{\rho}_{31_D} + \tilde{\rho}_{32_D}\right) - \frac{R}{T}\alpha\left(1 - e^{i\phi}\right) = 0$$

where C is the so called bistability coefficient [8], R is the reflectivity and T is the transmittivity of the mirrors and ϕ is the cavity detuning parameter given by

$$\phi = \frac{\omega - \omega_{cav}}{F.S.R.}; \quad (F.S.R. = \text{free spectral range})$$

where ω_{cav} defines the frequency of maximum transmission through the cavity. The density matrix elements have been integrated over the Doppler profile, i.e.

$$\tilde{\rho}_{31_D} = \int_{-\infty}^{\infty} \frac{\tilde{\rho}_{31}(v)\, \exp\left(-\frac{v^2}{V_0^2}\right)}{V_0\sqrt{\pi}}\, dv$$

where $v_0 = \frac{2kT}{m}$

is the most probable velocity when the atoms are in thermal equilibrium.

C is estimated from the expression

$$C = \frac{\mu^2 \omega T_2 \rho}{4 \hbar \varepsilon_0 \kappa}$$

where ρ is the atomic density and κ is the inverse of the cavity lifetime. The estimation is complicated by the fact that we do not have a strictly three-level system and that the induced dipole moment between hyperfine levels and sub-levels varies from transition to transition. For example, for linearly polarized light interacting with the D_2 transition the dipole moment is calculated [9] to be 1.65×10^{-29} cm for the $F=2 \rightarrow F^1=3$, M=0 scheme but only 0.61×10^{-29} cm for the $F=2 \rightarrow F^1=2$, M=1 scheme. Thus, at 150°C, C is estimated to lie in the range from 500 to 3800.

The density matrix equations and field equation are solved numerically for values of input field as a function of output field for a given laser detuning. If bistability exists, a characteristic S-curve will be obtained. Values of output field for a given input field are then found by a numerical routine that finds the intersection of the output-input function and the line representing the input field desired. This is repeated for a selected range of injected field frequencies and so a picture of cavity transmitted intensity versus laser frequency can be obtained for comparison with the experimental data.

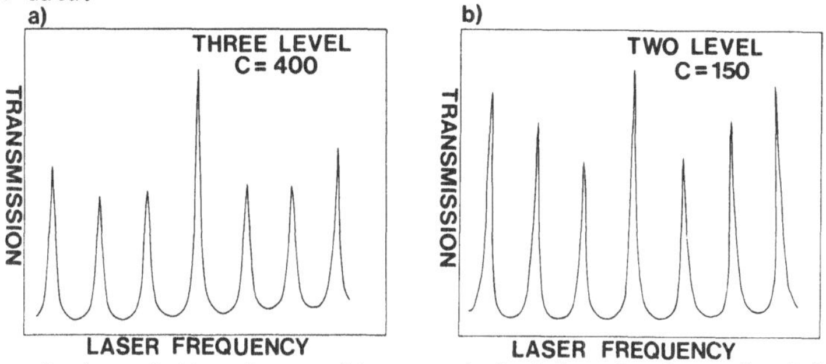

Fig. 7 Transmission of the ring cavity computed at low C values for Rabi frequency of 100 MHz a) three-level, C = 400 and b) two-level, C = 150.

The computed transmission intensity versus laser frequency for the ring cavity for an injected field Rabi frequency of 100 MHz and C of 400 is shown in Fig. 7 (a). Comparison with the experimental data in Fig. 3 (b) shows some general agreement. However, the overall absorption profile appears wider in the theoretical calculation than in that observed and, unlike in the experimental data, the first peak from the centre is larger than the second.

Variation of the parameters Rabi frequency, C, τ and cavity detuning do not enhance agreement. The problem appears to be that by using a sufficiently high value of C to obtain the necessary absorption leads to a profile that is too broad.

The transmission of a ring cavity containing an ensemble of two-level systems was also calculated. The element ρ_{21} was calculated from the steady-state equations, viz.

$$\frac{\rho_{22}}{T_1} + 2\ \mathrm{Im}\ (\alpha\tilde{\rho}_{12}) = 0$$

$$\left(\Delta-\frac{i}{T_2}\right)\tilde{\rho}_{12} - \alpha(\rho_{22}-\rho_{11}) = 0$$

$$\rho_{11} + \rho_{22} = 1$$

$$\tilde{\rho}_{12} = \tilde{\rho}_{21}{}^{*}$$

where

$$\Delta = \omega - kv - \omega_{21}.$$

Fig. 7 (b) shows the results of such a calculation for a Rabi frequency of 100 MHz and C of 150. There appears to be closer resemblance to the experimental results than the three-level model. This is probably due to the fact that more absorption is attainable at lower C values than for the three-level system so that the absorption profile is narrower. Intuitively, these sodium transitions should not be behaving as two-level systems under these conditions. A Rabi frequency of 100 MHz would not induce power broadening of the magnitude necessary to mix the two ground states. And there is no large homogeneous broadening. In fact, collisions should be negligible at our pressures (mean free path ≃ 5 cm). And lastly, there is the non-absorption resonance feature in the Fabry-Perot data which requires at least a three-level system.

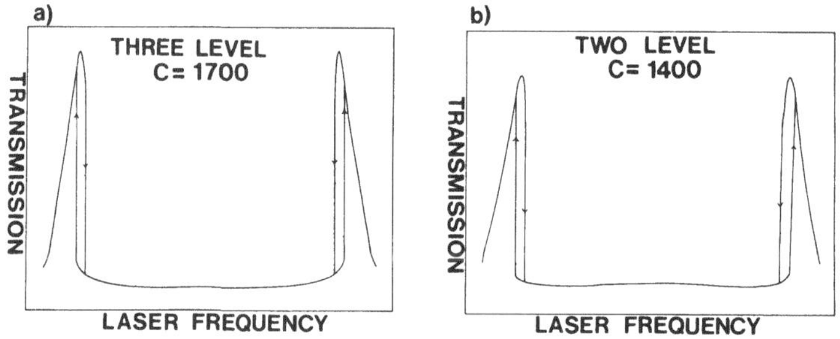

Fig. 8 Computed transmission of the ring cavity showing bistability a) three-level, C = 1700, Rabi frequency = 1 GHz b) two-level, C = 1400, Rabi frequency = 1.1 GHz.

The three-level model will produce bistable effects. Fig. 8 a) shows the transmitted intensity of the ring with an injected field of 1 GHz Rabi frequency, C = 1700 and the cavity detuned 443 MHz (one half a free spectral range) from the frequency mid-way between the transitions. This plot should be compared with the experimental data of Fig. 2 b). There is general agreement with the form of the hysteresis and switching. What is not in evidence is the absorption. The switching is always to almost 100% transmission in the calculations.

Also, in disagreement with the experimental results is the difference in power level required to obtain the effects of Fig. 7 a) and Fig. 8 a). The 100 MHz Rabi frequency used earlier is below threshold for bistability in this model. And 1 GHz Rabi frequency plots at the lower C number shows no absorption. Yet the experimental data shown in Fig. 2 (b) and Fig 3 (b) were recorded at almost the same injected field intensity. Only the temperature was varied.

Fig. 8 b) is the computer plot of transmission for the two-level model. C is slightly lower and the Rabi frequency higher than for the three-level model but the plot is almost the same.

There are several other avenues that have yet to be followed in any detail but which have the potential for introducing the necessary absorption effects. Firstly, there is the removal of the mean field approximation and including in the model the absorption in the cell. This implies allowing for a decreasing field as the beam progresses through the cell. Second, is the introduction of unequal dipole moments for the two transitions.

Added absorption can also be introduced by decreasing T_1 and T_2 but this is not considered realistic for two reasons. One, spontaneous emission should dominate collisions as the relaxation process under our conditions of operation. Hence there should be no increase in the homogeneous line width. Two, T_1 could be affected by allowing for atoms entering the laser beam in the ground state. However, since we only observe the centre of the transmitted beam, all atoms contributing to the data should have reached a steady state.

Acknowledgements

The work reported here has been very much a team effort. I would particularly like to acknowledge the contributions of Margaret Marshman and Max Standage. Most of all, I would like to express appreciation to Werner Schulz who has undertaken most of the grind involved in data collection and programme writing. The work has been funded by the Australian Research Grants Scheme.

References

1) W.J. Sandle and A. Gallagher, Phys. Rev. A24 (1981) 2017.
2) H.M. Gibbs, S.L. McCall and T.N.C. Venkatesan, Phys. Rev. Lett. 36 (1976) 1135.
3) W.E. Schulz, W.R. MacGillivray and M.C. Standage, submitted Optics Commun.
4) G. Alzetta, L. Moi and G. Orriols, Il Nuovo Cimento 52 (1979) 209.
5) G. Orriols, Il Nuovo Cimento 53 (1979) 1.
6) J.E. Bjorkholm and A. Ashkin, Phys. Rev. Lett. 32 (1974) 629.
7) T.W. Hänsch, I.S. Shakin and A.L. Schawlow, Phys. Rev. Lett. 27 (1971) 707.
8) D.F. Walls and P. Zoller, Optics Commun. 34 (1980) 260.
9) W.R. MacGillivray and M.C. Standage, Optics Commun. 36 (1981) 189.

POLARIZATION SWITCHING AND OPTICAL BISTABILITY WITH RESONANTLY DRIVEN J=½ TO J=½ ATOMS IN A RING CAVITY

W.J.SANDLE and M.W.HAMILTON

Physics Department, University of Otago, P.O. Box 56, Dunedin, New Zealand

Introduction.

We have three main objectives in this seminar:

- to introduce in a tutorial vein the use of the standard irreducible tensor approach for the calculation of atomic polarization;
- to illustrate the method by application to a specific problem of interest: the steady-state behaviour of J=½ to J=½ atoms in a ring cavity with a weak magnetic field applied transverse to the propagation direction;
- to present a selection of our experimental results on polarization switching[1] of cavity output field with linearly polarized excitation.

These experimental results are the first which show the complete polarization switching sequence - from linearly polarized output at low input power to elliptically polarized output at intermediate input power and back to linear output at high laser power - and which report purely absorptive polarization switching[2].

Standard irreducible tensor methods.

These methods[3], although now widely applied in the atomic physics community, are only beginning to be appreciated in quantum optics. They enable a rigorous and complete solution to atomic behaviour while making maximum use of symmetries present[3].

The density operator ρ is expanded in a basis of operators $T_q^{(k)}$ (which we can more usefully describe as basis vectors in product (Liouville) space) which have the same transformation properties under rotations as a simple ket $|\alpha FM\rangle$ (in Hilbert space) with angular momentum quantum number $F(=k)$ and component $M(=q)$. (The label α describes remaining quantum numbers needed to fully specify the state).

Thus, restricting discussion to a two level case with degeneracies $2J_u+1$ (upper) and $2J_\ell+1$ (lower) we expand

$$\rho = \sum_{\substack{\alpha,\beta=u,\ell \\ k,q}} \rho_q^k(\alpha\beta)\; T_q^{(k)}(\alpha\beta) \tag{1}$$

in terms of standard irreducible components $\rho^k_q(uu)$, i.e. upper state population (k=0), orientation (k=1) alignment (k=2) etc; $\rho^k_q(\ell\ell)$, lower-state population, orientation, alignment; and the off-diagonal terms $\rho^k_q(u\ell)$, $\rho^k_q(\ell u)$ which are sometimes called the "atomic" or "optical" coherences to distinguish them from the "Zeeman" coherences (off-diagonal terms within the same level). The exact form of the $\rho^k_q(\alpha\beta)$ depends of course on the phase choice in the basis $T^{(k)}_q$; it is usual and convenient to make the standard choice from angular momentum coupling theory:

$$T^{(k)}_q(\alpha\beta) = \sum_M (-)^{G-N} \langle FGM{-}N|kq\rangle |\alpha FM\rangle\langle\beta GN| \tag{2}$$

where $\langle FGM{-}N|kq\rangle$ is the usual vector coupling (Clebsch-Gordan) coefficient (Messiah's notation [4]).

Atomic evolution for $J_\ell=\frac{1}{2}$ to $J_u=\frac{1}{2}$ in a transverse field

The equations for the evolution of the density operator components ρ^k_q under an applied electric field

$$\underset{\sim}{E}(\underset{\sim}{r},t) = E_0(\underset{\sim}{r})\{\hat{\underset{\sim}{\varepsilon}}\exp(-i\omega t) + \hat{\underset{\sim}{\varepsilon}}^*\exp(i\omega t)\} \tag{3}$$

of frequency ω with amplitude $2E_0(r)$ and polarization $\hat{\underset{\sim}{\varepsilon}}$ are given by Ducloy[5]. They are lengthy and will not be repeated here in their general form. For the special case of $J_\ell=\frac{1}{2}$ connected to $J_u=\frac{1}{2}$ by exactly resonant radiation propagating along the quantization axis (so only σ_+, σ_- transitions occur), neglect of all atom velocity effects (homogeneously broadened transition) and with the assumptions that inelastic collisional rates are zero, that the lower level is the ground level and that there are no incoherent pumping processes, we find the following equations

$$\dot{\rho}^0_0(\ell) = \gamma\rho^0_0(u) - iv\{\varepsilon_1\tilde{\rho}^1_1(\ell u)+\varepsilon_{-1}\tilde{\rho}^1_{-1}(\ell u)\} + iv^*\{\varepsilon^*_1\tilde{\rho}^{1*}_1(\ell u)+\varepsilon^*_{-1}\tilde{\rho}^{1*}_{-1}(\ell u)\} \tag{4a}$$

$$\dot{\rho}^0_0(u) = -\dot{\rho}^0_0(\ell) \tag{4b}$$

$$\begin{aligned}\dot{\rho}^1_0(\ell) = {} & -\Gamma_1(\ell)\rho^1_0(\ell) - \tfrac{1}{3}\gamma\rho^1_0(u) - \tfrac{i\omega_L}{\sqrt{2}}\{\rho^1_1(\ell)+\rho^1_{-1}(\ell)\} \\ & -iv\{\varepsilon_1\tilde{\rho}^1_1(\ell u)-\varepsilon_{-1}\tilde{\rho}^1_{-1}(\ell u)\} + iv^*\{\varepsilon^*_1\tilde{\rho}^{1*}_1(\ell u)-\varepsilon^*_{-1}\tilde{\rho}^{1*}_{-1}(\ell u)\}\end{aligned} \tag{4c}$$

$$\begin{aligned}\dot{\rho}^1_1(\ell) = {} & -\Gamma_1(\ell)\rho^1_1(\ell) - \tfrac{1}{3}\gamma\rho^1_1(u) - \tfrac{i\omega_L}{\sqrt{2}}\rho^1_0(\ell) \\ & +iv\varepsilon_{-1}\{\tilde{\rho}^0_0(\ell u)+\tilde{\rho}^1_0(\ell u)\} + iv^*\varepsilon^*_1\{\tilde{\rho}^{0*}_0(\ell u)-\tilde{\rho}^{1*}_0(\ell u)\}\end{aligned} \tag{4d}$$

$$\dot{\rho}^1_{-1}(\ell) = -\Gamma_1(\ell)\,\rho^1_{-1}(\ell) - \tfrac{1}{3}\gamma\,\rho^1_{-1}(u) - \frac{i\omega_L}{\sqrt{2}}\,\rho^1_0(\ell)$$
$$+iv\,\varepsilon_1\{\tilde{\rho}^0_0(\ell u) - \tilde{\rho}^1_0(\ell u)\} + iv^*\varepsilon^*_{-1}\{\tilde{\rho}^{0\,*}_0(\ell u) + \tilde{\rho}^{1\,*}_0(\ell u)\} \qquad (4e)$$

$$\dot{\rho}^1_0(u) = -\Gamma_1(u)\,\rho^1_0(u) - iv\{\varepsilon_1\tilde{\rho}^1_1(\ell u) - \varepsilon_{-1}\tilde{\rho}^1_{-1}(\ell u)\} + iv^*\{\varepsilon^*_1\tilde{\rho}^{1*}_1(\ell u) - \varepsilon^*_{-1}\tilde{\rho}^{1*}_{-1}(\ell u)\} \qquad (4f)$$

$$\dot{\rho}^1_1(u) = -\Gamma_1(u)\,\rho^1_1(u) - iv\,\varepsilon_{-1}\{\tilde{\rho}^0_0(\ell u) - \tilde{\rho}^1_0(\ell u)\} - iv^*\varepsilon^*_1\{\tilde{\rho}^{0\,*}_0(\ell u) + \tilde{\rho}^{1\,*}_0(\ell u)\} \qquad (4g)$$

$$\dot{\rho}^1_{-1}(u) = -\Gamma_1(u)\,\rho^1_{-1}(u) - iv\,\varepsilon_1\{\tilde{\rho}^0_0(\ell u) + \tilde{\rho}^1_0(\ell u)\} - iv^*\varepsilon^*_{-1}\{\tilde{\rho}^{0*}_0(\ell u) - \tilde{\rho}^{1*}_0(\ell u)\} \qquad (4h)$$

$$\dot{\tilde{\rho}}^0_0(\ell u) = -\Gamma_0(\ell u)\,\tilde{\rho}^0_0(\ell u) - iv^*\left(\varepsilon^*_1\{\rho^1_{-1}(u) - \rho^1_{-1}(\ell)\} + \varepsilon^*_{-1}\{\rho^1_1(u) - \rho^1_1(\ell)\}\right) \qquad (4i)$$

$$\dot{\tilde{\rho}}^1_0(\ell u) = -\Gamma_1(\ell u)\,\tilde{\rho}^1_0(\ell u) - iv^*\left(\varepsilon^*_1\{\rho^1_{-1}(u) + \rho^1_{-1}(\ell)\} - \varepsilon^*_{-1}\{\rho^1_1(u) + \rho^1_1(\ell)\}\right) \qquad (4j)$$

$$\dot{\tilde{\rho}}^1_1(\ell u) = -\Gamma_1(\ell u)\,\tilde{\rho}^1_1(\ell u) + iv^*\varepsilon^*_1\{\rho^0_0(u) - \rho^0_0(\ell) - \rho^1_0(u) - \rho^1_0(\ell)\} \qquad (4k)$$

$$\dot{\tilde{\rho}}^1_{-1}(\ell u) = -\Gamma_1(\ell u)\,\tilde{\rho}^1_{-1}(\ell u) + iv^*\varepsilon^*_{-1}\{\rho^0_0(u) - \rho^0_0(\ell) + \rho^1_0(u) + \rho^1_0(\ell)\} \qquad (4\ell)$$

Here, we have expressed the electric field polarization in standard components,

$$\hat{\underset{\sim}{\varepsilon}} = \varepsilon_1\hat{\underset{\sim}{e}}^*_1 + \varepsilon_{-1}\hat{\underset{\sim}{e}}^*_{-1} \qquad (5)$$

where

$$\hat{\underset{\sim}{e}}_{\pm 1} = \mp\frac{1}{\sqrt{2}}(\hat{\underset{\sim}{x}} \pm i\hat{\underset{\sim}{y}}) \quad , \qquad (6)$$

we have written $\rho^k_q(\ell\ell)$ as $\rho^k_q(\ell)$ (similarly for u), we have abbreviated $d_{\ell u}E_o/\hbar\sqrt{6}$ (where $d_{\ell u}$ is the reduced dipole matrix element for the transition) as v, we have employed a rotating frame at frequency ω to supress rapid time dependence of the atomic coherences, and we have expressed as $\Gamma_k(\ell)$, $\Gamma_k(u)$, $\Gamma_k(\ell u)$ the relaxation rates of $\rho^k_q(\ell)$, $\rho^k_q(u)$ and $\rho^k_q(\ell u)$ respectively. (Note that one of the fundamental reasons for using an irreducible tensor basis for our treatment is that radiative decay and collisions do not give rise to mixing of irreducible components with different k's and q's. If we had used a Zeeman representation, $\rho_{mm'}(\ell)$, $\rho_{mm'}(u)$, $\rho_{mm'}(\ell u)$, such mixing would have to have been taken into account). We shall be concerned with the case where $\Gamma_k(u)$ and $\Gamma_k(\ell u)$, which include the effect of radiative decay and collisional destruction of upper-state-Zeeman and atomic coherences, are much larger than $\Gamma_k(\ell)$. Finally, we have included terms $-(i\omega_L/\sqrt{2})\,\{\rho^1_1(\ell)+\rho^1_{-1}(\ell)\}$ and $-(i\omega_L/\sqrt{2})\rho^1_0(\ell)$ to account for the evolution of $\rho^1_0(\ell)$ and $\rho^1_{\pm 1}(\ell)$ due to the transverse magnetic field B_x (Fig.1) ($\omega_L = -g_{J_\ell}\,\mu_B B_x/\hbar$ = Larmor frequency). [These terms follow from evaluating within the lower (ℓ) manifold $\frac{1}{i\hbar}[\mathcal{H}_B,\rho] = -i\omega_L[J_x,\rho] = -i\frac{\omega_L}{2}[J_+ + J_-,\rho]$, employing $[J_\pm, T^{(k)}_q] = \sqrt{k(k+1)-q(q\pm 1)}\;T^{(k)}_{q\pm 1}$ (ref.4) and using the orthogonality of the $T^{(k)}_q$ (ref.1). We note that Larmor precession terms are retained only for the lower state; it is supposed that $\omega_L \sim \Gamma_1(\ell)$, but that $\omega_L \ll \Gamma_1(u)$ or $\Gamma_1(\ell u)$.]

The macroscopic polarization.

For density N of $J_\ell = \frac{1}{2}$, $J_u = \frac{1}{2}$ atoms in volume V, the polarization is ($\sigma+$, $\sigma-$ transitions only)

$$\underset{\sim}{P} = \frac{N}{V}\frac{d_{u\ell}}{3}\left[\,\underset{\sim}{\hat{e}}_1(\rho^1_1(\ell u) + \rho^1_{-1}(\ell u)^*) + \underset{\sim}{\hat{e}}_{-1}(\rho^1_{-1}(\ell u) + \rho^1_1(\ell u)^*)\,\right] . \tag{7}$$

Upon solving equations (4a) - (4ℓ) for $\rho^1_1(\ell u)$ and $\rho^1_{-1}(\ell u)$ in the steady state (in the rotating frame) for the case of elliptic polarization of high eccentricity with major axis along B_x, we find for the susceptibility

$$\chi_\pm = \frac{i\alpha c}{\omega}\,\eta_\pm \quad , \tag{8}$$

involving the weak field absorption coefficient

$$\alpha = \frac{N\pi c^2}{\omega_a^{\,2} V}\;\frac{\gamma}{\Gamma_1(\ell u)} \tag{9}$$

and the function $\eta_\pm$ given by

$$\eta_\pm = \frac{n_\pm}{D} \tag{10a}$$

where

$$n_\pm = [1+2\beta_L(x_+^{\,2}+x_-^{\,2})][1+4\beta_L x^2_\mp] + \frac{\omega_L^{\,2}}{\Gamma_1(\ell)^2}[1+2\beta_H(x_+^{\,2}+x_-^{\,2})][1+4\beta_H x^2_\mp] \tag{10b}$$

$$D = [1+2\beta_L(x_+^{\,2}+x_-^{\,2})][1+(4+2\beta_L)(x_+^{\,2}+x_-^{\,2})+32\beta_L x_+^{\,2}x_-^{\,2}] + \frac{\omega_L^{\,2}}{\Gamma_1(\ell)^2}[1+2\beta_H(x_+^{\,2}+x_-^{\,2})]$$
$$\times\,[1+(4+2\beta_H)(x_+^{\,2}+x_-^{\,2})+32\beta_H x_+^{\,2}x_-^{\,2}] \tag{10c}$$

$$\beta_L = \frac{\gamma}{\Gamma_1(\ell)\Gamma_1(u)}\left[\Gamma_1(\ell)+\Gamma_1(u)-\frac{\gamma}{3}\right] \simeq \frac{\gamma}{\Gamma_1(\ell)} \quad \text{for } \Gamma_1(\ell) \ll \gamma \tag{11a}$$

$$\beta_H = \frac{\gamma}{\Gamma_1(u)} \tag{11b}$$

and

$$x_\pm = |E_0\varepsilon_\mp|/E\,\text{sat} \tag{12a}$$

where $$E\,\text{sat} = \hbar[6\gamma\Gamma_1(\ell u)]^{\frac{1}{2}}/d_{u\ell} \tag{12b}$$

Special Cases.

This general expression for the susceptibility is complicated, but in two cases of interest we obtain the following simple form for $\eta_{\pm}$;

$$\eta_{\pm} = \{1+8x_{\pm}^{2} + \frac{(2\beta-4)(x_{\pm}^{2}-x_{\mp}^{2})}{1+4\beta x_{\mp}^{2}}\}^{-1} \quad . \tag{13}$$

These cases are:

(i) where we have low transverse magnetic field, $\omega_L \ll \Gamma_1(\ell)$, when β (eq.(13)) is to be identified with β_L (eq.(11a)), and,

(ii) for "high" transverse field $\omega_L \gg \Gamma_1(\ell)$ (but $\omega_L \ll$, $\Gamma_1(u)$, $\Gamma_1(\ell u)$) when the same equation for η (eq.(13)) applies but where now β is to be identified with β_H (eq.(11b)). (Note: This requires $x_+ \sim x_-$ with polarization major axis along B_x).

We observe that these two values of β (β_L and β_H) are widely dissimilar. In essence, the application of the transverse magnetic field has the effect of modifying β. When the transverse field is small, β is essentially (eq.(11a)) determined by the ratio of natural decay rate to collisional transfer rate in the lower state. When on the other hand the transverse field is large enough that $\omega_L \gg \Gamma_1(\ell)$, the lower-state collisional rate becomes immaterial and β is determined by collisional and radiative relaxation in the upper state.

The importance of the conclusions reached above will become clear following the next section when we see that the predicted switching behaviour in an optical bistability/polarization switching experiment is expected to be strongly dependent on β, and thus should be strongly modified by the presence of a magnetic field.

Polarization switching state equations.

Given the susceptibilities χ_+ and χ_- (eq.(8)), state equations for the arrangements as in fig.1 are obtained by the usual method upon invoking the mean-field approximation (see, e.g. ref.(6)). We suppose that one or other of the limits $\omega_L \ll \Gamma_1(\ell)$ or $\omega_L \gg \Gamma_1(\ell)$ applies, and we remember that the case of large ellipticity of the polarization (cavity field) with major axis along B_x has been assumed. Then for the cavity tuned to resonance

$$y_{\pm} = x_{\pm}\left[1 + \frac{2C}{1+8x_{\pm}^{2} + \frac{(2\beta-4)(x_{\pm}^{2}-x_{\mp}^{2})}{1+4\beta x_{\mp}^{2}}}\right] \tag{14}$$

where the scaled input field amplitudes are

$$y_{\pm} = E^{in}\,\varepsilon^{in}_{\mp 1} / [E\,\mathrm{sat}\,(1-R)^{\frac{1}{2}}] \qquad (15)$$

and cooperativity

$$C = \frac{\alpha LR}{4(1-R)} \quad ; \qquad (16)$$

x_+ and x_- are scaled output field amplitudes defined in eq.(12). The state equations (14) are identical to those obtained in ref.(6) (case of exact resonance and zero transverse field) where no constraint was placed upon the polarization ellipticity.[7]

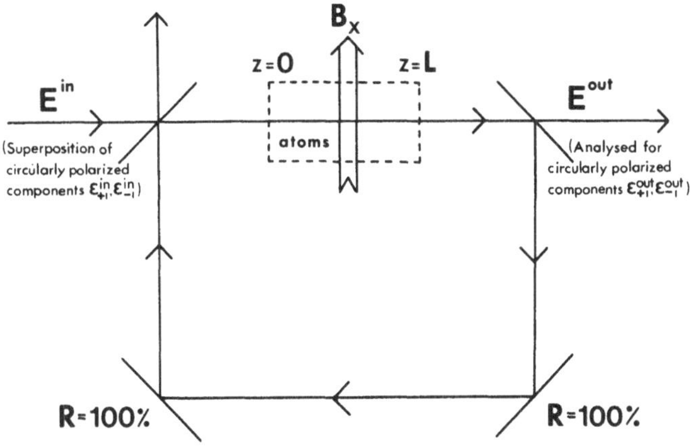

Fig.1: Ring cavity configuration that is analysed theoretically.

Thus for $\omega_L = 0$ eqs.(14) are exact (within the mean field limit) for all input fields y_+, y_-, and we shall first explore the form of the solutions to eqs.(14) as β is changed, imagining that control of β is via the values of $\Gamma_1(\ell)$, $\Gamma_1(u)$, γ inserted in eq.(11a) and that the magnetic field is being kept at zero.

Dependence on β for $y_+ = y_-$.

Figs.2(a) and (b) show steady state solutions to eq.(14) (x_+ versus y_+ and x_+ versus x_-) for $C = 10$, $\beta = 8.8$. Unstable solutions are shown as dotted lines. We note the existence of symmetric (i.e. linearly polarized with $x_+ = x_-$) and asymmetric (elliptically polarized $x_+ \neq x_-$) solutions. The symmetric solutions obey (from eq. (14))

$$y_+ = x_+ \left[1 + \frac{2C}{1+8x_+^{\,2}} \right] , \tag{17}$$

which is the form of the two-state optical bistability equation as expected[8].

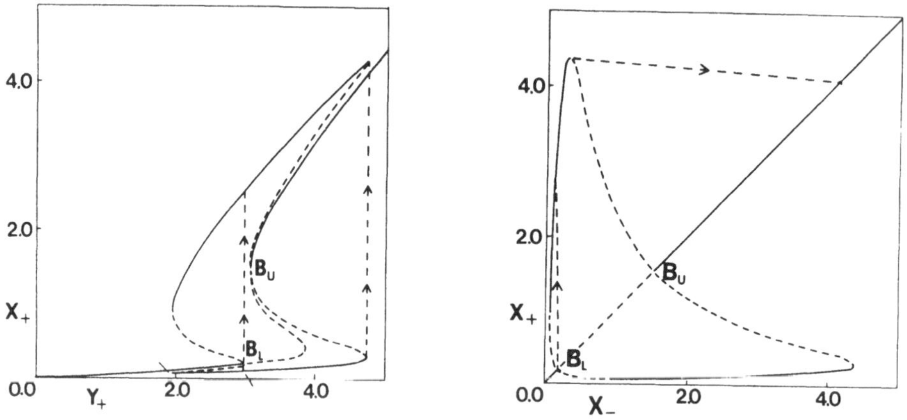

(a) *(b)*

Fig. 2: Predicted behaviour for β>2. The arrows indicate one of the two possibilities for polarization switching (where x_+ *switches first).*

However, as y_+ ($=y_-$) is increased from zero, the symmetric solution is followed only to the lower bifurcation point B_L, beyond which asymmetric solutions appear. The abrupt change in the cavity output polarization (for example as described by the arrows in fig. (2)) is what we term polarization switching. At sufficiently high input field, saturation of the transition occurs and the output field reverts to linearly polarized. Clearly on reducing the input field there are regions of hysteresis with (e.g.) switching back to the asymmetric state at the upper bifurcation point B_u.

An analytic examination[6] of eqs. (14) shows that polarization switching is expected for $\beta>2$; the switching will be abrupt (with hysteresis) if $C>\frac{3}{2}+\frac{5}{\beta}$ and will evolve smoothly (no hysteresis) if $1+\frac{6}{\beta}<C<\frac{3}{2}+\frac{5}{\beta}$. If $\beta<2$, a situation such as shown in fig.3 will occur[6] where the asymmetric solutions are contained within the unstable part of the symmetric branch; normal optical bistability with no polarization switching of the output field should be observed as the input field amplitude is varied. Note: $C=1+\frac{6}{\beta}$ gives minimum power for polarization switching, but switching can be observed for $C>\frac{1}{2}+\frac{3}{\beta}+\left(\frac{4}{\beta}+\frac{8}{\beta^2}\right)^{\frac{1}{2}}$.

Importance of a transverse field.

Now the importance of B_x becomes clear. If B_x is sufficiently large ($\omega_L \gg \Gamma_1(\ell)$) then the value of β to be used in eq. (14) is necessarily less than two, since

$\beta = \beta_H = \gamma/\Gamma_1(u) < 1$. Thus we expect normal optical bistability but not polarization switching to occur; the approximation of large ellipticity leading to eqs.(10) is justified, and eq.(14) with $\beta = \beta_H$ can be applied.

Alternatively, for $B_x \simeq 0 (\omega_L << \Gamma_1(\ell)$ the value of β to be used is β_L, which may be greater than two provided $\Gamma_1(\ell)$ is sufficiently small. This is the case in the experiments upon which we shall now report.

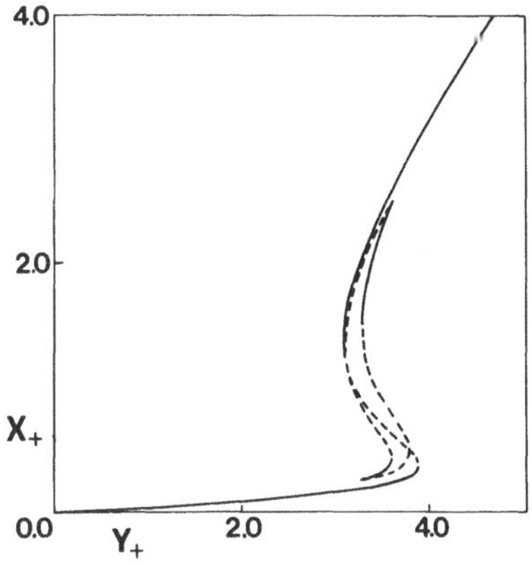

Fig.3: Predicted behaviour for $\beta = 1.33$, $C = 10$.

Experiment.

Fig.4 gives a schematic of the experimental arrangement. Up to 500 mW of linearly polarized single mode stabilised laser radiation is mode matched into a near-concentric Fabry-Perot cavity (250 MHz FSR, R = 98%) containing an L = 10 cm Na cell with 50 torr Ar buffer gas. The pyrex cell windows are set perpendicular to the beam and single layer antireflection coated on the outer surface. During the course of the experiments Na attack on the windows slowly reduced the overall finesse; the value appropriate to the experiments reported is 11 (for zero Na density).

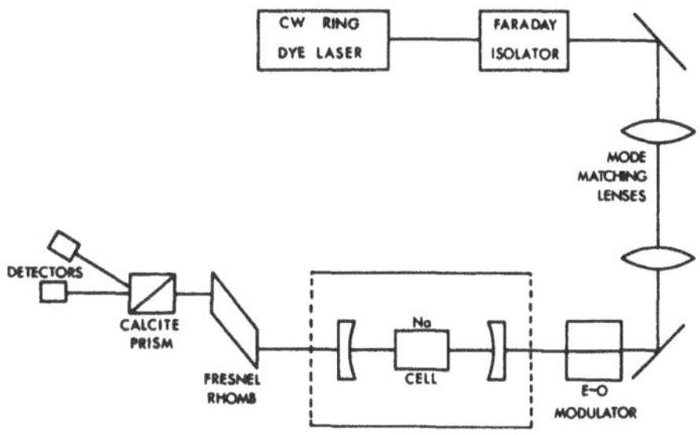

Fig.4: Experimental arrangement. The cavity is a linear Fabry-Perot mounted in a mild steel evacuable tank. The laser beam following the electro-optic modulator is linearly polarized.

The cell and etalon are mounted inside an evacuable tank which serves the dual purposes of reducing schlieren effects and screening magnetic fields. Fine control of magnetic field is achieved via coils in the tank. The output from the cavity is analysed for polarization character by use of a Fresnel rhomb and calcite crystal to direct $\sigma+$ and $\sigma-$ beams into separate detectors.

Figs.5(a) and (b) illustrate, for the absorptive case with zero transverse field,polarization switching with and without hysteresis respectively. In fig.5(a) the Na density is $\sim 1.3 \times 10^{11}$ cm^{-3}, and only the lower switching to the asymmetric state is accessible with our relatively poor cavity gain. In fig.5(b) the Na density has been reduced to approximately 4×10^{10} cm^{-3} in order to show the upper switch

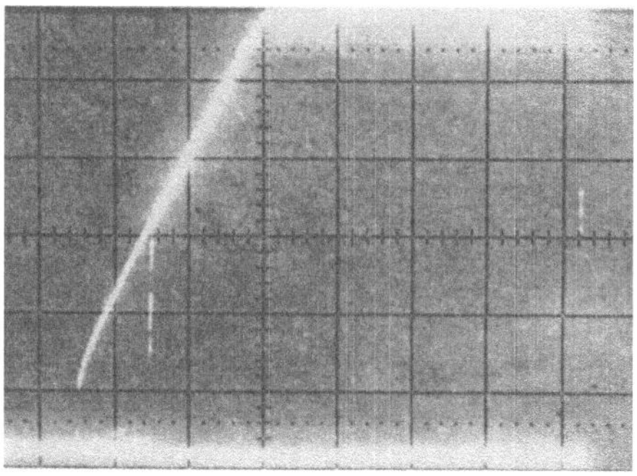

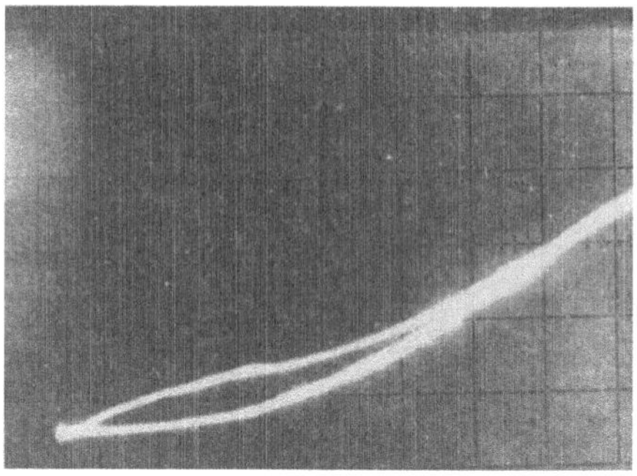

(a) (b)

Fig.5: Oscilloscope displays of σ_+ (lower trace) and σ_- (upper trace) intensities versus input laser power.
(a) horizontal trace 0-165 mW. The flat portion at the top is detector saturation.
(b) horizontal trace 0-180 mW.

non-abrupt) back to the symmetric state as saturation occurs. We confirmed that application of a transverse magnetic field of ~1 Gauss was sufficient to move the threshold for purely absorptive polarization switching outside our range of observation. We were not able to check that this field gave absorptive optical bistability (because of low cavity gain) and we refer (e.g.) to ref.9 for experimental evidence for this case.

Finally in fig.6 we show (again with B_x=0) the complete polarization switching sequence from symmetric at low powers to asymmetric at intermediate power back to symmetric at high laser power. In order to see this, we detuned the laser frequency 1.5 GHz below line centre and required the full available power at the cavity of 500 mW.

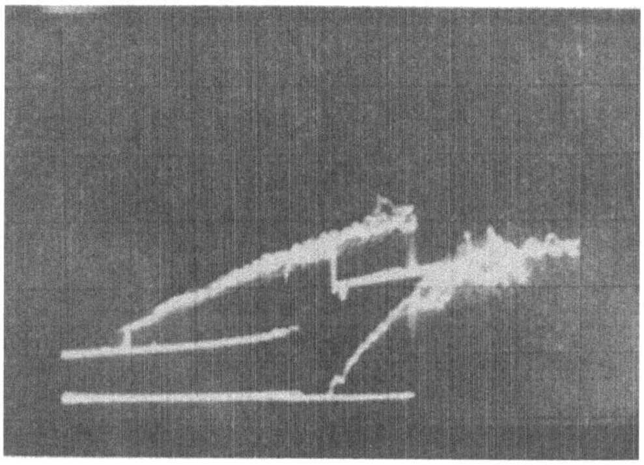

Fig.6: Oscilloscope displays of σ_+ (lower trace) and σ_- (upper trace) intensities for the dispersive case. The traces are separated vertically for clarity. Horizontal trace: laser power (0-500 mW) Na density 9 $\times 10^{10}$ cm^{-3}

Acknowledgements.

We wish to thank Drs R.J.Ballagh and D.M.Warrington for valuable assistance. We acknowledge financial support provided by the Research Committees of the University Grants Committee and the University of Otago.

References.

1. Polarization switching (optical tristability) was first discussed by M.Kitano, T.Yabuzaki and T.Ogawa, Phys.Rev.Lett., 46, 926 (1981) for the dispersive regime invoking a 3-state model not including saturation.

2. Previous observation of optical tristability as predicted in ref.1 (dispersive case, no saturation) has been reported by S.Cecchi, G.Giusfredi, E.Petriella and P.Salieri, Phys.Rev.Lett., 49, 1928 (1982).

3. For an excellent review of irreducible tensor methods, see A.Omont, Prog.Quant. Electr., 5, 69 (1977).

4. A.Messiah, Quantum Mechanics, Vol.II, (English Edition) (North Holland: Amsterdam) (1963).

5. M.Ducloy, Phys.Rev.A., 8, 1844 (1973).

6. M.W.Hamilton, R.J.Ballagh and W.Sandle, Z.Physik B, 49, 263 (1982).

7. Identical equations (to within a reparametrization) have also been obtained for a 3-level (Λ) case with zero coherence assumed between the lower levels by C.M.Savage, H.J.Carmichael and D.F.Walls, Opt.Commun., 42, 211 (1982), and recently (same authors) for a 4-level ($J=\frac{1}{2} \rightarrow J=\frac{1}{2}$) case. (Phys.Rev.Lett., 50, 163 (1983)).

8. R.J.Ballagh, J.Cooper and W.J.Sandle, J.Phys.B: At.Mol.Phys., 14, 3881 (1981) and J.Phys.B: At.Molec.Phys., 14, 4941 (1981).

9. W.J.Sandle and A.Gallagher, Phys.Rev.A., 24, 2017 (1981). See also W.J.Sandle, Laser Physics, Eds.D.F.Walls and J.D.Harvey (Academic Press: Sydney) p.225 (1980).

CHAOS IN NONLINEAR OPTICAL SYSTEMS

H.J. Carmichael†
Physics Department,
University of Waikato,
Hamilton, New Zealand.

ABSTRACT

Predictions of self-oscillation and chaos in optical bistability in a ring cavity are reviewed. Each is derived as a special case from a single unifying stability analysis. Three other systems which also show self-oscillations and chaos are discussed. Each comprises two ring-cavity modes interacting via a nonlinear medium. Nonlinear couplings are provided by a $J = \frac{1}{2}$ to $J = \frac{1}{2}$ transition, a two-photon transition, and a second-order nonlinear susceptibility, respectively. A new type of period-doubling to chaos which occurs in the first of these models, and also in the Lorenz equations, will be described.

INTRODUCTION

As evidenced by other articles in this volume, optical bistability is of interest both for the fundamental physics involved in its study, and for its potential use in the construction of practical optical devices. It has been an important development from both perspectives to learn that bistable systems may become unstable, and for a CW input produce an oscillating output(1-12). These oscillations may be periodic, suggesting obvious applications, or aperiodic with a broadband power spectrum, a phenomenon known as chaos. In these lectures I will be concerned with the theoretical analysis which predicts these possible behaviours in ordinary absorptive and dispersive optical bistability, as well as in bistable and multistable systems designed around other optical nonlinearities.

While periodic oscillations clearly have more direct device applications, it has been the possible chaotic behaviour of bistable systems which has captured the imagination of so many workers. Instabilities leading to periodic self-oscillation were first discussed by McCall(1) and Bonifacio and Lugiato(2). The current rapid growth of interest in optical chaos followed Ikeda's prediction of chaotic oscillations in the output from a ring cavity with dispersive nonlinearity(3,4). Since Ikeda's work experimental observations of chaos have been made(9,12), and further predictions of chaos in bistable or multistable systems have appeared(10,11,13,14). Throughout these lectures I will be concerned only with passive optical systems. The subject of self-oscillation and chaos in laser systems is covered elsewhere in this volume.

†On leave from Physics Department, University of Arkansas, Fayetteville, AR72701, U.S.A.

It is not my intention to discuss the general theory and nature of chaos beyond mentioning here some of the most elementary ideas. For thorough discussions of chaos and examples in various scientific disciplines there are a number of reviews available[15-20].

We are concerned with nonlinear dissipative systems. Physically a cavity mode dissipates energy through partially reflective mirrors or an atomic medium dissipates energy via fluorescence. Mathematically the system evolves in a multi-dimensional phase space where the volume of phase space is not conserved but shrinks to zero for infinite times. For such a system we expect the long-time behaviour to evolve, after transients, to something simple, and persistent - e.g. a steady state (a fixed point in phase space) or a periodic oscillation (a limit cycle). These are the attracting solutions, or simply attractors. The long-time behaviour need not be simple however. Another possible attractor is a quasiperiodic oscillation, decomposable into a finite number of noncomensurate frequencies and their harmonics (motion on an n-torus). Such an oscillation is not periodic and may be very complicated, even appear noisy - though still the Fourier spectrum is discrete. Then there is the possibility of chaos. The long-time behaviour may be an aperiodic oscillation associated with evolution in phase space on a strange attractor. Here the Fourier spectrum contains broadband features and the evolution is unpredictable in so much as nearby points on the attractor evolve into the future along diverging trajectories.

As the physical parameters (laser intensity, etc.) in a system are altered, attracting solutions may become unstable and be replaced by new attractors. This is the process of bifurcation. In optical bistability, the 'bistability' refers to the coexistence of two stable steady states. For certain experimental configurations one or both of these steady states may become unstable and be replaced by a periodic oscillation. In some cases a sequence of bifurcations to more complicated oscillations occurs with the eventual appearance of chaos. In the first part of these lectures I will discuss the stability of the steady state for optical bistability using the familiar model of a two-level homogeneously broadened medium in a ring cavity. For this system seven instabilities have been reported in the literature. Bonifacio and Lugiato[2], and Carmichael et al.[31], have reported instabilities in absorptive systems leading to periodic self-oscillation. Instabilities reported by Ikeda[3], Ikeda et al.[4], Ikeda and Akimoto[10], and Lugiato et al.[11], in systems with dispersion, lead eventually to chaos. I have not seen a detailed study of the oscillatory states that arise from the instability reported for dispersive systems by Lugiato.[5]

Since optical bistability by nonlinear absorption and dispersion was first proposed,[21] it has been realised that bistable or multistable systems may be devised using virtually any optical nonlinearity. Polarisation switching has been predicted for two ring-cavity modes with opposite circular polarisation interacting via a $J = \frac{1}{2}$ to $J = \frac{1}{2}$ transition [22]. This has recently been observed[23,24]. Also, two-photon optical bistability has been predicted[25] and observed in Rubidum vapour[26]. In the second part of these lectures I will discuss recent predictions of chaos in both of these systems[13,14], and in a system comprising a ring-cavity mode and its second harmonic interacting via a second-order nonlinear susceptibility. The latter has been shown to exhibit periodic self-oscillation by McNeil et al.[27] In my discussion of polarisation switching I will look carefully at the sequence of bifurcations leading to chaos as the incident laser intensity is changed (the route to chaos). A period-doubling sequence of a new type has been found (Carmichael et al. Ref.13). This sequence also exists in the Lorenz equations and appears to correspond to a sequence reported recently for one dimensional mappings with two extrema[28].

In Section 2 I review the instabilities predicted for ordinary optical bistability in a two-level homogeneously broadened medium in a ring cavity. Sections 3, 4 and 5 are devoted to chaos in polarisation switching, two-photon bistability, and second harmonic generation, respectively.

2. ABSORPTIVE AND DISPERSIVE BISTABILITY IN A RING CAVITY

The ring cavity model with homogeneous broadening is by now a familiar vehicle for theoretical studies of optical bistability[29]. Laser light of frequency ω_0 is incident at the input of a ring cavity containing a two-level homogeneously broadened medium with resonant frequency ω_a. The cavity input and output mirrors both have reflection and transmission coefficients R and $T = (1 - R)$, with associated phase changes ϕ_R and ϕ_T, respectively. All remaining mirrors are perfect reflectors. The internal cavity field propagates in the z-direction through a medium with density N_V extending from $z = 0$ to $z = L$. The full round-trip distance in the cavity is $L + 1$.

In what follows transverse effects will be neglected and the cavity field is expanded as a plane wave:

$$\vec{E}(z,t) = \vec{e}_0 E(z,t) e^{-i(\omega_0 t - k_0 z)} + \text{c.c.}, \qquad (2.1)$$

where $\vec{e}_0$ is a polarisation vector, $\vec{k}_0 = \omega_0/c$, and $E(z,t)$ is a slowly varying complex field amplitude. As indicated, spatial variation of the cavity-field amplitude in the direction of propagation will be included. Boundary conditions for the cavity may conveniently be referred to the ends of the medium. Then at $z = 0$,

$$E(0,t) = (1 - R)^{\frac{1}{2}} e^{i\phi_T} E_I + RE(L, t - \frac{1}{c})\, e^{-i\theta}, \qquad (2.2)$$

and at $z = L$,

$$E_T(t) = (1 - R)^{\frac{1}{2}} e^{i\phi_T} E(L, t - \frac{1'}{c}), \qquad (2.3)$$

where E_I and $E_T(t)$ are incident and transmitted field amplitudes, respectively; $\theta = -[k_0(L + 1) + 2\phi_R]_{\text{mod}2\pi}$, $-\pi < \theta < \pi$, is the empty cavity detuning; and $1'$ is the distance from the end of the medium at $z = L$ to the output mirror. The mathematical formulation is completed by Maxwell-Bloch equations describing the propagation of $E(z,t)$ through the medium. In the slowly varying amplitude approximation Maxwell's equations give

$$\frac{\partial E}{\partial z} + \frac{1}{c}\frac{\partial E}{\partial t} = i\,\frac{\mu k_0}{2\varepsilon_0}\,P, \qquad (2.4)$$

where $\vec{\mu}$ ($\mu = \vec{e}_0.\vec{\mu}$) is the dipole moment. The polarisation P (μP in units of Cm^{-2}) and inversion density D satisfy the Bloch equations

$$\frac{\partial P}{\partial t} = -i\,\frac{\mu}{\hbar}\,ED - \gamma_{\perp}(1 + i\Delta)P,$$

$$\frac{\partial D}{\partial t} = -2i\,\frac{\mu}{\hbar}\,(E^*P - EP^*) - \gamma_{\parallel}(D + N_v), \qquad (2.5)$$

where $\gamma_{\parallel}$ and $\gamma_{\perp}$ are longitudinal and transverse relaxation rates, and $\Delta = (\omega_a - \omega_0)/\gamma_{\perp}$ is a dimensionless detuning.

It is convenient to introduce the dimensionless field amplitudes

$$\tilde{E}_I = (1-R)^{\frac{1}{2}} e^{i\phi_T}(2\mu/\hbar)[\gamma_{\parallel}\,\gamma_{\perp}(1+\Delta^2)]^{-\frac{1}{2}}E_I,$$

$$\tilde{E}(z,t) = (2\mu/\hbar)[\gamma_{\parallel}\,\gamma_{\perp}(1+\Delta^2)]^{-\frac{1}{2}}\,E(z,t) \qquad (2.6)$$

$$\tilde{E}_T(t) = (1-R)^{-\frac{1}{2}} e^{-i\phi_T}(2\mu/\hbar)[\gamma_{\parallel}\,\gamma_{\perp}(1+\Delta^2)]^{-\frac{1}{2}}E_T(t),$$

and the average inversion seen by a wavefront propagating from $z = 0$ to z (during $t - z/c$ to t):

$$-N_v\tilde{D}(z,t) = \frac{1}{z}\int_0^z dz'D(z',t - \frac{z-z'}{c}). \qquad (2.7)$$

Then from Eqs. (2.4) and (2.5) the steady-state solution for the cavity-field amplitude, $0 \leqslant z \leqslant L$, is[29]

$$\tilde{E}_{ss}(z) = \tilde{E}_{ss}(0)\,e^{-\alpha z/2(1-i\Delta)\tilde{D}_{ss}(z)}, \qquad (2.8)$$

with

$$\tilde{D}_{ss}(z) - 1 = -(\alpha z)^{-1}|\tilde{E}_{ss}(0)|^2(1 - e^{-\alpha z\tilde{D}_{ss}(z)}), \qquad (2.9)$$

where

$$\alpha = (1+\Delta^2)^{-1}(\mu^2 N_v k_0/\varepsilon_0\hbar\gamma_{\perp})$$

is the off-resonant absorption coefficient. The relationship between incident and transmitted fields follows from the boundary conditions (Eqs. (2.2) and (2.3)):

$$\tilde{E}_T = \tilde{E}_I e^{-\alpha L/2(1-i\Delta)\tilde{D}_{ss}(L)}[1 - Re^{-\alpha L/2(1-i\Delta)\tilde{D}_{ss}(L)}e^{-i\theta}]^{-1} \qquad (2.10)$$

where

$$\tilde{D}_{ss}(L) - 1 = -(\alpha L)^{-1}|\tilde{E}_T|^2(e^{\alpha L\tilde{D}_{ss}(L)} - 1). \qquad (2.11)$$

Our concern is with the stability of this steady state. In principle the question of stability is addressed in the same manner as for simple systems of ordinary differential equations. We linearise Eqs. (2.4) and (2.5) about the steady state and look for solutions governed by an exponential growth (unstable) or decay (stable). In practice, since Eq. (2.4) is a partial differential equation, the calculations are a little more complex. I will omit the tedious detail and jump directly from a formulation of the problem to its solution.

A general solution to Eqs. (2.4) and (2.5) is written

$$\vec{V}(z,t) = \vec{V}_{ss}(z) + \sum_{\lambda} \vec{A}_{\lambda}(z) e^{\lambda t} , \tag{2.12}$$

where $\vec{V} \equiv (E,E^*,P,P^*,D)$, and $\vec{A}_{\lambda}(z)$ describes a spatially dependent mode corresponding to the eigenvalue λ. After linearisation Eqs. (2.4) and (2.5) are solved for

$$\begin{pmatrix} A_{E,\lambda}(L) \\ A_{E^*,\lambda}(L) \end{pmatrix} = M(\lambda) \begin{pmatrix} A_{E,\lambda}(0) \\ A_{E^*,\lambda}(0) \end{pmatrix} \tag{2.13}$$

where $M(\lambda)$ is a 2×2 matrix and $A_{E,\lambda}$ and $A_{E^*,\lambda}$ are the first two components of $\vec{A}_{\lambda}$. However, boundary conditions (Eqs. (2.2) and (2.3)) require

$$\begin{pmatrix} A_{E,\lambda}(0) \\ A_{E^*,\lambda}(0) \end{pmatrix} = Re^{-1/c\lambda} \begin{pmatrix} e^{-i\theta} & 0 \\ 0 & e^{i\theta} \end{pmatrix} \begin{pmatrix} A_{E,\lambda}(L) \\ A_{E^*,\lambda}(L) \end{pmatrix} \tag{2.14}$$

Then together Eqs. (2.13) and (2.14) define a homogeneous algebraic system, which, for a nontrivial solution, requires

$$\det\left[Re^{-1/c\lambda} \begin{pmatrix} e^{-i\theta} & 0 \\ 0 & e^{i\theta} \end{pmatrix} M(\lambda) \right] = 0 \tag{2.15}$$

The eigenvalues λ are solutions to this characteristic equation. For absorptive bistability $(\Delta = \theta = 0)$ Eq. (2.15) reads[30]

$$\left\{1 - Re^{-\lambda\tau_r} \left(\frac{\tilde{E}_{ss}(L)}{\tilde{E}_{ss}(0)}\right)^{\gamma_\perp/(\lambda+\gamma_\perp)} \left[\frac{(\lambda+\gamma_\parallel)(\lambda+\gamma_\perp) + \gamma_\parallel \gamma_\perp \tilde{E}_{ss}(0)^2}{(\lambda+\gamma_\parallel)(\lambda+\gamma_\perp) + \gamma_\parallel \gamma_\perp \tilde{E}_{ss}(L)^2}\right]^{\frac{1}{2}\frac{\lambda+2\gamma_\perp}{\lambda+\gamma_\perp}}\right\}$$

$$\times \left[1 - Re^{-\lambda\tau_r} \left(\frac{\tilde{E}_{ss}(L)}{\tilde{E}_{ss}(0)}\right)^{\gamma_\perp/(\lambda+\gamma_\perp)}\right] = 0 \tag{2.16}$$

where $\tau_r = (L + l)/c$ is the cavity round trip time. For systems with dispersion I do not have a general analytical solution for $M(\lambda)$; however, after an adiabatic elimination of the polarisation Eq. (2.15) reads[31]

$$1 + R^2 e^{-2\lambda\tau_r} \frac{|\tilde{E}_{ss}(L)|^2}{|\tilde{E}_{ss}(0)|^2} \frac{\lambda + \gamma_\parallel (1 + |\tilde{E}_{ss}(0)|^2)}{\lambda + \gamma_\parallel (1 + |\tilde{E}_{ss}(L)|^2)} - Re^{-\lambda\tau_r}$$

$$\times \left\{ \left[\frac{\lambda + \gamma_{\parallel}\,(1 + |\tilde{E}_{ss}(0)|^2)}{\lambda + \gamma_{\parallel}\,(1 + |\tilde{E}_{ss}(L)|^2)} + 1 \right] \mathrm{Re}\left(\frac{\tilde{E}_{ss}(L)}{\tilde{E}_{ss}(0)} e^{-i\theta} \right) \right.$$

$$\left. + \Delta \left[\frac{\lambda + \gamma_{\parallel}\,(1 + |\tilde{E}_{ss}(0)|^2)}{\lambda + \gamma_{\parallel}\,(1 + |\tilde{E}_{ss}(L)|^2)} - 1 \right] \mathrm{Im}\left(\frac{\tilde{E}_{ss}(L)}{\tilde{E}_{ss}(0)} e^{-i\theta} \right) \right\} = 0 \qquad (2.17)$$

The seven instabilities I will now review may all be identified in the solutions of either Eq. (2.16) or Eq. (2.17). Of course, these equations are too complicated to be solved generally. Each instability is associated with a set of simplifying assumptions used to cast the characteristic equation into a solvable form. These assumptions fall into two broad categories; those based on time scales, and the familiar assumption of the mean-field limit.

There are three time scales explicitly evident in Eq. (2.16), each represented by a characteristic time appearing in combination with λ - the terms $\lambda\tau_r$, $(\lambda + \gamma_{\parallel})$, and $(\lambda + \gamma_{\perp})$. The size of λ compared with τ_r^{-1}, $\gamma_{\parallel}$, and $\gamma_{\perp}$ determines the relative importance of these terms. For example, Eq. (2.17) eliminates the polarisation a priori. Consequently, for $\Delta = \theta = 0$, Eq. (2.16) agrees with Eq. (2.17) if we look for solutions with $\lambda \ll \gamma_{\perp}$. Assuming $\lambda \ll \gamma_{\perp}$ simplifies Eq. (2.16) considerably.

The mean-field limit has been a popular device for simplifying calculations on many aspects of optical bistability. It assumes that round-trip losses and phase shifts are small - strictly;[(29)]

$$\alpha L \to 0,\ (1 - R) \to 0,\ \theta \to 0,\ |E_I|^2 \to 0,\ |E_T|^2 \to 0$$

with

$$C = \frac{\alpha L}{4(1-R)}, \quad \phi = \frac{\theta}{1-R}, \quad Y = \frac{|\tilde{E}_I|^2}{1-R}, \quad X = |\tilde{E}_T|^2$$

all constant. In Eqs. (2.16) and (2.17) the medium is represented by terms which differ from unity only to the extent that $\tilde{E}_{ss}(z)$ is not uniform, and in taking the mean-field limit this deviation from uniformity is of the order $(1 - R)$:

$$\frac{|\tilde{E}_{ss}(L)|}{|\tilde{E}_{ss}(0)|} = 1 - (1 - R)(1 - i\Delta)\frac{2C}{1+X} + O[(1 - R)^2]. \qquad (2.18)$$

It follows that Eq. (2.16), for example, may be written as

$$(e^{\lambda\tau_r} - 1)^2[(\lambda + \gamma_{\parallel})(\lambda + \gamma_{\perp}) + \gamma_{\parallel}\gamma_{\perp}X] + O[(1 - R)] = 0 \qquad (2.19)$$

Without the term $O[(1 - R)]$ the solutions are eigenvalues for an empty cavity with perfect reflectors and a two-level medium driven by the intensity X - there is no coupling between the cavity and the medium. With the term $O[(1 - R)]$ corrections to these eigenvalues may be calculated by perturbation theory.

Although it is often overlooked, something should be said about time scales in the mean-field limit. I will distinguish two versions of this limit according to the behaviour of $\gamma\tau_r \equiv \gamma_{\parallel}\tau_r$ or $\gamma_{\perp}\tau_r$. This parameter characterises the relationship between the homogeneous width for the medium and the frequency spacing of the longitudinal cavity modes. I will define the <u>mean-field multimode limit</u> by, in addition to the above,

$$\gamma\tau_r \text{ constant}, \quad \frac{\kappa}{\gamma} = \frac{(1-R)}{\gamma\tau_r} \to 0;$$

and the <u>mean-field single-mode limit</u> by

$$\gamma\tau_r \to 0, \quad \frac{\kappa}{\gamma} = \frac{(1-R)}{\gamma\tau_r} \text{ constant.}$$

Here κ is the cavity decay time. The perturbative solution of the characteristic equation is carried out differently for the two cases[32]. Usually, by the mean-field limit, it is meant the mean-field single-mode limit.

In what follows the stability analysis as presented only indicates the existence of an instability. The behaviour which replaces the unstable steady state, be it precipitation to a coexisting steady state, periodic self-oscillation, or period-doubling to chaos, has in each case been determined by additional work which goes beyond a linear stability analysis.

INSTABILITIES IN THE MEAN-FIELD MULTIMODE LIMIT

1. Absorptive Systems - Ref.2

The two factored equations following from Eq. (2.16) determine independently the eigenvalues governing the evolution of in phase and in quadrature fluctuations from the steady state. To first order in $(1 - R)$ these read, respectively,

$$1 - e^{\lambda\tau_r} - (1-R)\left[1 + \frac{2C}{1+X^2}\gamma_{\perp}\frac{\lambda + \gamma_{\parallel}(1-X)}{(\lambda+\gamma_{\parallel})(\lambda+\gamma_{\perp}) + \gamma_{\parallel}\gamma_{\perp}X}\right] = 0$$

$$1 - e^{\lambda\tau_r} - (1-R)\left[1 + \frac{2C}{1+X^2}\gamma_{\perp}\frac{1}{\lambda+\gamma_{\perp}}\right] = 0 \tag{2.20}$$

I will scale time in units of $\gamma_{\parallel}^{-1}$ and write $\hat{\lambda} = \lambda\gamma_{\parallel}^{-1}$ and $\Gamma = \gamma_{\perp}/\gamma_{\parallel}$. Then the solutions for $\hat{\lambda}$ in the limit $(1 - R) \to 0$ are

$$\hat{\lambda}^{(0)}_{1,n} = \hat{\lambda}^{(0)}_{2,n} = -i\alpha_n = -i2\pi n/\gamma_{\parallel}\tau_r, \qquad n = 0, \pm 1, \pm 2, \ldots$$

$$\hat{\lambda}^{(0)}_{3,4} = -\frac{1}{2}\{1 + \Gamma \pm [(1-\Gamma)^2 - 4\Gamma X]^{\frac{1}{2}}\}, \tag{2.21}$$

$$\hat{\lambda}^{(0)}_{5} = -\Gamma,$$

and in the multimode limit their perturbative corrections are

$$\hat{\lambda}_{1,n} = -i\alpha_n - \frac{\kappa}{\gamma_{\parallel}}\left[1 + \frac{2C}{1+X}\Gamma\frac{1 - X - i\alpha_n}{(1 - i\alpha_n)(\Gamma - i\alpha_n) + \Gamma X}\right], \qquad n = 0, \pm 1, \pm 2, \ldots$$

$$\hat{\lambda}_{2,n} = -i\alpha_n - \frac{\kappa}{\gamma_{\parallel}}\left[1 + \frac{2C}{1+X}\Gamma\frac{1}{\Gamma - i\alpha_n}\right] \qquad n = 0, \pm 1, \pm 2, \ldots$$

$$\hat{\lambda}_{3,4} = \hat{\lambda}^{(0)}_{3,4} - (1 - R)\frac{2C}{1+X}(1 + \Gamma + 2\hat{\lambda}^{(0)}_{3,4})^{-1}\Gamma\frac{1 - X + \hat{\lambda}^{(0)}_{3,4}}{\exp(\hat{\lambda}^{(0)}_{3,4}\gamma_{\parallel}\tau_r) - 1} \tag{2.22}$$

$$\hat{\lambda}_5 = \hat{\lambda}^{(0)}_5 - (1 - R)\frac{2C}{1+X}\Gamma\frac{1}{\exp(\hat{\lambda}^{(0)}_5\gamma_{\parallel}\tau_r) - 1}$$

These are the "dressed" eigenvalues[32, 33]. There are two for each cavity mode (complex field amplitude) and three for the medium (inversion and complex polarisation). The terms propertional to $\kappa/\gamma_{\parallel}$ and $(1 - R)$ vanish in the limit; however, the cavity eigenvalues $\hat{\lambda}_{1,n}$ and $\hat{\lambda}_{2,n}$ are then pure imaginary. Stability for the cavity modes is then determined by the perturbative corrections. Bonifacio and Lugiato[2] found that the real part of $\lambda_{1,n}$ may be positive for certain off-resonant modes $(n \neq 0)$ along part of the upper branch in absorptive bistability. The resulting instability leads sometimes to sustained periodic oscillations and sometimes to transient oscillations with eventual precipitation to the lower stable branch. No chaos has been reported.

2. Dispersive systems - Ref. 5

Perturbative solutions to Eq. (2.17) may be found in a similar fashion. Since this equation does not factor, its form in the mean-field limit is rather complicated. I will simply give the solutions:

$$\hat{\lambda}'_{2,n} = -i\alpha_n - \frac{\kappa}{\gamma_{\parallel}}\left\{1 + \frac{2C}{1+X}\frac{1 - i\alpha_n}{1 - i\alpha_n + X} \mp \left[(1 + \Delta^2)\left(\frac{2C}{1+X}\frac{X^2}{1 - i\alpha_n + X}\right)^2 - \left(\phi - \Delta\frac{2C}{1+X}\frac{1 - i\alpha_n}{1 - i\alpha_n + X}\right)^2\right]^{\frac{1}{2}}\right\} \tag{2.23}$$

$$n = 0, \pm 1, \pm 2, \ldots$$

As the polarisation has been eliminated a priori in Eq. (2.17) (requiring $\gamma_{\perp} \gg \kappa, \gamma_{\parallel}$) no solutions for $\hat{\lambda}_4$ and $\hat{\lambda}_5$ are found. The solution for $\hat{\lambda}_3$ is the same as given in Eqs. (2.22) (but with $\Gamma \to \infty$). Lugiato[5] did not eliminate the polarisation, but the instability he found remains in Eq. (2.23). The real part of $\hat{\lambda}_{1,n}$ may be positive for certain off-resonant modes along part of the upper branch in dispersive bistability, and also in systems which are not bistable. I have not seen a study of the oscillations which arise from this instability.

INSTABILITIES IN THE MEAN-FIELD SINGLE-MODE LIMIT

1. Dispersive systems - Ref. 11

In Eqs. (2.22) the perturbative corrections to $\hat{\lambda}_{3,4}$ and $\hat{\lambda}_5$ involve a denominator $\exp(\hat{\lambda}^{(0)}\gamma_{\parallel}\tau_r) - 1$. The mean-field single-mode limit has $\gamma_{\parallel}\tau_r \sim (1 - R) \to 0$, and therefore $[\exp(\hat{\lambda}^{(0)}\gamma_{\parallel}\tau_r) - 1] \sim (1 - R) \to 0$. This cancels the explicit $(1 - R)$ dependence which in the mean-field multimode limit justified, self-consistently, the perturbative method used to calculate $\hat{\lambda}_{3,4}$ and $\hat{\lambda}_5$. This perturbative method can no longer be applied[32]. Rather, we may introduce the mean-field limit in the characteristic equation as before, and also write

$$(e^{\lambda\tau_r} - 1) = (e^{\hat{\lambda}\gamma_{\parallel}\tau_r} - 1) \sim \hat{\lambda}\gamma_{\parallel}\tau_r,$$

but beyond that make no further simplification. We consider only the resonant mode as $\alpha_n \sim (1 - R)^{-1} \to \infty$ corresponding to the migration of all off-resonant modes to infinity. From Eq. (2.16) [via Eqs. (2.20)] we obtain a cubic and a quadratic in $\hat{\lambda}$ giving the five eigenvalues for the resonant cavity mode coupled to the medium. From Eq. (2.17) we obtain the cubic equation

$$\hat{\lambda}^3 + \hat{\lambda}^2\left[1 + X + 2\frac{\kappa}{\gamma_{\parallel}}\left(1 + \frac{2C}{1+X}\right)\right] + \hat{\lambda}\frac{\kappa}{\gamma_{\parallel}}\left[2(1 + X)\left(1 + \frac{2C}{1+X}\,\frac{1-X}{1+X}\right) + \frac{\kappa}{\gamma_{\parallel}}\,\frac{Y}{X}\right] + \left(\frac{\kappa}{\gamma_{\parallel}}\right)^2(1 + X)\frac{dY}{dX} = 0 \qquad (2.24)$$

where Y/X and dY/dX refer to the familiar steady-state input-output relationship for the mean-field limit[29,33]. Again two eigenvalues are not found because Eq. (2.17) eliminates the polarisation a priori. Equation (2.24) corresponds to the eigenvalue equation for the model studied by Lugiato et al.[11]. Applying the Hurwitz stability criterion[34] these authors found an instability along part of the upper branch in dispersive bistability. They report sustained periodic oscillations, transient oscillations with eventual precipitation to the lower stable branch, and period-doubling to chaos.[35]

2. Dispersive systems, the dispersive limit - Ref.10

In the dispersive limit the incident laser is tuned far from the medium resonance so that the nonlinear absorption can be neglected. This implies the use of intensities low enough so as not to power broaden the homogeneous line significantly compared to this detuning. For $X \ll 1$ Eq. (2.24) reads

$$\hat{\lambda}^3 + \hat{\lambda}^2(1 + 2b) + \hat{\lambda}[2b + b^2 + (\bar{X} - z_0)^2] + b^2 + (\bar{X} - z_0)(3\bar{X} - z_0) = 0 \tag{2.25}$$

where I have defined

$$b = \frac{\kappa}{\gamma_{\parallel}}(1 + C), \quad z_0 = \frac{\kappa}{\gamma_{\parallel}}(C\Delta - \phi), \quad \bar{X} = \frac{\kappa}{\gamma_{\parallel}} C\Delta X .$$

For this equation the Hurwitz stability criterion reproduces Eq. (4) from Ref. 10 (in Ref. 10 the > should read <). There Ikeda and Akimoto found an instability leading to period-doubling and chaos along part of the upper branch in dispersive bistability. This and the instability found by Lugiato et al.[11] are probably not two distinct instabilities, i.e. they probably exist in a single connected region of parameter space.

INSTABILITIES OUTSIDE THE MEAN-FIELD LIMIT

1. Dispersive systems - Ref. 3

Each of the final three instabilities is found with both the polarisation and inversion adiabatically eliminated (but without the mean-field limit so we require $\gamma\tau_r \gg 1$, $\gamma \equiv \gamma_{\parallel}$ or $\gamma_{\perp}$). Setting $\lambda \ll \gamma_{\parallel}$ in Eq. (2.17) gives a quadratic in $\exp(\lambda\gamma_{\parallel}\tau_r)$:

$$1 - 2BSe^{-\hat{\lambda}\gamma_{\parallel}\tau_r} + B^2 e^{-2\hat{\lambda}\gamma_{\parallel}\tau_r} = 0 , \tag{2.26}$$

where

$$B = R\,\frac{|\tilde{E}_{ss}(L)|}{|\tilde{E}_{ss}(0)|}\left(\frac{1 + |\tilde{E}_{ss}(0)|^2}{1 + |\tilde{E}_{ss}(L)|^2}\right)^{\frac{1}{2}} , \tag{2.27}$$

$$S = \frac{R}{2B}\left[\left(\frac{1 + |E_{ss}(0)|^2}{1 + |\tilde{E}_{ss}(L)|^2} + 1\right)\mathrm{Re}\left(\frac{\tilde{E}_{ss}(L)}{\tilde{E}_{ss}(0)}\,e^{-i\theta}\right)\right.$$

$$\left. + \Delta\left(\frac{1 + |\tilde{E}_{ss}(0)|^2}{1 + |\tilde{E}_{ss}(L)|^2} - 1\right)\mathrm{Im}\left(\frac{\tilde{E}_{ss}(L)}{\tilde{E}_{ss}(0)}\,e^{-i\theta}\right)\right]$$

Solutions are

$$\hat{\lambda}_{\frac{1}{2},n} = -i\alpha_n - (\gamma_{\parallel}\tau_r)^{-1}\ln\frac{1}{B}\left[S \pm (S^2 - 1)^{\frac{1}{2}}\right] , \tag{2.28}$$

$$n = 0, \pm 1, \pm 2, \ldots$$

The requirement for stability is that the real part of the logarithm be positive. This reproduces the condition for stability obtained from a description of cavity dynamics in terms of a mapping of the complex cavity field amplitude over successive round trips[3,31]. Ikeda[3] found that this mapping may be unstable along part of the upper branch in dispersive bistability. In terms of Eq. (2.28) the real part of the logarithm becomes negative and all of the modes corresponding to $\hat{\lambda}_{1,n}$, $n = 0, \pm 1, \pm 2, \ldots$, are simultaneously unstable. The instability leads to periodic oscillation, period-doubling, and chaos[3,35].

2. Dispersive systems, the dispersive limit - Ref. 4

In the dispersive limit, with $|\tilde{E}_{ss}(z)|^2 \ll 1$ and $|\tilde{E}_{ss}(L)|^2 = \exp(-\alpha L)|\tilde{E}_{ss}(0)|^2$ (only linear absorption remains), the eigenvalues given by Eq. (2.28) have

$$B = Re^{-\alpha L/2},$$

$$S = e^{\alpha L/2}\left[\mathrm{Re}\left(\frac{\tilde{E}_{ss}(L)}{\tilde{E}_{ss}(0)}e^{-i\theta}\right) + \frac{1}{2}\Delta|\tilde{E}_{ss}(0)|^2(1-e^{-\alpha L})\,\mathrm{Im}\left(\frac{\tilde{E}_{ss}(L)}{\tilde{E}_{ss}(0)}e^{-i\theta}\right)\right] \tag{2.29}$$

Again, from the equivalent mapping, Ikeda et al.[4] found regions of instability in dispersive bistability leading to periodic oscillation, period-doubling, and chaos[35]. This prediction of chaos has lead to experimental observations in a hybrid system[9], and, more recently, in an all optical system[12]. As with the instabilities reported in Refs. (10) and (11), those reported in Refs. (3) and (4) are probably not distinct.

3. Absorptive systems - Ref. 31

For a purely absorptive medium $(\Delta = 0)$ in a detuned cavity $(\theta \neq 0)$ B is as given in Eq. (2.27) and

$$S = \frac{R}{2B}\left(\frac{1 + |\tilde{E}_{ss}(0)|^2}{1 + |\tilde{E}_{ss}(L)|^2} + 1\right)\cos\theta . \tag{2.30}$$

A detailed analysis of stability based on the eigenvalues in Eq. (2.28)[31] shows that for $|\theta| < \frac{\pi}{2}$ (in particular $\theta = 0$, a tuned cavity) absorptive bistability is possible; always with both the upper and lower branches stable. For $\frac{\pi}{2} < |\theta| < \pi$ (in particular $|\theta| = \pi$, a fully detuned cavity) there is no bistability, but the real part of the logarithm in Eq. (2.28) may become negative. The resulting instability leads to a periodic oscillation with period $2\tau_r$ where the medium oscillates between saturated and unsaturated states on successive round trips. There is no chaos.

3. OPTICAL TRISTABILITY AND POLARISATION SWITCHING

I will now turn the discussion from ordinary optical bistability to a system of two ring-cavity modes coupled via a nonlinear medium. In what I have called the mean-field single-mode limit (there are now two single modes), and with the medium adiabatically eliminated, complex cavity field amplitudes E_1 and E_2 obey the coupled equations,

$$K^{-1}\dot{E}_{1,2} = (1 - R)^{-\frac{1}{2}} e^{i\phi_T} E_I^{1,2} - E_{1,2}\left[1 + i\phi_{1,2} - i(1 - R)^{-1} \frac{k_{1,2}L}{2} \chi_{1,2}\right] \tag{3.1}$$

where χ_1 and χ_2 are nonlinear susceptibilities and all other quantities are obvious generalisations of those previously defined. There are now two cavity outputs

$$E_T^{1,2} = (1 - R)^{\frac{1}{2}} e^{i\phi_T} E_{1,2} \tag{3.2}$$

Each of the systems discussed in this and the following two sections uses a different nonlinear medium and therefore different χ's.

In this first example the nonlinear interaction takes place via a $J = \frac{1}{2}$ to $J = \frac{1}{2}$ transition (Fig. 1). A linearly polarised field of frequency ω_0 and amplitude E_I ($E_I^1 = E_I^2 = E_I/\sqrt{2}$) is incident at the cavity input and the cavity field is composed of two circularly polarised components:

$$\vec{E}(z,t) = \frac{1}{\sqrt{2}}\left[E_1(t)\vec{e}_+ + E_2(t)\vec{e}_-\right] e^{-i(\omega_0 t - k_0 z)} + \text{c.c.}\,, \tag{3.3}$$

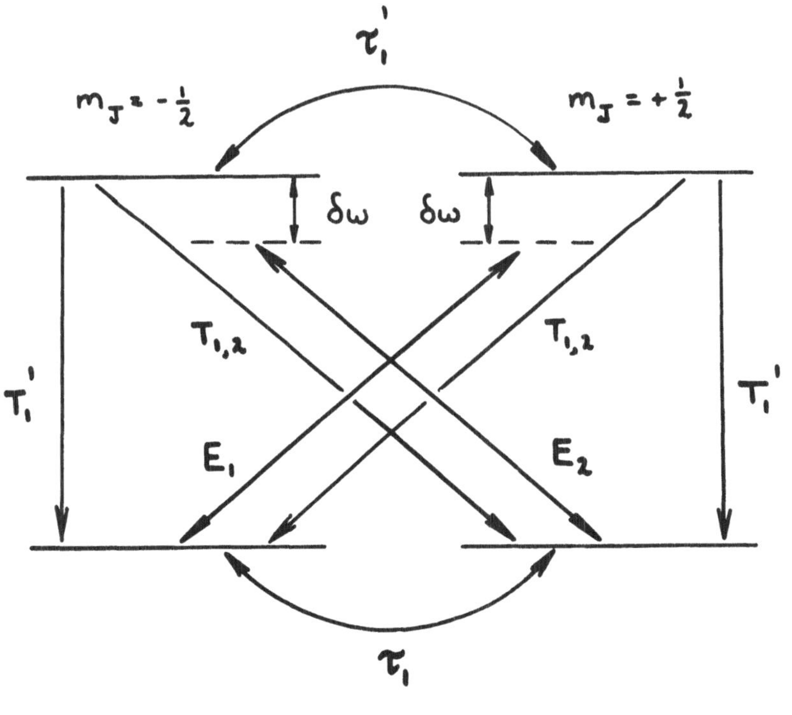

Figure 1: Energy level diagram for $J = \frac{1}{2}$ to $J = \frac{1}{2}$ transition.

where $\vec{e}_\pm = (\vec{X} \pm i\vec{y})/\sqrt{2}$ are unit vectors. In Fig. 1 T_1 and T_1' are relaxation times for the excited state population; T_2 is the relaxation time for the atomic dipole coherences τ_1 and τ_1' are relaxation times for the population differences between ground and excited state magnetic sublevels; and $\delta\omega = \omega_a - \omega_0 = \Delta/T_2$. Atomic density matrix element equations for this system are solved by Sandle and Hamilton elsewhere in this volume, and Hamilton et al.[42] have analysed the resulting steady states in the purely absorptive case. I will go directly to the steady state result. It will be convenient to introduce the dimensionless field amplitudes

$$\tilde{E}_{1,2} = (2\sqrt{\tfrac{2}{3}}\,|\mu|/\hbar)\left[T_2\left(\frac{1}{T_1} + \frac{1}{T_1'}\right)^{-1}/(1+\Delta^2)z\right]^{\frac{1}{2}} \frac{E_{1,2}}{\sqrt{2}}$$

$$\tilde{E}_I = (1-R)^{-\frac{1}{2}} e^{i\phi_T}(2\sqrt{\tfrac{2}{3}}\,|\mu|/\hbar)\left[T_2\left(\frac{1}{T_1} + \frac{1}{T_1'}\right)^{-1}/(1+\Delta^2)z\right]^{\frac{1}{2}} \frac{E_I}{\sqrt{2}} \qquad (3.4)$$

where μ is the reduced atomic dipole matrix element and

$$z = \left[\left(\frac{1}{T_1} + \frac{1}{T_1'}\right)^{-1}/\frac{\tau_1}{2}\right]\left(\frac{1}{\tau_1} + \frac{1}{\tau_1'} + \frac{1}{T_1'}\right)\left(\frac{2}{\tau_1'} + \frac{1}{T_1} + \frac{1}{T_1'}\right)^{-1} \qquad (3.5)$$

There are two nonzero atomic coherences which define the polarisations $P_{1,2}$ and the susceptibilities $\chi_{1,2} = \sqrt{\tfrac{2}{3}}\,\mu\, P_{1,2}/\varepsilon_0 E_{1,2}$. In addition to the population differences $D_{1,2}$ between the radiatively coupled levels, if the cavity field is not linearly polarised $(|\tilde{E}_1|^2 \neq |\tilde{E}_2|^2)$, nonzero population differences D_g and D_e arise between the ground and excited state magnetic sublevels. In the steady state

$$D_{1,2} = -\frac{N_V}{2}(1 + |\tilde{E}_{2,1}|^2)\, S(|\tilde{E}_1|^2, |\tilde{E}_2|^2)^{-1},$$

$$D_g = \frac{N_V}{2}\frac{1}{\tau_1}\left(\frac{1}{\tau_1} + \frac{1}{\tau_1'} + \frac{1}{T_1'}\right)^{-1}(|\tilde{E}_2|^2 - |\tilde{E}_1|^2)S(|\tilde{E}_1|^2, |\tilde{E}_2|^2)^{-1}, \qquad (3.6)$$

$$D_e = \frac{N_V}{2}\left(\frac{1}{\tau_1'} + \frac{1}{T_1'}\right)\left(\frac{1}{\tau_1} + \frac{1}{\tau_1'} + \frac{1}{T_1'}\right)^{-1}(|\tilde{E}_2|^2 - |\tilde{E}_1|^2)S(|\tilde{E}_1|^2, |\tilde{E}_2|^2)^{-1},$$

and

$$\chi_{1,2} = i(k_0L)^{-1}\frac{\alpha L}{2}(1 - i\Delta)(1 + |\tilde{E}_{2,1}|^2)S(|\tilde{E}_1|^2, |\tilde{E}_2|^2)^{-1}, \qquad (3.7)$$

where

$$S(|\tilde{E}_1|^2, |\tilde{E}_2|^2) = 1 + \frac{1}{2}(1+z)(|\tilde{E}_1|^2 + |\tilde{E}_2|^2) + z|\tilde{E}_1|^2|\tilde{E}_2|^2 \qquad (3.8)$$

The coupled field equations are then (with time measured in units of κ^{-1})

$$\dot{\tilde{E}}_{1,2} = \tilde{E}_I - \tilde{E}_{1,2}\left[1 + i\phi + C(1 - i\Delta)(1 + |\tilde{E}_{2,1}|^2)S(|\tilde{E}_1|^2, |\tilde{E}_2|^2)^{-1}\right]. \qquad (3.9)$$

Steady-state solutions to Eq. (3.9) are given in Fig. 2 where the mode intensities $X_{1,2} = |\tilde{E}_{1,2}|^2$ are plotted against the incident intensity $Y = |\tilde{E}_I|^2$ for fixed values of C, Δ, ϕ, and z. The three-dimensional plots are an aid to understanding the projections $X_{1,2}$ v Y. The special case z = 1 (Fig. 2(a)) is particularly instructive. Here $S(|\tilde{E}_1|^2, |\tilde{E}_2|^2)$ factorises and Eqs. (3.9) decouple to give two independent state equations

$$Y = X_{1,2}\left[\left(1 + \frac{C}{1 + X_{1,2}}\right)^2 + \left(\phi - \frac{C\Delta}{1 + X_{1,2}}\right)^2 \right] \tag{3.10}$$

Each mode satisfies the familiar cubic bistability equation[32,36] and the projections $X_{1,2}$ v Y are the familiar S-shaped curves. However, in three dimensions, what is a bistable region for X_1 and X_2 individually is in fact a region of quadrastability. There are two stable symmetric branches (linear polarisation) with X_1 and X_2 both in either high transmission or low transmission states, and two stable asymmetric branches (elliptical polarisation) with X_1 in high transmission and X_2 in low transmission, or visa versa. In Fig. 2(a) the loop of asymmetric solutions is such that it overlaps the symmetric curve in projections on the $(X_{1,2}, Y)$ planes. Fig. 2(b) shows how this loop moves as z is changed so that both the symmetric and asymmetric branches are apparent in these projections. In this figure the point C marks the bifurcation to optical tristability predicted by Kitano et al.[22] and recently observed[23,24] - one stable symmetric branch and two stable asymmetric branches. Our inclusion of saturation in the present model brings the added structure for higher incident intensities.

With the inclusion of saturation it is also possible for the asymmetric branches to become unstable[13]. I will not discuss the stability analysis itself. It simply involves the application of standard techniques to a set of four ordinary differential equations (Eqs. (3.9)) and the numerical solution of a quartic equation for the eigenvalues. Figure 3 shows the unstable region as a function of Y for C = 4, Δ = 5, ϕ = 15, and z = 0.03. Over a significant range of Y both the symmetric and asymmetric steady states are unstable and some form of oscillation must occur. I will spend the rest of my time on this system discussing the oscillations which are found by solving Eqs. (2.9) numerically.

The solutions to Eqs. (2.9) may be traced as trajectories in a four-dimensional phase space. I will illustrate the solutions found in the long-time limit by projections of these trajectories onto a plane; plotting the intensity $X_2(t)$ versus the intensity $X_1(t)$ to obtain a curve parametrised by t. A periodic oscillation $X_{1,2}(t + T) = X_{1,2}(t)$ is represented by a closed curve (limit cycle), and an aperiodic oscillation by a curve which never retraces itself. In four dimensions the trajectories do not cross, but there is no such restriction on their projections. In the range $145 \lesssim Y \lesssim 170$ periodic oscillations period-doubling, and chaos are all observed[13]. For the upper part of this range these are illustrated in Figs. 4-7

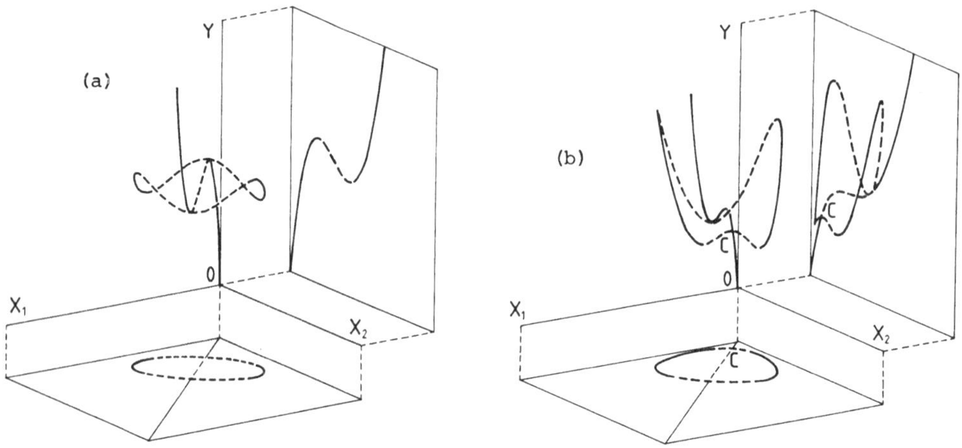

Figure 2: Steady-state solutions to Eqs. (3.9):
(a) z = 1, (b) z = 0.3 Solid (dashed) curves are stable (unstable)

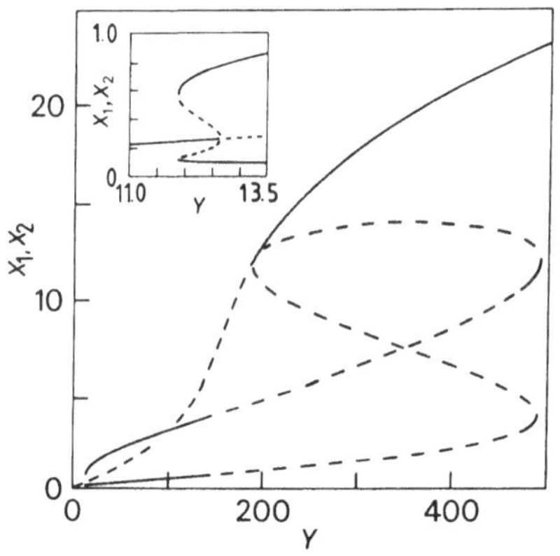

Figure 3: Steady-state solutions to Eqs. (3.9) for C = 4, Δ = 5, φ = 15 and z = 0.03. Solid (dashed) curves are stable (unstable). The inset shows the region of tristability.[22]

and I intend to concentrate on these examples. However, first I will briefly describe the situation as Y is increased from $Y < 145$ where initially the system will be on one or other of the two stable asymmetric branches (Fig. 3).

At $Y \simeq 145$ both of the asymmetric branches become unstable via a Hopf bifurcation - the real part of a pair of complex eigenvalues changes sign, from negative to positive. Technically, it should be proved whether these are supercritical or subcritical bifurcations. That is to say, does a stable limit cycle exist beyond the bifurcation point which contracts onto the asymmetric steady state as Y decreases, or does an unstable limit cycle exist before the bifurcation point which contracts onto the asymmetric steady state as Y increases? The mathematical

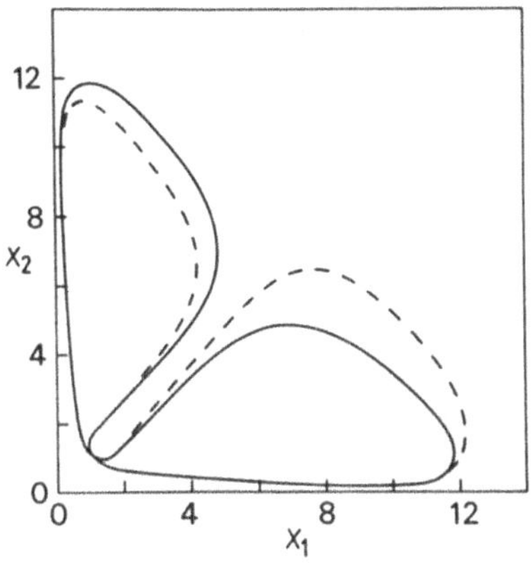

Figure 4: *A symmetric limit cycle (solid curve) at Y = 170 has bifurcated to a pair of asymmetric cycles (dashed curve and its reflection about $X_1 = X_2$) at Y = 166.*

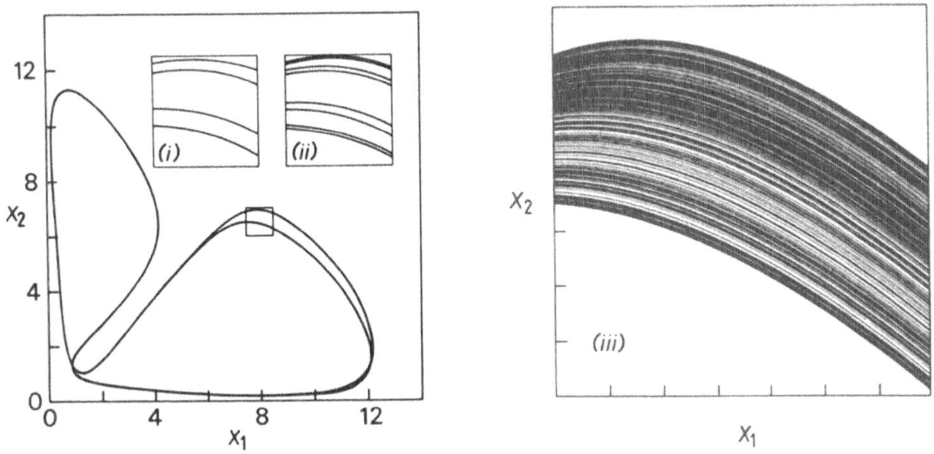

Figure 5: *Each asymmetric cycle period-doubles to chaos: period two at Y = 165, i) period four at Y = 164.9, ii) period eight at Y = 164.88, iii) chaos at Y = 164.8.*

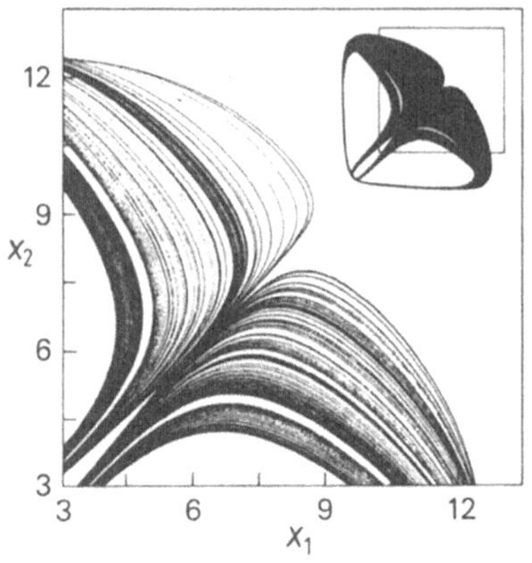

Figure 6: A symmetric chaotic attractor at Y = 163.7.

techniques which enable us to answer this question are described and illustrated by Mandel elsewhere in this volume[36]. I will not carry out such an analysis for this example but simply rely in what follows on numerical evidence derived from a long-time integration of Eqs. (2.9). Beyond the bifurcation point ($Y \simeq 145$) this evidence suggests that a stable limit cycle exists around each of the unstable asymmetric steady states. These cycles grow with increasing Y, until for $Y \simeq 152$ they become unstable and the long-time trajectories wind onto a 'chaotic figure eight' encircling both unstable asymmetric steady states. A chaotic attractor is observed qualitatively similar to that illustrated in Fig. 6. If Y is now decreased the chaotic behaviour persists until at $Y \simeq 149$ the long-time trajectories return to one or other of the stable limit cycles. In summary, for $149 \lesssim Y \lesssim 152$ there is a hysteresis involving the coexistence of stable limit cycles about each of the unstable asymmetric steady states and a chaotic attractor encircling both. This is reminiscent of a similar hysteresis observed as a function of Rayleigh number in the Lorenz equations[37].

Returning now to Figs. 4-7, these illustrate the interesting bifurcation structure which unfolds as Y is decreased from $Y = 170$. This initial value of Y is beyond the value over which chaos is observed and a stable limit cycle exists in the shape of a symmetric figure eight (Fig. 4). The bifurcations to chaos from this symmetric cycle are illustrated in Figs. 4-6:

(1) The symmetric cycle becomes unstable and is replaced by two asymmetric cycles (Fig. 4).

(2) The asymmetric cycles period-double to chaos to form coexisting asymmetric chaotic attractors (Fig. 5).

(3) The two asymmetric chaotic attractors merge on a single symmetric chaotic attractor (Fig. 6).

This behaviour has also been seen in the Lorenz equations as the Rayleigh number is decreased[38]. It appears however, that it is not the last word on this route to chaos. As Y is decreased further a new stable limit cycle appears, looking very like a superposition of the two asymmetric cycles previously observed. This new symmetric cycle bifurcates to chaos as before. Then in a second periodic window, and a third, the same bifurcations from a symmetric limit cycle to chaos occur. These observations have led myself, Savage, and Walls[13] to suggest that this is the beginning of an infinite sequence of periodic windows based on a sequence of symmetric cycles which are related via a period-doubling of a new type. The first four cycles in this sequence are plotted in Fig. 7. We expect this new period-doubling sequence to occur in other systems which possess a symmetry like the reflection symmetry in Eqs. (2.9). The Lorenz equations provide one such example, and indeed, there the same sequence does occur. The symmetric cycles corresponding to those in Fig. 7 are plotted for the Lorenz equations in Fig. 8. In further support of our suggestion I conclude this section with reference to a recent paper where what appears to be the same period-doubling sequence has been analysed in a one-dimensional mapping with two extrema[28].

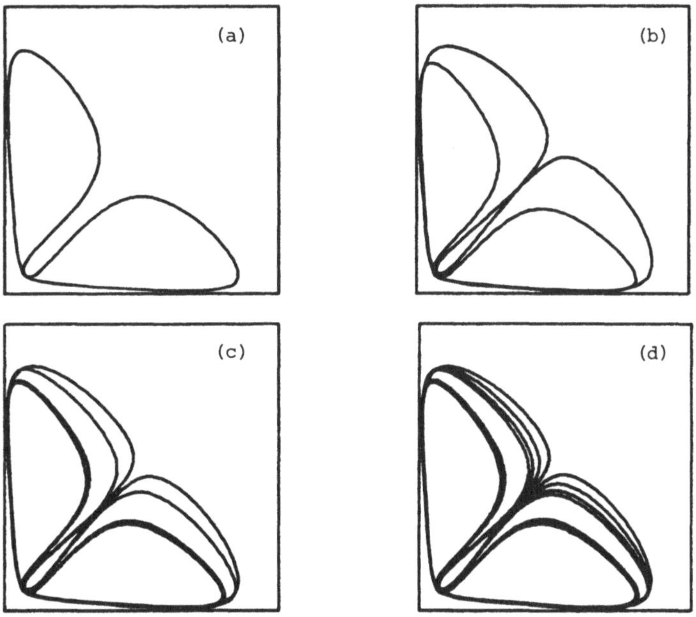

Figure 7: Four symmetric cycles in the proposed new period-doubling sequence: (a) Y = 170, (b) Y = 163.6, (c) Y = 161.0, (d) Y = 160.88.

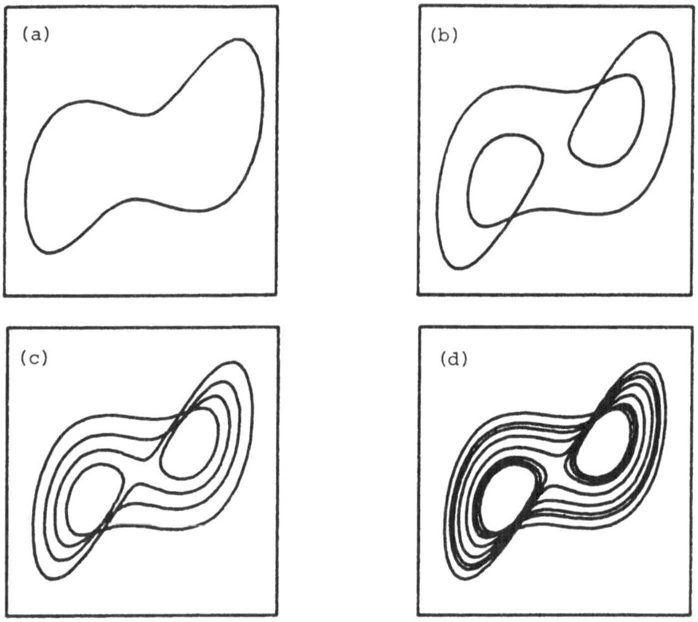

Figure 8: Four symmetric cycles from the Lorenz equations corresponding to the sequence in Fig. 7. $\sigma = 5$, $b = 1$: (a) $r = 250$, (b) $r = 126.1$, (c) $r = 105.5$, (d) $r = 103.2$.

4. TWO-PHOTON OPTICAL BISTABILITY AND TRISTABILITY

In this second example of Eqs. (3.1) two ring-cavity modes with degenerate frequencies ω_0 interact via a two-photon transition with resonant frequency ω_a. Both modes are excited by incident fields with an amplitude E_I. It is convenient to introduce the dimensionless amplitudes

$$\begin{aligned}\tilde{E}_{1,2} &= (2|K|/\hbar)^{\frac{1}{2}}(\gamma_{\parallel}\,\gamma_{\perp})^{-\frac{1}{4}}\,E_{1,2}\\ \tilde{E}_I &= (1-R)^{\frac{1}{2}}\,e^{i\phi_T}(2|K|/\hbar)^{\frac{1}{2}}(\gamma_{\parallel}\,\gamma_{\perp})^{-\frac{1}{4}}\,E_I,\end{aligned} \tag{4.1}$$

where $\gamma_{\parallel}$ and $\gamma_{\perp}$ are phenomenological damping rates, and

$$K = \frac{2}{\hbar}\sum_j \mu_{gj}\mu_{je}/(\omega_{jg} - \omega_0)$$

where μ_{jg}, μ_{je} and ω_{jg} are dipole moments and resonant frequencies for single-photon transitions between the ground and excited states and the intermediate state $|j\rangle$. From a standard calculation of two-photon susceptibilities[39] it follows that the mode amplitudes satisfy the equations (time is measured in units of κ^{-1})

$$\tilde{E}_{1,2} = \tilde{E}_I - \tilde{E}_{1,2}\left\{1 + i\phi + 2C\left[1 - i\delta(|\tilde{E}_1|^2, |\tilde{E}_2|^2)\right]|\tilde{E}_{2,1}|^2 S(|\tilde{E}_1|^2, |\tilde{E}_2|^2)^{-1}\right.$$

$$\left.-i\xi 2C\Gamma\left[1 + \delta(|\tilde{E}_1|^2, |\tilde{E}_2|^2)^2\right] S(|\tilde{E}_1|^2, |\tilde{E}_2|^2)^{-1}\right\}, \qquad (4.2)$$

with

$$\delta(|\tilde{E}_1|^2, |\tilde{E}_2|^2) = \Delta + \frac{\xi}{2}(|\tilde{E}_1|^2 + |\tilde{E}_2|^2)$$

$$S(|\tilde{E}_1|^2, |\tilde{E}_2|^2) = 1 + \delta(|\tilde{E}_1|^2, |\tilde{E}_2|^2) + |\tilde{E}_1|^2|\tilde{E}_2|^2 \qquad (4.3)$$

where ϕ and Γ are as previously defined; $\Delta = (\omega_a - \omega_0)/\gamma_\perp$; C is defined as before in terms of the resonant two-photon absorption coefficient

$$\alpha = (k_0 N_v |k|/\varepsilon_0 \Gamma^{\frac{1}{2}}) ;$$

and

$$\xi = (h|K|)^{-1}\Gamma^{-\frac{1}{2}} \sum_j (|\mu_{jg}|^2 \frac{\omega_{jg}}{\omega_{jg}^2 - \omega_0^2} + |\mu_{je}|^2 \frac{\omega_{je}}{\omega_{je}^2 - \omega_0^2}) \qquad (4.4)$$

is the Stark coefficient.

Without Stark effects ($\xi = 0$) Hermann[40] has predicted periodic self-oscillations in this system, but no chaos. With the inclusion of Stark terms an instability leading to periodic oscillation, period-doubling, and chaos has been predicted by Parriger et al.[14] Figures 9 and 10 are due to these authors. In Fig. 9 steady-state intensities $X_{1,2}$ are plotted against the incident intensity Y for fixed values of C, Δ, ϕ, and Γ, and for both a small Stark coefficient $\xi = 0.1$ (Fig. 9(a)) and a large Stark coefficient $\xi = 1.4$ (Fig. 9(b)). For both values of ξ a pitchfork bifurcation to a pair of asymmetric branches occurs. In Fig. 9(a) the asymmetric branches are always stable, while in Fig. 9(b) they become unstable via a Hopf bifurcation in the manner of the previous example. Figure 10 illustrates the occurence of period-doubling and chaos.

5. SECOND HARMONIC GENERATION

In a third and final example of Eqs. (3.1) I consider a ring cavity mode with frequency ω_0 interacting with its second harmonic via a second order susceptibility $\chi^{(2)}$. Only the fundamental frequency is incident on the cavity input and I will assume that both the fundamental and the second harmonic are equally detuned from the cavity resonance. If E_1 and E_2 are the complex amplitudes for fundamental and second harmonic waves, respectively, with an appropriate scaling they satisfy the equations

$$\dot{\tilde{E}}_1 = \tilde{E}_I - \tilde{E}_1(1 + i\phi) + \tilde{E}_1^*\tilde{E}_2 ,$$

$$\dot{\tilde{E}}_2 = -\tilde{E}_2(1 + i\phi) - 2\tilde{E}_1^2. \qquad (5.1)$$

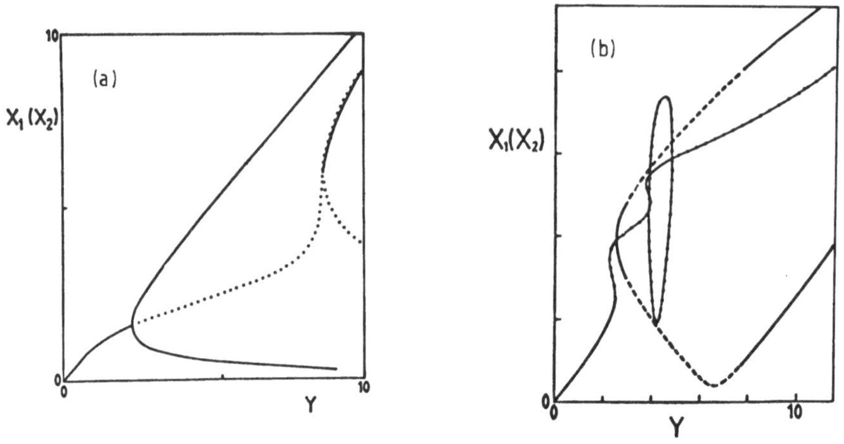

Figure 9: *Steady-state solutions to Eqs. (4.2) for $C = 2$, $\Delta = -10$, $\phi = 0$, $\Gamma = 0.5$, and (a) $\xi = 0.1$, (b) $\xi = 1.4$. Solid curves are stable. Dashed and dotted curves are unstable.*

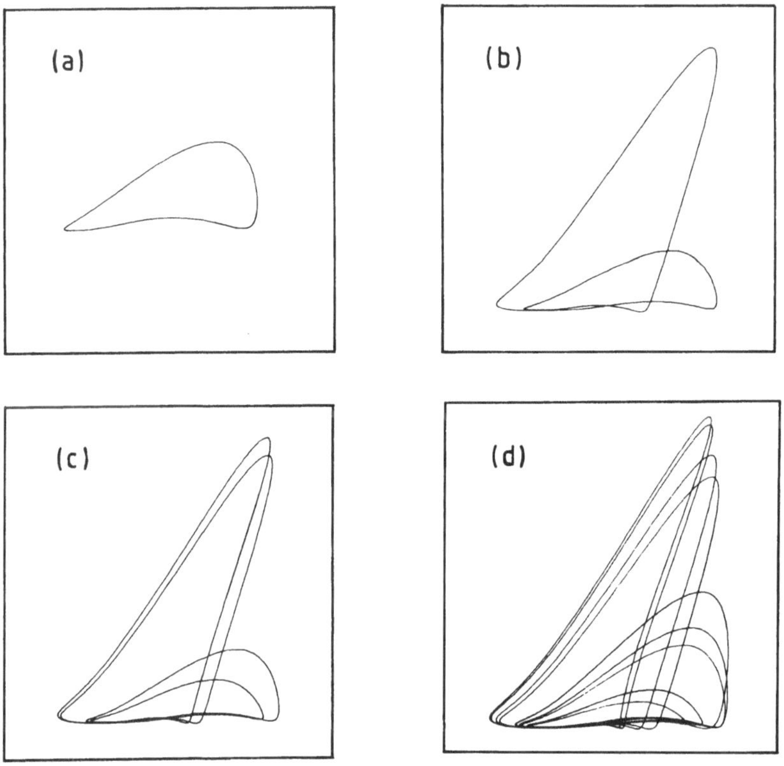

Figure 10: *Period-doubling in Eqs. (4.2) for the parameters of Fig. 9: (a) $Y = 8.2$, (b) period two at $Y = 8.4$, (c) period four at $Y = 8.32$, (d) chaos at $Y = 8.25$.*

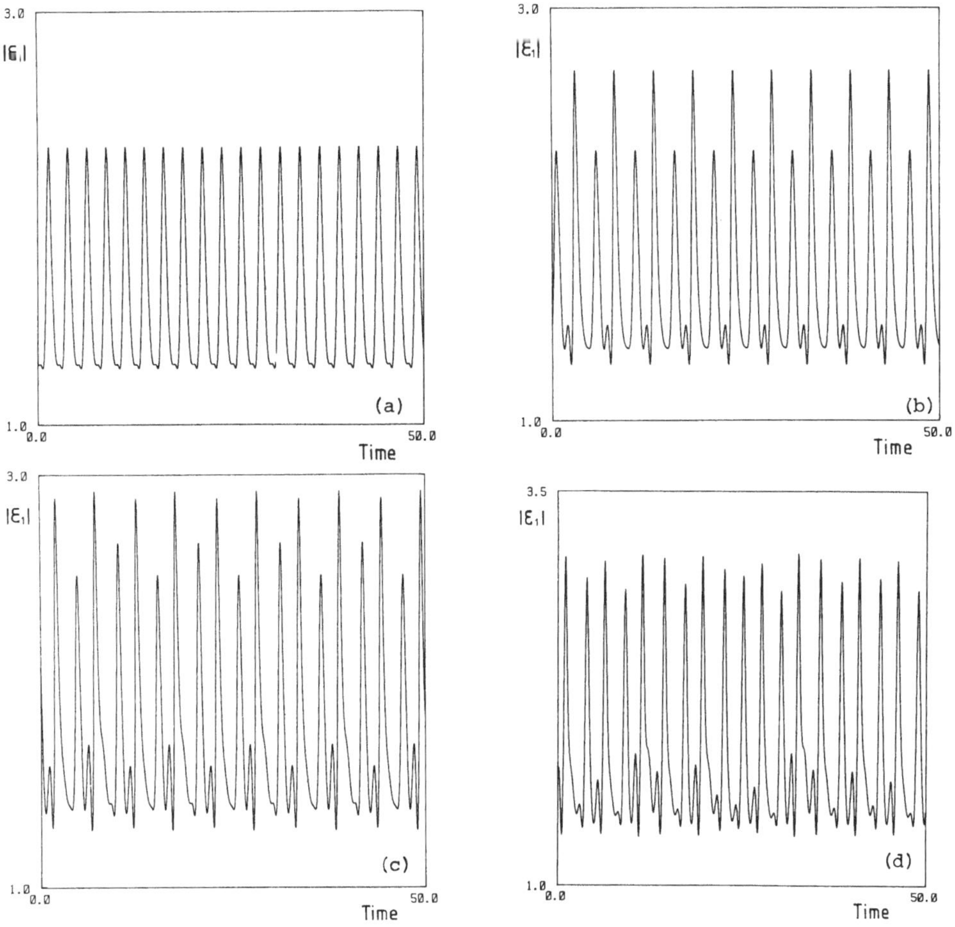

Figure 11: Period-doubling in Eqs. (5.1) for $\phi = 1$: (a) $Y = 6.0$, (b) period two at $Y = 6.5$, (c) period four at $Y = 6.95$, (d) chaos at $Y = 7.5$

McNeil et al.[27] have shown that with $\phi = 0$ Eqs. (5.2) may produce periodic oscillations, and Mandel and Erneux[36] have shown these to be stable. However, with $\phi = 0$ period-doubling and chaos have not been found. With the inclusion of a nonzero cavity detuning period-doubling to chaos can occur[41] as illustrated in Fig. 11.

ACKNOWLEDGEMENT

The research reported here has been supported in part by the United States Army through its European Research Office.

REFERENCES

1. S.L. McCall, Appl.Phys.Lett., 32, 284, (1978).

2. R. Bonifacio and L.A. Lugiato, Lett.Nuovo.Cimento, 21, 510, (1978); R. Bonifacio, M. Gronchi and L.A. Lugiato, Optics Commun., 30, 129, (1979); M. Gronchi, V. Benza, L.A. Lugiato, P. Meystre and M. Sargent III, Phys.Rev., A24, 1419, (1981).

3. K. Ikeda, Optics Commun., 30, 257, (1979).

4. K. Ikeda, H. Daido and O. Akimoto, Phys.Rev.Lett., 45, 709, (1980).

5. L.A. Lugiato, Optics Commun., 33, 108, (1980).

6. F. Casagrande, L.A. Lugiato and M.L. Asquini, Optics Commun., 32, 492, (1980).

7. M. Sargent III, Kvant.Elektron. (Moscow) 10, 2151, (1980) (Sov.J.Quantum Electron., 10, 1247, (1980)).

8. W.J. Firth, Optics Commun., 39, 343, (1981).

9. H.M. Gibbs, F.A. Hopf, D.L. Kaplan and R.L. Shoemaker, Phys.Rev.Lett., 46, 474, (1981); F.A. Hopf, D.L. Kaplan, H.M. Gibbs and R.L. Shoemaker, Phys.Rev., A25, 2172, (1982).

10. K. Ikeda and O. Akimoto, Phys.Rev.Lett., 48, 617, (1982).

11. L.A. Lugiato, L.M. Narducci, D.K. Brandy and C.A. Pennise, Optics Commun., 43, 281, (1982).

12. H. Nakatsuka, S. Asaka, H. Itah, K. Ikeda and M. Matsuoka, Phys.Rev.Lett., 50, 109, (1983).

13. C.M. Savage, H.J. Carmichael and D.F. Walls, Optics Commun., 42, 211, (1982); H.J. Carmichael, C.M. Savage and D.F. Walls, Phys.Rev.Lett., 50, 163, (1983).

14. C. Parigger, P. Zoller and D.F. Walls, Optics Commun., 44, 213, (1983).

15. R.M. May, Nature, 261, 459, (1976).

16. A.S. Monin, Usp.Fiz.Nauk., 125, 97, (1978) (Sov.Phys.-Usp., 21, 429, (1978)).

17. M.I. Rabinovich, Usp.Fiz.Nauk., 125, 123, (1978) (Sov.Phys.-Usp., 21, 443, (1978)).

18. J.P. Exkmann, Rev.Mod.Phys., 53, 643, (1981).

19. E. Ott, Rev.Mod.Phys., 53, 655, (1981).

20. K. Tomita, Phys.Rep., 86, 115, (1982).

21. H. Seidel, U.S. Patent No. 3 610 731 (5 October 1971); A. Szoke, V. Daneu, J. Goldhar and N. Kurnit, Appl.Phys.Lett., 15, 376, (1969); S.L. McCall, Phys.Rev., A9, 1515, (1974); H.M. Gibbs, S.L. McCall and T.N.C. Venkatesan, Phys.Rev.Lett., 36, 1135, (1976).

22. M. Kitano, T. Yabuzaki and T. Ogawa, Phys.Rev.Lett., 46, 926, (1981).

23. S. Cecchi, G. Giusfredi, E. Petriella and P. Salieri, Phys.Rev.Lett., 49. 1928, (1982).

24. W.J. Sandle and M.W. Hamilton, this volume.

25. F.T. Arecchi and A. Politi, Lett.Nuovo.Cim., 23, 65, (1978); G.P. Agrawal and C. Flytzanis, Phys.Rev.Lett., 44, 1058, (1980).

26. A. Giacobino, M. Devaud, F. Biraben and G. Brynberg, Phys.Rev.Lett., 45, 434, (1980).

27. K.J. McNeil, P.D. Drummond and D.F. Walls, Optics Commun., 27, 292, (1978).

28. J. Coste and N. Peyraud, Physica, 5D, 415, (1982).

29. For the ring cavity model with spatial effects see Ref. 3 and the following: R. Bonifacio and L.A. Lugiato, Lett.Nuovo Cimento., 21, 505, (1978); R. Bonifacio, L.A. Lugiato and M. Gronchi, in Laser Spectroscopy IV, Proceedings of the Fourth Conference on Laser Spectroscopy, 1979, edited by H. Walther and W.K. Rothe (Springer, Berlin, 1979); R. Roy and M.S. Zubairy, Phys.Rev., A21, 274, (1980); H.J. Carmichael and J.A. Hermann, Z.Phys., 38B, 365, (1980).

30. Only the first factor is obtained if the perturbation from the steady state is assumed real. This corresponds to the characteristic equation derived in Ref. 2; see Ref. 34.

31. H.J. Carmichael, R.R. Snapp and W.C. Schieve, Phys.Rev., A26, 3408, (1982).

32. H.J. Carmichael, Phys.Rev.A., to be published.

33. G.P. Agrawal and H.J. Carmichael, Phys.Rev.A., 19, 2074, (1979).

34. H. Leipholz, Stability Theory, (Academic, New York, 1970), p.33.

35. The period-doubling sequences predicted in Refs. 3 and 4 are studies in some detail in Ref. 34 where general references on period-doubling are given - see also; R.R. Snapp, H.J. Carmichael and W.C. Schieve, Optics Commun., 40, 68, (1981).

36. P. Mandel and T. Erneux, Optica Acta, 29, 7, (1982).

37. R. Gilmore, Catastrophe Theory for Scientists and Engineers, (Wiley, New York, 1981), pp. 553-62.

38. K.A. Robbins, SIAM J.Appl.Math., 36, 457, (1979).

39. See for example; J.N. Elgin, G.H.C. New and K.E. Orkney, Optics Commun., 18. 250, (1976).

40. J.A. Hermann, Optics Commun., 44, 62, (1982).

41. C.M. Savage and D.F. Walls, Optica Acta, (in press).

42. M.W. Hamilton, R.J. Ballagh, W. Sandle, Z.Physik B., 49, 263, (1982).

SPONTANEOUS PULSATIONS IN LASERS

Lee W. Casperson
School of Engineering and Applied Science
University of California, Los Angeles, California 90024

I. INTRODUCTION

Although lasers and related devices have now been in wide use for over two decades, it cannot yet be said that all aspects of laser operation are well understood. New laser lines and designs are often found by trial and error, and the literature abounds with interesting experimental observations relating to lasers that have never been interpreted convincingly in terms of theoretical models. Conversely, some of the theoretical literature that ostensibly relates to lasers can never be confirmed by means of laboratory experiments. The purpose of this paper is to provide an overview of a specific laser-related topic which has been discussed in the literature since the earliest days of laser physics, but which has been attracting greatly increased attention within the last year. This topic involves the spontaneous coherent pulsations that sometimes occur in laser oscillators even when all of the usual destabilizing mechanisms (saturable absorption, multiple modes, etc.,) are believed to be absent. Besides providing an overview of this topic, the paper also reports the first quantitative agreement that has been obtained between theory and experiment.

Because research on the topic of spontaneous coherent pulsations in lasers has occurred at a low level over such a long period of time, it is easy for a researcher to be unaware of previous studies in this field. However, the subject has now matured sufficiently that it is important to establish a proper historical perspective. Section 2 of this paper includes a brief historical survey of the early research relating to pulsations in homogeneously broadened lasers. The subject is continued in Section 3 with a review of experimental and theoretical developments that relate to inhomogeneously broadened lasers including especially the most recent theoretical results. The author is fully aware of the futility of trying to prepare a complete discussion of earlier work on this or any other topic, and he apologizes in advance for any references that are neglected or misrepresented. For conciseness only the previous work on semiclassical instabilities in ordinary single-mode lasers is considered, and the vast related literature on rate equation instabilities, mode-locking, bistability, chaos, etc., is largely ignored.

II. HOMOGENEOUSLY BROADENED LASERS

The study of spontaneous coherent pulsations in laser oscillators is almost as old as the study of lasers themselves. Thus, for example, a classic prophetic study of the requirements and properties of cw laser oscillators was conducted by Schawlow and Townes and published in 1958 [1]. But at the same time, other researchers were already investigating the possibility of coherent instabilities when a two level system interacts with an electromagnetic field. In this section and the next, we review briefly some of the studies since 1958 that relate to the concept of spontaneous pulsations in lasers.

Among the earliest studies of maser transients were those carried out by the Russian authors Khaldre and Khokhlov (1958) [2], Gurtovnik (1958) [3], and Orayevskiy (1959) [4]. These authors derived time dependent equations governing the interaction of a system of molecules with the electromagnetic cavity mode of a maser oscillator. The equations were then examined for various kinds of transient solutions, and the conditions for spontaneous instabilities were derived [3]. The immediate purpose of those studies seems to have been to explain and predict any time dependent phenomena that might be observed in molecular beam masers such as the ammonia maser [5].

The next major advance toward the understanding of laser instabilities was, of course, the experimental demonstration of the first laser. Operation of the first laser was reported by Maiman in 1960 at Hughes Research Laboratory in California [6]. That laser was based on a flashlamp pumped ruby light amplifier, and this choice of material provided an early impetus for the study of spontaneous pulsations. Even with very gently varying pump pulses ruby lasers tend to produce their outputs in the form of a train of spikes. Usually the spikes appear to be quite random in nature, but under some conditions regular damped pulsations can also be observed. Theoreticians were busy for several years trying to identify the cause of this spiking behavior, and one idea figuring prominently in their discussions involved the possibility of some fundamental instability in the laser equations. The pulsations were, of course, not unique to the ruby system, and similar effects were observed with the majority of solid lasers.

In 1960 Statz and DeMars developed a set of rate equations governing the interaction of the laser atoms with an electromagnetic field [7]. These equations were found to predict damped relaxation oscillations, and similar results were obtained by Dunsmuir in 1961 [8]. However, it soon became apparent that the rate equation approach to laser oscillation could not actually predict instabilities, and this point was also studied in detail by Makhov [9] and by Sinnett [10] in 1962. It was then tempting to assume that the spiking must somehow be a consequence of the transient nature of the pumping. But in 1962 Nelson and Boyle reported the observation of undamped pulsations in a ruby laser system with cw pumping [11], and since that time there have been dozens of reports of similar pulsations in cw pumped solid, liquid, and gas laser systems. Hence, it became obvious that the Statz and DeMars rate equations by themselves were not capable of explaining many of the observed pulsation phenomena.

Singer and Wang in 1961 were the first to point out explicitly that the observed laser spiking might be due to fundamental coherent nonlinearities in the laser equations [12]. Related arguments were formulated by Pao in 1962 [13]. A comprehensive study of laser transients was given by Tang in 1963 using essentially the same techniques and semiclassical density matrix equations that are still used in rigorous laser studies [14]. In a simplified and normalized form, these equations can be written:

$$\frac{dP}{dt} = - \gamma(P + Ar), \tag{1}$$

$$\frac{dr}{dt} = - \gamma_1(r - r_o - AP), \tag{2}$$

$$\frac{dA}{dt} = - \gamma_c(A + P), \tag{3}$$

where P is the normalized polarization (related to the off-diagonal elements of the density matrix), r is the threshold parameter measuring the population inversion compared to its value at threshold ($r = 1$), r_o represents the pump rate, A is the normalized electric field amplitude, and the normalizations have been chosen to include the numerous physical constants and coefficients of the laser medium and cavity. The polarization decay rate γ is the reciprocal of the coherence time (or transverse relaxation time), γ_1 is the inversion decay rate (or longitudinal relaxation time), and γ_c is the decay rate of the electric field in the laser cavity.

Tang showed that in the limit where the polarization decay rate is large compared to the cavity and inversion decay rates ($\gamma >> \gamma_c, \gamma_1$) the semiclassical equations reduce to the stable Statz and DeMars rate equations. In this limit the derivative in Eq.(1) may be neglected yielding $P = - Ar$, and with this substitution Eqs. (2) and (3) can be written:

$$\frac{dr}{dt} = - \gamma_1(r - r_o + rI), \tag{4}$$

$$\frac{dI}{dt} = - 2\gamma_c I(1 - r), \tag{5}$$

where $I = A^2$ is a normalized intensity. These are, of course, the usual rate equations, and their stability can be verified by linearization and use of Liapounoff's second method [15]. In a second limit, where the cavity decay rate is large compared to the polarization and inversion decay rates ($\gamma_c >> \gamma_1, \gamma$) the semiclassical equations reduce to another stable equation set. In this limit, the derivative in Eq. (3) may be neglected yielding $P = - A$, and with this substitution Eqs. (1) and (2) can be written:

$$\frac{dr}{dt} = - \gamma_1(r - r_o + I), \tag{6}$$

$$\frac{dI}{dt} = - 2\gamma I(1 - r). \tag{7}$$

The stability of these equations can also be readily verified. As discussed below, it was left for other researchers to discover that for cases intermediate between the limits described by Tang, Eqs. (1) to (3) actually do possess a region of instability.

Following the ideas of Singer and Wang, Uspenskiy was able in 1963 to demonstrate an instability in a somewhat simplified model of the ruby laser [16]. This model was superior to the rate equation treatments in the sense that it permitted the polarization to not depend instantaneously on the value of the electric field. Thus it is a coherent model and the first that could be applied specifically to the problem of spontaneous pulsations in lasers. However, because of the somewhat obsolete formulation of the problem, Uspenskiy's stability criteria are not reproduced here.

The next important development in the subject of semiclassical laser instabilities was the publication in 1964 of a series of much more general analyses by Korobkin, Uspenskiy, Grasyuk, and Orayevskiy [17-20]. In these papers, the authors have extended the earlier analyses of coherent instabilities in maser oscillators and tested the results against experimental data obtained with ruby lasers. Stability criteria were also derived which enable one to determine in advance whether a laser characterized by particular pump rates and decay lifetimes will exhibit spontaneous coherent pulsations. The most fundamental of these criteria can be written in the notation of Eqs. (1) to (3) as:

$$\gamma_c > \gamma_1 + \gamma. \tag{8}$$

If this condition is not satisfied, no instability is possible.

Korobkin, Uspenskiy, Grasyuk, and Orayevskiy also showed that even when Eq. (8) is satisfied there is still a minimum level of pumping that is necessary in order for the instability to be observed. In the notation of Eqs. (1) to (3) the condition they derived is:

$$r_o > 1 + \frac{(\gamma + \gamma_1 + \gamma_c)(\gamma + \gamma_c)}{\gamma(\gamma_c - \gamma - \gamma_1)}. \tag{9}$$

To avoid confusion, it may also be noted that in the extreme limit of Eq. (8) ($\gamma_c >> \gamma_1 + \gamma$), which is essentially one of the cases found by Tang to be stable [14], the pump level necessary to observe the instability becomes infinite.

Another major development reported in 1964 was the first numerical solution of the semiclassical laser equations by Buley and Cummings [21]. These authors found that under low loss conditions the equations predict damped pulsations. On the other hand, with high cavity loss the equations were found to yield undamped pulsations, and this behavior agrees with the instabilities predicted in the earlier analytical treatments. It is especially notable that the undamped pulsations found by Buley and Cummings were of two basic types. Under some conditions the pulses were periodic, while for operation far above threshold a series of "almost random" spikes was obtained. These were the first direct theoretical observations of periodic and chaotic pulsations in homogeneously broadened lasers. Much of the research in this field since 1964 can be regarded as clarifications, physical interpretations, and extensions of the basic stability criteria and pulsation effects which had by then already been discovered.

Detailed studies of the effects of spontaneous emission noise on laser intensity and phase fluctuations were reported in 1966 by Haken [22] and by Risken, Schmid, and Weidlich [23]. Among other things these studies showed that for some operating conditions undamped pulsations should be obtainable, and the resulting stability criteria were similar to those obtained by Korobkin, Uspenskiy, Grasyuk, and Orayevskiy. The exact forms in which Eqs. (8) and (9) are written were first given by Risken, Schmid, and Weidlich.

It is of interest to regard the cavity loss rate γ_c in Eq. (9) as an adjustable parameter and to derive the minimum possible value of the pump rate for instabilities to be observed. If the derivative of r_o with respect to γ_c is set equal to zero, one finds that the optimum value of the cavity loss rate is:

$$\gamma_c = \gamma + \gamma_1 + (4\,\gamma^2 + 6\,\gamma\,\gamma_1 + 2\,\gamma^2)^{1/2}, \tag{10}$$

and the corresponding value of the minimum pump rate is:

$$r_o > 5 + 3\,\frac{\gamma_1}{\gamma} + 2\left[4 + 6\frac{\gamma_1}{\gamma} + 2\,\left(\frac{\gamma_1}{\gamma}\right)^2\right]^{1/2} \tag{11}$$

The stability criterion given in Eq. (11) was derived in other ways by Risken and Nummedal [24,25] and by Graham and Haken [26], but this result was not necessarily considered to be dependent on the earlier stability conditions.

As mentioned above, Buley and Cummings had remarked briefly in 1964 that for operation far above threshold the semiclassical laser equations predict the possibility of undamped chaotic pulsations. This fact wasn't fully appreciated until 1975 when Haken demonstrated the mathematical equivalence of the laser equations and the simplified Boussinesq equations of fluid mechanics [27]. That fluid convection possesses chaotic solutions is obvious to anyone who has heated a pot of water, and the mathematical conditions relating to such solutions had been investigated by Lorenz [28].

During the late seventies, there were several studies of laser stability, but for the most part, these works emphasized spontaneous mode locking [29-32]. Detailed numerical solutions of the laser equations were reported by Mayr, Risken, and Vollmer in 1981 including especially a discussion of chaotic solutions [33]. Other studies have related laser instabilities and chaos to similar behavior in chemical reactions, thermodynamics, nonlinear optics, and more complex optical structures [34]. Since Carmichael will be discussing some of these topics in detail at this symposium it is unnecessary to pursue them further here.

Very recently there has been an increased level of research relating directly to laser instabilities, and though our consideration of this material will necessarily soon be obsolete, a few items may be mentioned. In 1981, Zorell extended the results of Risken and Nummedal to include the effects of tuning away from line center [35]. It was found that higher pumping levels are required to achieve spontaneous pulsations in a detuned laser. In 1982 Minden and Casperson [36] and Hendow and Sargent [37] showed explicitly that in the homogeneous limit a frequency domain stability test employed by Casperson [38,39] yields the same results as earlier perturbational methods. Hendow and Sargent also introduced the claim that pulsations of one of the three quantities of interest (polarization, population, field), namely the population, cause the laser instabilities [37]. It may be noticed, however, that these three quantities have equal status in Eqs.(1) to (3), and pulsations of all of them accompany any instability. (The reader may easily demonstrate that if any one of P, A, and r is replaced by its cw value in Eqs. (1) to (3) the instability is eliminated.) No corresponding suggestion has been made yet for a single dependent variable in the other physical systems which exhibit similar instabilities. Hendow and Sargent have also examined the effects on instabilities of tuning away from line center [40], but the relationship of these results to Zorrell's study remains to be clarified. In a surprising development, Lawandy has recently claimed that the stable Eqs. (4) and (5) can actually be unstable, and that the implications of this result are in agreement with the observed instabilities in CH_3F lasers [41].

As a final observation relating to spontaneous pulsations in homogeneously broadened lasers, it is important to indicate how these numerous theoretical studies relate to practical experiments. The original motivation behind the development of the semiclassical models was, after all, the experimental observation of spontaneous pulsations in ruby lasers and others. Unfortunately, the practical usefulness of the theoretical results is highly doubtful, and sometimes authors admit that "conditions are not realized for the existence of these solutions" [42]. First of all, very few lasers are known which satisfy Eq. (8), and a partial list is given in Section III of this paper. If the cavity loss rate in such a laser could be adjusted to satisfy Eq. (10), it is still doubtful that the pump rate threshold of Eq. (11) could be reached. Even in the most favorable circumstance that the inversion decay rate is much less than the polarization decay rate ($\gamma >> \gamma_1$) an unsaturated threshold parameter $r_o > 9$ is clearly required, and in the twenty years that these concepts have been known no laser has been found which satisfies the necessary criteria. Thus, for the time being it must be acknowledged that the fascinating models of semiclassical spontaneous pulsations in homogeneously broadened lasers do not have practical applications. As will be seen below, the situation is very different with inhomogeneous broadening.

III. INHOMOGENEOUS BROADENING

Most of the studies of laser instabilities that have been mentioned so far in this overview have been concerned primarily or exclusively with lasers that are homogeneously broadened. However, almost from the first studies of He-Ne lasers in 1961 [43], it has been recognized that the saturation characteristics of inhomogeneously broadened lasers might be substantially different and could lead to striking new types of laser behavior. In 1962, Bennett introduced to laser studies the concept of spectral hole burning and showed that the glitches on the dispersion curve that are associated with the gain holes can lead to important but marvelously complicated new spectral effects [44]. The first comprehensive semiclassical model of inhomogeneously broadened lasers was given by Lamb in 1964 [45]. However, a stability analysis of Lamb's model was not carried out until very recently.

The first investigation of spontaneous pulsations in inhomogeneously broadened lasers seems to have been that of Yakubovich in 1969 [46]. In that work it was shown that intensity pulsations are possible if the initial inhomogeneous spectral distribution has two peaks. In a somewhat related analysis, Idiatulin and Uspenskiy investigated the stability of a laser system containing two discrete spectral classes of atoms [47]. While these results are interesting, they do not correspond directly to any practical laser systems.

In 1969, in the course of his thesis research, the author discovered that under some conditions, a cw high-gain single-mode xenon laser operating at 3.51 microns produces its output in the form of an infinite train of pulses. In some circumstances, these pulses occur periodically with the individual pulses possessing a remarkably complex structure. Under other conditions, the pulses alternate in height or are aperiodic having a broadband noiselike frequency spectrum. Although considerable effort was expended, the author was unable to find an adequate explanation for the pulsation effects, and the first published descriptions of this work appeared in 1971 and 1972 [48,49]. The two principal theoretical models that had been separately explored involved coherence effects and the effects of inhomogeneous broadening. It was observed, however, that the models could be generalized to simultaneously "include both inhomogeneous broadening and a finite coherence time, but the mathematics would be more complicated" [48]. It wasn't discovered until later that the combination of inhomogeneous broadening and a long coherence time actually result in the experimentally observed instability.

As will be shown below, another related discovery of the same era was to play an important part in the eventual interpretation of spontaneous pulsation data. It was found that in a highly dispersive laser medium several frequencies can exist, all having the same wavelength, and this effect was termed "mode splitting" [48,50]. The author isn't aware that the pulsation effect attracted much attention from other quarters, but nevertheless he spent many leisure hours over the next several years investigating ever more complex and obscure models that might account for this instability. These efforts culminated with the discovery that Lamb's semiclassical equations have a low threshold instability and that a direct numerical solution produces pulses which in a qualitative way possess all of the properties of the experimentally observed pulses. The details of this work were published in 1978 [51].

Most of the author's experiments were carried out using a laser discharge that was 1.1 m in length and 5.5 mm in diameter with a gas pressure of about 5×10^{-3} torr. Some typical early experimental results are represented in Fig. 1. The curve in Fig. 1c is the frequency domain version of the 2.5 MHz repetition rate pulsations shown in Fig. 1.a. The main pulses are usually followed by weaker echo pulses, and the example in Fig. 1a is typical. Under some conditions, the successive dominant pulses alternate in height as shown in Fig. 1.b, and this effect has since come to be called period doubling. Under other conditions the output appears noiselike, and a typical frequency spectrum corresponding to this behavior is shown in Fig. 1d. The output is actually not random in this case, and in modern terminology, the results might be termed optical chaos.

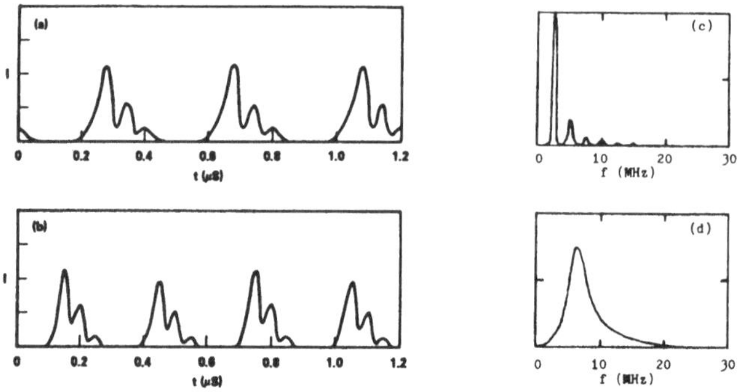

Fig. 1: Experimental plots of the pulsation instability for (a) a discharge current of 40 ma and (b) a discharge current of 50 ma. Period doubling is evident in (b) as the intensity alternates between successive bursts. Part (c) shows the frequency spectrum corresponding to part (a), and part (d) is the frequency spectrum for the chaotic output observed with a current of 70 ma.

The most general formalism that is usually needed to model the effects just described consists of the density matrix equations for the active medium coupled to Maxwell's wave equation for the electric field. These equations were discussed thoroughly by Lamb [45], and a useful simplification was introduced by Feldman and Feld [52]. The equations form a complete set from which the time and space dependences of the electric field can be determined, subject to the boundary conditions at the resonator mirrors. In their most general form, these equations allow for a distribution of velocities (Doppler broadening) and a distribution of natural center frequencies (isotope shifts, hyperfine splitting, Stark broadening, etc.), and hence they apply to quite arbitrary inhomogeneously broadened laser media [39]. In the past few years, various techniques have been used to investigate these equations as they apply to spontaneous pulsations in lasers. As mentioned above, direct numerical integration was first reported in 1978 [51]. A typical result for the amplitude and intensity fluctuations in a periodically pulsing standing wave xenon laser are shown in Fig. 2. It can be seen from the figure that each major pulse is followed by a damped train of secondary pulses, and this behavior is common to most laboratory and numerical experiments. Figure 2 also illustrates the effect known as period doubling in which succeeding pulses alternate in amplitude.

While the intensity is pulsing the atomic populations are also exhibiting rapid fluctuations [51]. In Fig. 3 is a plot of the frequency dependent population differences $\rho_{aa} - \rho_{bb}$ at the instant of time t = 0.55 μs in Fig. 2. This difference is normalized to a line center value of unity, and frequency is meas-

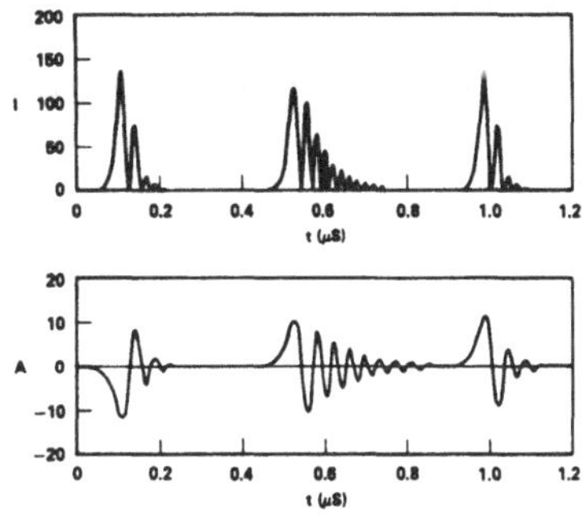

Fig. 2 : Theoretical plots of the pulsation instability in a xenon laser. The normalized field amplitude is represented by A, and $I = A^2$ is the normalized intensity. This is an example of period doubling, in which the pulse shape alternates between successive pulses.

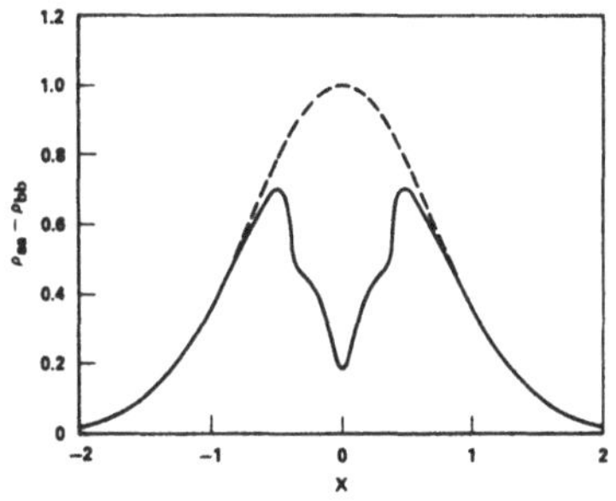

Fig. 3 : Normalized population difference as a function of normalized frequency at the time t = 0.55 microseconds for the population pulsations associated with Fig. 2. The dashed line is the population difference that would result in the absence of saturating fields.

ured in the normalized units $x = 2(\nu - \nu_o)/\Delta\nu_D$. This figure thus represents a snapshot of the pulsing population difference, and for comparison, the unsaturated Gaussian population is shown as a dashed line. The dips away from line center are associated with the Fourier components of the high frequency secondary pulsations. At increased excitation levels the population difference also develops a time dependent fine structure having minima at frequencies corresponding to harmonics of the fundamental repetition frequency.

While brute force numerical solutions are always useful for comparison with experiments, it was also considered worthwhile to look for simpler stability criteria which might at least indicate the approximate conditions under which spontaneous pulsations would be observed. Such criteria were first described in 1979 [53], and comprehensive treatments were given in 1980 and 1981 for Doppler and non-Doppler inhomogeneously broadened lasers respectively [38,39]. In the first of these treatments [38], the population fluctuations described above were not explicitly included (they aren't essential to the existence of the instability), while in the second discussion [39] a detailed procedure for including these pulsations was described. The stability criteria were derived using, in effect, the principles of mode-splitting [48,50]. In the mode-splitting discussions it had been shown that a strong frequency dependence of the index of refraction leads to the possibility of several frequencies having the same average wavelength (i.e., belonging to the same longitudinal mode) inside of a laser cavity. Beats between these frequency components could be detected as spontaneous pulsations.

In a saturating single-mode inhomogeneously broadened laser, the dispersion variations associated with the spectral hole burned in the gain profile have just the right properties to cause mode-splitting. Thus, a saturating mode can develop sidebands, and if these sidebands have net gain one concludes that the laser is pulsing spontaneously. The details of the resulting stability criteria depend on the values of the homogeneous and inhomogeneous lifetimes as well as on the values of the decay coefficients of the laser medium and cavity. Specific stability criteria for a xenon ring laser are shown in Fig. 4 as a function of the threshold parameter r and the cavity lifetime t_c [36]. Roughly speaking, with larger values of the homogeneous linewidth (higher pressure) higher pump rates are required to observe the instability. The approximations and numerical coefficients used in deriving the curves in Fig. 4 are exactly the same as those used in the direct numerical integrations discussed above. Although the detailed pulse shapes cannot be inferred from stability calculations, it is important to note that such calculations are usually less expensive to carry out than numerical time domain solutions, and it is possible to represent a wide range of operating conditions on a single plot. Recently, stability calculations similar to those described above have been performed by Hendow and Sargent [37,40].

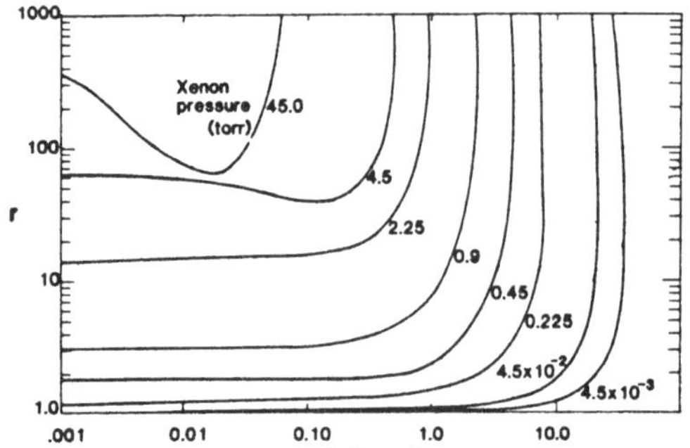

Fig. 4 : Stability thresholds for a xenon laser as a function of the cavity lifetime $t_c = 1/2\gamma_c$ for various values of pressure. In the xenon laser the maximum gain tends to decrease with increasing pressure, so only the lower curves are susceptible to experimental confirmation.

Those authors have used a simpler analytical model in which the population decays are characterized by a single lifetime, and the results are not so directly comparable to present experimental data. If one further assumes that the single population decay rate is equal to the coherence decay rate (i.e., there are no phase interrupting collisions), then the stability criteria for an inhomogeneously broadened laser can be obtained analytically, and this has been discussed by Mandel [54].

Theoretically speaking, the time domain and frequency domain solutions that have been briefly described here would seem to be quite complete and quite adequate. An experimentalist would, however, be far from satisfied. The numerical solutions represented by Fig. 2 only agree in a qualitative way with the pulsation waveforms obtained in xenon laser experiments. The magnitude of these and other discrepancies is sufficient to cast some doubt on the predictive ability of the model, especially regarding lasers other than xenon. Accordingly, it has seemed worthwhile to explore various generalizations of the basic semiclassical model to see whether better agreement with experiment might be obtainable.

One of the conspicuous deficiencies of the semiclassical models employed in previous stability studies is their characterization of relaxation processes within the laser medium. In the most general treatments spontaneous decay out of each of the lasing levels is governed by a single decay rate γ_a or γ_b, and the models don't allow for direct spontaneous decay from the upper laser state to the lower one. On the other hand, it is known that most of the actual spontaneous decays from the upper state in xenon do go directly to the lower state. Another defect of the previous models as applied to xenon lasers is their neglect of spectral cross relaxation due to velocity changing collisions. In our latest investigations, realistic decay processes and spectral cross relaxation integrals have been included in the semiclassical equations using the best available spectroscopic data relating to xenon. The effect of using a more accurate semiclassical model is illustrated in Fig. 5a, where a direct numerical solution of the equations is shown for a xenon ring laser at 5×10^{-3} torr pressure and line center tuning. The only nonspectroscopic parameters used in computing this result are the cavity lifetime $t_c = 1$ nanosecond and the threshold parameter $r = 1.7$. This theoretical pulse train may be seen to be in excellent agreement with the experimental curve in Fig. 1a, and the values of t_c and r seem consistent with the experimental conditions. This is a substantial improvement over our previous calculations, where no reasonable adjustment of parameters would produce significantly better agreement than that suggested by a comparison of Figs. 1a and 2.

Under a wide range of operating conditions xenon lasers produce chaotic pulsations, and this behavior is also shown clearly by the theoretical models. An example of a theoretical chaotic pulse train is given in Fig. 5b using the

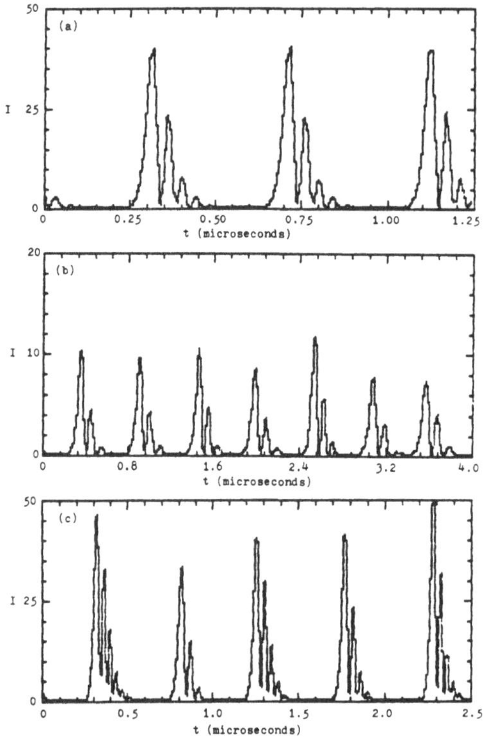

Fig. 5 : Numerical solutions using an improved model for relaxation processes in xenon. Part (a) shows exactly periodic pulsations for a center-tuned laser with a threshold parameter of r = 1.7. Part (b) shows the chaotic pulsations of lower amplitude and frequency that result when the threshold parameter is reduced to r = 1.3. Part (c) illustrates the combination of period doubling and chaos that occurs when the laser of part (a) is detuned by 5 $\Delta\nu_h$.

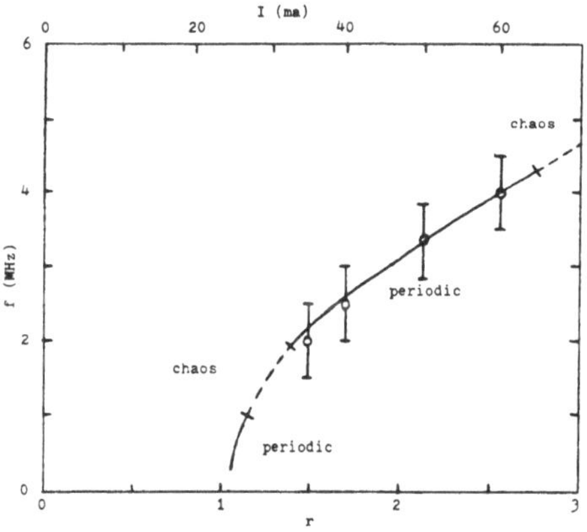

Fig. 6 : Theoretical and experimental pulsation frequencies as a function of threshold parameter (discharge current) for a laser tuned to line center. In the regions labeled chaos, the average frequency is given.

improved semiclassical model described above with the parameter values $t_c = 1$ nanosecond and r = 1.3. A series of similar results have been computed for various values of r, and the corresponding pulsation frequencies are plotted in Fig. 6 together with the experimental data they are intended to represent. For values of r that are close to the oscillation threshold the pulsation waveform is essentially a low frequency sinusoid. For increasing values of r the pulses develop more structure and more intensity, pass through a chaotic operating region, and enter the second periodic region, which was accessible to our experiments. For very large values of r a second region of chaos occurs, and this behavior is in agreement with our frequency domain experiments. Besides the theoretical time domain solutions like those shown in Fig. 5, Fourier transform frequency domain plots are also being obtained.

Another interesting aspect of the time domain solutions which we didn't specifically investigate previously [51] concerns the dependence of the pulsation waveforms on detuning. In a ring laser model a slight detuning from line center is found to result in the intensity not quite dropping to zero between pulses, and this fact should be considered along with detection limitations and spontaneous emission effects when interpreting the corresponding experimental results. In particular, the nonzero intensity between secondary pulses in Fig. 1a may be due in part to detuning, and Fig. 5c gives a recomputation of Fig. 5a including a detuning of $5\Delta\nu_h$. Detuning also modifies the regions of chaos, period doubling, etc., and the pulsation behavior changes significantly as a standing wave laser is tuned in the vicinity of the Lamb dip. However, for low pressure operation the Lamb dip is narrow and well removed from the frequency of maximum power because of dispersion focusing [48,55]. Thus, in a low pressure standing-wave xenon laser a ring laser theoretical model should be adequate for most purposes, and this conclusion is supported by the results that have been described above. Due to space limitations, a more comprehensive written description of this work will be given elsewhere [56].

In another recent development, Minden has extended the frequency domain stability analyses to include small but finite saturating sidebands on a central lasing mode [57]. With a knowledge of the phase and amplitude of the various frequency components it is possible to construct the time domain waveforms without actually carrying out numerical integrations of the differential equations. A typical result of this procedure is shown in Fig. 7, and it is clear that the general pulsation characteristics are similar to the experimental data and numerical simulations described above.

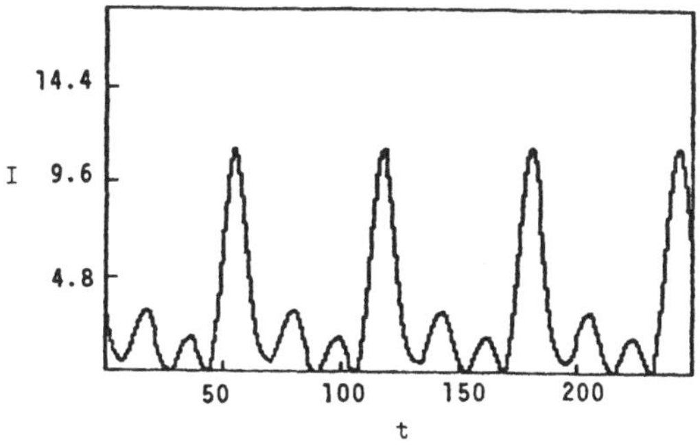

Fig. 7 : Typical pulsations predicted in a frequency domain solution of the semiclassical laser equations. The intensity and time scales are arbitrary

The preceding paragraphs have dealt primarily with recent theoretical developments relating to spontaneous pulsations in xenon lasers. At the same time, Abraham and his associates have been obtaining important new experimental data concerning mode-splitting, spontaneous pulsations, and chaos in helium-xenon lasers [58,59]. Since Abraham will be describing his results at this meeting, it is inappropriate to say any more here. Instead we will conclude by listing some other representative laser types in which spontaneous pulsations might be observable or may already have been seen. Among the requirements that such a laser should satisfy is Eq.(8). In particular, the polarization decay rate should be small compared to the cavity decay rate. In order for a laser to operate with a large cavity loss rate it must have a high gain, and Eq.(8) can be approximated by the useful alternate form:

$$cg/(2\pi\Delta\nu_h) > 1, \qquad (18)$$

where g and c represent respectively the average small signal gain coefficient and speed of light, and $\Delta\nu_h$ is the homogeneous linewidth ($\Delta\nu_h = \gamma/\pi$). Besides satisfying Eq. (18), it should be possible to operate the laser under conditions of inhomogeneous broadening (low pressure in gas lasers).

Several high gain lasers are listed in Table 1 together with an estimate of the value of the ratio $cg/(2\pi\Delta\nu_h)$ for comparison with Eq. (18) [60]. It is clear from this table that there are some lasers which are candidates for the spontaneous pulsation effect, and the reader will probably think of others. Also shown is an indication of whether spontaneous pulsations have yet been observed. At this writing it is only in the case of xenon and helium-xenon lasers that spontaneous pulsations have been clearly shown to be caused by the semiclassical instability, and in the case of the HF and He-Ne lasers only very noisy outputs have been observed without clear evidence of periodic pulsations.

TABLE 1 : LASER CANDIDATES FOR THE SEMICLASSICAL SPONTANEOUS PULSATION INSTABILITY

LASER	$cg/(2\pi\Delta\nu_h)$	INSTABILITY KNOWN?
Xe, He-Xe	150	yes [48, etc.]
CH_3F	50	yes [61]
HF	20	noise only
CO_2	9	yes [62,63]
GaAs	5	yes [64]
He-Ne (3.39 μm)	5	noise only
CH_3I	3	yes [65]
Rhodamine B	3	no

IV. CONCLUSION

In this study, we have briefly reviewed some of the historical developments leading to the present understanding of semiclassical spontaneous pulsations in lasers. Even before the first demonstration of an optical frequency laser, it was known that the semiclassical maser equations possessed instabilities. These ideas were quite well developed by 1964, and periodic pulsations and chaos had been seen in numerical solutions of the homogeneously broadened laser equations. This subject has attracted increased attention in recent years with the growing awareness that the laser equations are similar to the nonlinear equations of other unstable physical systems. However, no homogeneously broadened laser has yet been shown to satisfy the severe conditions for realizing the instability. Spontaneous pulsations in xenon lasers were reported in 1971, and it was shown in 1978 that these pulsations correspond to an instability of the Lamb equations for inhomogeneously broadened lasers. The conditions for observing instabilities in inhomogeneously broadened lasers are much more easily met than the corresponding conditions for homogeneously broadened lasers, and it may be expected that several more spontaneously pulsing inhomogeneously broadened lasers will be found.

There is no single easy physical explanation for why the pulsations occur. It is safest to simply observe that whenever one has a physical system which is governed by several nonlinear equations instabilities and pulsations are a liklihood. Numerous such systems are known ranging in size between nuclear and stellar dimensions. Mode-splitting interpretations are helpful, and it is also possible to derive the stability criteria using Liapounoff's method. Recently, more realistic semiclassical models have been developed, and it can now be said that the pulsation characteristics of at least the low pressure xenon lasers are reasonably well understood. The time may be nearing when all of the highly complex laser pulsation data that is being reported can be rigorously modelled. While spontaneously pulsing lasers are now mainly in the category of laboratory curiosities, perhaps one day they will even find some practical applications.

ACKNOWLEDGEMENT

The author is pleased to acknowledge valuable discussions and correspondence with M.L. Minden and N.B. Abraham.

REFERENCES

[1] A.L. Schawlow and C.H. Townes, Physical Review, Vol. 112, p. 1940 (1958).

[2] K.Y. Khaldre and R.V. Khokhlov, Izvestiya Vuzov MVO SSSR (Radiofiszika), Vol. 1, p. 60 (1958), (in Russian).

[3] A.G. Gurtovnik, Izvestiya Vuzov MVO SSSR (Radiofizika), Vol. 1, p. 83 (1958), (in Russian).

[4] A.N. Orayevskiy, Radio Engineering and Electronic Physics, Vol. 4, p.228 (1959).

[5] See for example, K. Shimoda, T.C. Wang, and C.H. Townes, Physical Review, Vol. 102, p. 1308 (1956).

[6] T.H. Maiman, Nature, Vol. 187, p. 493, (1960).

[7] H. Statz and G. DeMars, in Quantum Electronics edited by C.H. Townes, (Columbia University Press, New York, 1960), p. 530.

[8] R. Dunsmuir, Journal of Electronics and Control, Vol. 10, p. 453 (1961).

[9] G. Makhov, Journal of Applied Physics, Vol. 33, p. 202 (1962).

[10] D.M. Sinnett, Journal of Applied Physics, Vol. 33, p. 1578 (1962).

[11] D.F. Nelson and W.S. Boyle, Applied Optics, Vol. 1, p. 181 (1962).

[12] J.R. Singer and S. Wang, Physical Review Letters, Vol. 6, p. 351 (1961).

[13] Y.-H. Pao, Journal of the Optical Society of America, Vol. 52, p. 871 (1962).

[14] C.L. Tang, Journal of Applied Physics, Vol. 34, p. 2935 (1963).

[15] See for example, W. Kaplan, Ordinary Differential Equations, (Addison-Wesley, Reading, Massachusetts, 1958), p. 433.

[16] A.V. Uspenskiy, Radio Engineering and Electronic Physics, Vol. 8, p. 1145 (1963).

[17] V.V. Korobkin and A.V. Uspenskiy, Soviet Physics JETP, Vol. 18, p. 693 (1964).

[18] A.V. Uspenskiy, Radio Engineering and Electronic Physics, Vol. 9, p. 605 (1964).

[19] A.Z. Grasyuk and A.N. Orayevskiy, Radio Engineering and Electronic Physics, Vol. 9, p. 424 (1964).

[20] A.Z. Grasyuk and A.N. Orayevskiy, in Quantum Electronics and Coherent Light, edited by P.A. Miles, (Academic, New York, 1964), p. 192.

[21] E.R. Buley and F.W. Cummings, Physical Review, Vol. 134, p. 1454 (1964).

[22] H. Haken, Zeitschrift fur Physik, Vol. 190, p. 327 (1966).

[23] H. Risken, C. Schmid, and W. Weidlich, Zeitschrift fur Physik, Vol. 194, p. 337 (1966).

[24] H. Risken and K. Nummedal, Physics Letters, Vol. 26A, p. 275 (1968)

[25] H. Risken and K. Nummedal, Journal of Applied Physics, Vol. 39, p. 4662 (1968).

[26] R. Graham and H. Haken, Zeitschrift fur Physik, Vol. 213, p. 420 (1968).

[27] H. Haken, Physics Letters, Vol. 53A, p. 77 (1975).

[28] E.N. Lorenz, Journal of the Atmospheric Sciences, Vol. 20, p. 130 (1963).

[29] H. Haken and H. Ohno, Optics Communications, Vol. 16, p. 205 (1976).

[30] H. Ohno and H. Haken, Physics Letters, Vol. 59A, p. 261 (1976).

[31] H. Haken and H. Ohno, Optics Communications, Vol. 26, p. 117 (1978).

[32] P.R. Gerber and M. Buttiker, Zeitschift fur Physik B, Vol. 33, p. 219 (1979).

[33] M. Mayr, H. Risken, and H.D. Vollmer, Optics Communications, Vol. 36, p. 480 (1981).

[34] See for example, V. Degiorgio, in NATO advanced Study Institute "Nonlinear Phenomena at Phase Transitions and Instabilities", edited by T. Riste, (Plenum, New York, 1981).

[35] J. Zorell, Optics Communications, Vol. 38, p. 127 (1981).

[36] M.L. Minden and L.W. Casperson, IEEE Journal of Quantum Electronics, Vol. QE-18, p. 1952 (1982).

[37] S.T. Hendow and M. Sargent III, Optics Communications, Vol. 40, p. 385 (1982).

[38] L.W. Casperson, Physical Review A, Vol. 21, p. 911 (1980).

[39] L.W. Casperson, Physical Review A, Vol. 23, p. 248 (1981).

[40] S.T. Hendow and M. Sargent III, Optics Communications, Vol. 43, p. 59 (1982).

[41] N.M. Lawandy, IEEE Journal of Quantum Electronics, Vol. QE-18, p. 1992 (1982).

[42] E.M. Belenov, V.N. Morozov, and A.N. Orayevskiy in Quantum Electronics in Lasers and Masers, part 2, edited by D.V. Skobel'tsyn (Plenum, New York, 1972), p. 217.

[43] A. Javan, W.R. Bennett, Jr., and D.R. Herriott, Physical Review Letters Vol. 6, p. 106 (1961).

[44] W.R. Bennett, Jr., Physical Review, Vol. 126, p. 580 (1962).

[45] W.E. Lamb, Jr., Physical Review, Vol. 134, p. 1429 (1964).

[46] E.I. Yakubovich, Soviet Physics JETP, Vol. 8, p. 160 (1960).

[47] V.S. Idiatulin and A.V. Uspenskiy, Radio Engineering and Electronic Physics, Vol. 18, p. 422 (1973).

[48] L.W. Casperson, Modes and Spectra of High Gain Lasers, Ph.D. thesis (California Institute of Technology, Pasadena California 1971), (University

Microfilms, Ann Arbor, Michigan, order no. 72-469).

[49] L.W. Casperson and A. Yariv, IEEE Journal of Quantum Electronics, Vol. QE-8, p. 69 (1972).

[50] L.W. Casperson and A. Yariv, Applied Physics Letters, Vol. 17, p. 259 (1970).

[5]] L.W. Casperson, IEEE Journal of Quantum Electronics, Vol. QE-14, p. 756 (1978).

[52] B.J. Feldman and M.S. Feld, Physical Review A, Vol. 1, p. 1375 (1970).

[53] L.W. Casperson and K.C. Reyzer, Journal of the Optical Society of America, Vol. 69, p. 1430 (1979).

[54] P. Mandel, Optics Communications, to be published (1983).

[55] L.W. Casperson and A. Yariv, Applied Optics, Vol. 11, p. 462 (1972).

[56] L.W. Casperson, to be submitted for publication.

[57] M.L. Minden, Mode Splitting and Coherent Instabilities in High Gain Lasers, Ph.D. thesis (University of California, Los Angeles, California (1982).

[58] J. Bentley and N.B. Abraham, Optics Communications, Vol. 41, p. 52 (1982).

[59] M. Maeda and N.B. Abraham, Physical Review A, Vol. 26, p. 3395 (1982).

[60] A similar table is given by Minden, ref. 57

[61] N.M. Lawandy and G.A. Koepf, IEEE Journal of Quantum Electronics, Vol. QE-16, p. 701 (1980).

[62] G.B. Jacobs, Applied Optics, Vol. 5, p. 1960 (1966).

[63] P.L. Hanst, J.A. Morreal, and W.J. Henson, Applied Physics Letters, Vol. 12, p. 58 (1968).

[64] See for example, T.L. Paoli, IEEE Journal of Quantum Electronics, Vol. QE-13, p. 351 (1977).

[65] I.H. Hwang, M.H. Lee and S.S. Lee, IEEE Journal of Quantum Electronics Vol. QE-18, p. 148 (1982).

EXPERIMENTAL EVIDENCE FOR SELF-PULSING AND CHAOS IN CW-EXCITED LASERS

N.B. Abraham, T. Chyba, M. Coleman, R.S. Gioggia, N.J. Halas, L.M. Hoffer
S.-N. Liu, M. Maeda, and J.C. Wesson

Department of Physics
Bryn Mawr College
Bryn Mawr, PA 19010

1. INTRODUCTION

As Casperson [1-4] and, more recently, others [6-8] have shown, the inhomogeneously-broadened laser may be particularly susceptible to instabilities even in single mode operation. Before this was understood, such lasers (notably HeNe .6328) had been intensely studied because the combination of Doppler broadening and the resulting anomalous dispersion led to a variety of interesting phenomena [9,10] including hole-burning, Lamb dips, and mode-pulling in single-mode lasers. In several-mode lasers, additional phenomena were observed including mode-mode interactions leading to "mode-pushing", phase-locking, or frequency locking, among others. Thus it is that when we came to study the predominantly inhomogeneously broadened, high-gain transition at 3.51 microns in xenon a wealth of these phenomena were apparent.

In addition, several rather esoteric-seeming predictions had been made for inhomogeneously-broadened, standing-wave lasers. Bennett [11] showed that for Doppler-broadened standing-wave lasers the overlapping of the two holes burned in the gain profile would lead to more than the Lamb dip in the power output versus detuning. In particular, the mode-pulling of the operating frequency toward the atomic resonance frequency could become more complicated, including a region where the operating frequency was triple-valued for a given detuning of the laser cavity. Knowing what we do today from studies of optical bistability [12-14] and multistability [13-15], we might ask whether the stability of the laser in this region had been checked. The answer is no from both theoreticians and experimentalists. In part this is historical, the multi-valued stationary solutions appeared before the current interest in instabilities. Then too, the prediction was for lasers operating far above threshold which may have seemed unreachable experimentally.

Somewhat later, Casperson and Yariv [16] suggested that inhomogeneously-broadened lasers could support more than one frequency of oscillation for a single cavity mode. This phenomenon would result in several stationary states for a single set of operating parameters, leaving the stability open to question again. Although this "mode-splitting" effect would require rather extreme mode-pulling, the fact that these multiple resonant frequencies (as determined by the unsaturated anomalous dispersion) would be relatively unaffected by each other for extreme inhomogeneous broadening suggests that the separate frequencies of oscillation might be observed to coexist in a laser in much the same way as several frequencies of different mode structure exist simultaneously.

The laser of choice for studying all of these anomalies is the xenon 3.51 micron system. This is particularly true because (unlike the HeNe laser which requires the helium for nonresonant, inelastic pumping of the neon transition) the xenon can provide a high gain medium for helium pressures ranging from 0 to 6 Torr, giving a range of inhomogeneous to homogeneous linewidth ratios of from 26:1 to 1:1. (The xenon natural linewidth is about 4.6 MHz (FWHM) and the helium pressure broadening is 18.6 MHz/Torr [5]).

As we will see, this laser is susceptible to a variety of instabilities. These may provide highly stable pulse trains that are even somewhat tunable in the pulsing frequency. We also find a variety of transitions from stable to "unstable" behavior. Included among these instabilities are a single pulsing frequency, more than one pulsing frequency, subharmonics of a pulsing frequency, and chaotic behavior. It is the chaos that expands the company of those interested from laser physicists to a broad interdisciplinary group. (Experimentally, condensed matter has gotten most of the press and results in physics [17], with nonlinear electronic devices [18] and opto-electronic devices [13,19] close seconds.) Recently chaos has been observed in "all optical" systems including optical bistability [20] and in other laser systems [21,22].

In these lectures we will review experimental results on instabilities and chaos in simple laser systems, covering three particular topics. The first will be instabilities and chaotic behavior in single-mode standing wave lasers using the 3.51 micron line in xenon. The second will be similar studies of a self-pulsing ring laser also using the xenon transition. The third will be studies of mode-mode interaction and couplings in two-mode and three-mode lasers making transitions from free-running to mode-locked operation.

2. INSTABILITIES IN SINGLE-MODE FABRY-PEROT LASERS

The design of a single-mode laser to show these interesting effects requires several considerations. 1) The cavity linewidth must exceed the homogeneous linewidth of the laser medium. 2) The gain must bring the laser above threshold and, if possible, the laser must be above threshold even for detunings of the operating frequency of up to one and a half times the Doppler halfwidth. 3) The laser must continue to oscillate on only one mode throughout this detuning.

The first requirement means that $2\gamma t_c < 1$ must be satisfied, where γ for the xenon line is about 14×10^6 sec^{-1} and can be pressure-broadened to about 300×10^6 sec^{-1}. This sets limits on the quality of the cavity which is generally dominated by the mirror reflectivity but is affected to a lesser extent by window absorption and diffraction losses due to apertures.

The second requirement means that the laser medium must provide relatively high gain per unit length to offset the cavity losses and to provide sufficient gain for the detuned operation. This will require that at line center the laser be between two and four times above the lasing threshold.

The third condition is the one which severely cramps the design because, for example, single-mode operation under the 120 MHz Doppler-broadened gain medium will require that successive longitudinal modes operate of order 150-200 MHz apart. Considering that mode-pulling for high gain conditions of interest may be a much as a factor of 4-6, the free-spectral range of a prototype single-mode unstable laser must be of order 1000 MHz. This squeezes the laser cavity down to about 15 cm, leaving barely 10 cm of active region for the laser medium. Hence the high atomic weight of the xenon atoms which gives a relatively small Doppler broadening and the high gain of the 3.51 micron transition (20-200 dB/meter for various pressures and currents) are nearly essential. In addition apertures are required to reject the transverse modes.

The design we settled on for our studies of instabilities in a single-mode, Fabry Perot laser is shown in Figure 1. The 16.5 cm long cavity is formed by a spherical mirror and a 50% reflecting output coupling wedge. Vertical polarization was ensured by the use of windows at Brewster's angle.

A full characterization of the behavior of this laser requires recording of the power output, the intensity fluctuations, and the optical frequency of the laser as the gain or detuning is varied. The first two can be obtained by suitable intensity monitors, at least one of which needs a 100 MHz bandwidth to pass various pulsation frequencies. The measurement of the optical frequency, which cannot be inferred from the laser cavity length because of extreme and often nonlinear modepulling, is achieved by heterodyning the laser under study with a single-mode, frequency-stabilized, reference laser. In order to achieve maximum sensitivity, a two-laser reference system as shown in Figure 2 was used.

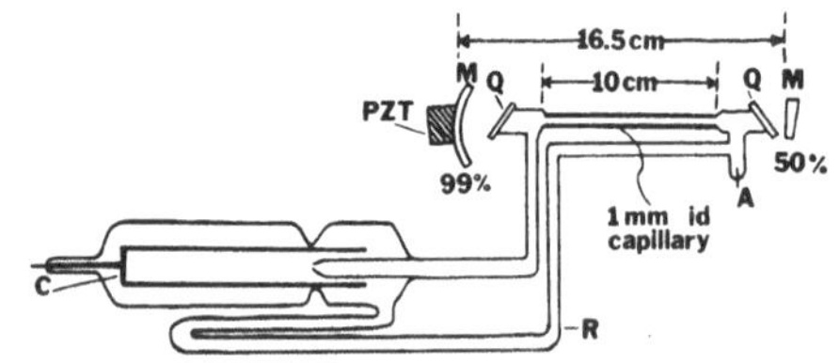

Figure 1: *Single-mode, high-gain laser (top). PZT, piezoelectric crystal; M, mirror; Q, quartz window; R, nonexcited return path for pressure equalization; A, anode; C, cold cathode. Depending on intracavity apertures, t_c ranged from .2 to 2 ns.*

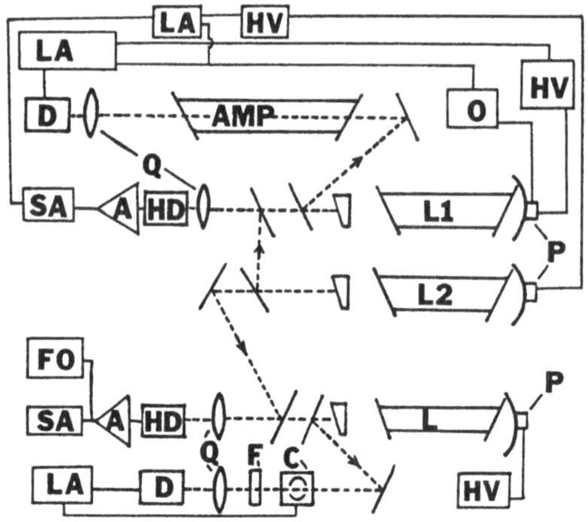

Figure 2: *Optical bench and electronics layout (right). L1, L2, and L, lasers; AMP, laser amplifier; P, piezoelectric crystals; HV, high voltage op amp; D, InAs detector; HD, high frequency InAs detector; LA, lockin amplifier; SA, spectrum analyzer; FO, fast oscilloscope; A, amplifier; C, chopper; Q, quartz lens; O, oscillator (500 Hz); F, line filter.*

The first laser was feedback-stabilized to the peak of the gain of a He-Xe discharge tube providing a frequency reference that had long term stability of a few MHz per day and a short-term stability of about 1/4 MHz. The stabilization was achieved by modulating the length of the laser, causing a frequency variation over a range of 2 MHz. The resulting power fluctuations were detected by a phase-sensitve detector (lock-in amplifier) which provided an error signal to keep the laser-amplifier combination operating at the peak output. The long term drift in the laser frequency was irrelevant to these measurements as only relative frequency variation with detuning of the laser was of interest. A second reference laser was used as the actual reference by heterodyning the two lasers and stabilizing the length of the second laser to provide a particular beat frequency with the first laser. This resulted in a stabilized laser which was not modulated and thus had a linewidth of about 100 kHz with a stability of 250 kHz. This second laser had the additional advantage of being stabilized to an arbitrary frequency which could be selected for optimum visibility of the beat frequencies. Both reference lasers were operated using natural xenon which resulted in a shift of about 50 MHz for their peak output from the peak of the laser under study which was filled with single-isotope enriched Xe-136 to avoid complications from the mix of isotopes that contribute to the gain in the natural xenon 3.51 micron peak.

In Figure 3 we show a sequence of graphs of the inter-laser beat frequency versus cavity detuning for different admixtures of helium.

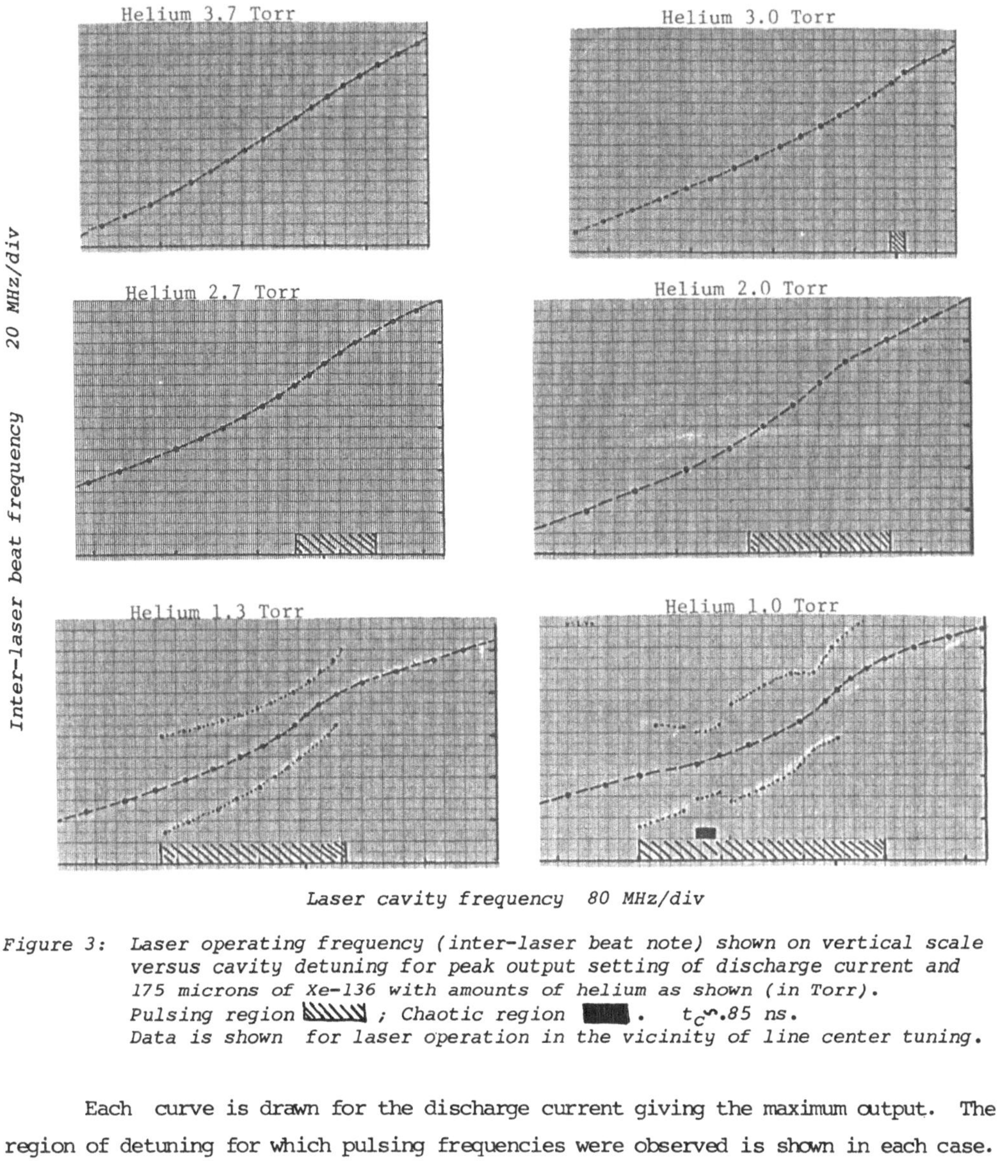

Figure 3: *Laser operating frequency (inter-laser beat note) shown on vertical scale versus cavity detuning for peak output setting of discharge current and 175 microns of Xe-136 with amounts of helium as shown (in Torr). Pulsing region* ▨ *; Chaotic region* ■ . $t_c \sim .85$ *ns. Data is shown for laser operation in the vicinity of line center tuning.*

Each curve is drawn for the discharge current giving the maximum output. The region of detuning for which pulsing frequencies were observed is shown in each case. For 3.7 Torr of helium (or more) no instabilities were observed, but near line center frequency a small kink in the plot of the operating frequency was noticeable. This can be attributed to reduced mode-pulling in the region of overlap of the two holes burned in the inhomogeneously broadened gain profile [11]. Overall, however, these plots are relatively straight near linecenter. For example, the 3.7 Torr case shows the laser frequency varying 22 MHz for cavity detuning of 80 MHz. As such things go in other lasers, this factor of four is "extreme" mode pulling [see 23-25], though not nearly the record factor of 34 observed for He-Xe under special conditions [26].

As the helium pressure is reduced we see the appearance of a region of instability near linecenter, though actually more correlated with the peak output. An exact determination of the line center frequency is complicated by dispersive focussing effects which lead to asymmetries in the power output versus detuning curves [27,28]. The helium added to increase the homogeneous broadening is also known to provide a pressure shift of about 3 MHz/Torr [29]. Thus the "true" line center frequency is obscured by complicated power output and Lamb dip shifts and asymmetries. It may well be that the observed kink in the detuning is the most sensitive indicator of the location of the frequency of the peak intensity and the chaos region may indicate the frequency of zero velocity atoms. The original Bennett prediction of a kink in such plots was for stable cw lasing and we are not able to make detailed connections with it because of the clearly present instabilities and asymmetries in our system.

The region of observed instability widens as the helium pressure is reduced. The lower homogeneous linewidth reduces the threshold gain required for the instability [4,7,8], thus widening the region over which it is observed. With decreasing helium pressure we also find more complicated pulsing phenomena, including the appearance of sidebands to the inter-laser beat note and a region of chaotic power spectra near the asymmetric Lamb dip.

Figure 4 shows results for the lowest helium pressure studied (1 Torr) at a lower gain setting. The pulsing is observed near line center begining at about 20 Mhz with harmonics appearing as the strength of the pulsations grows. The pulsing

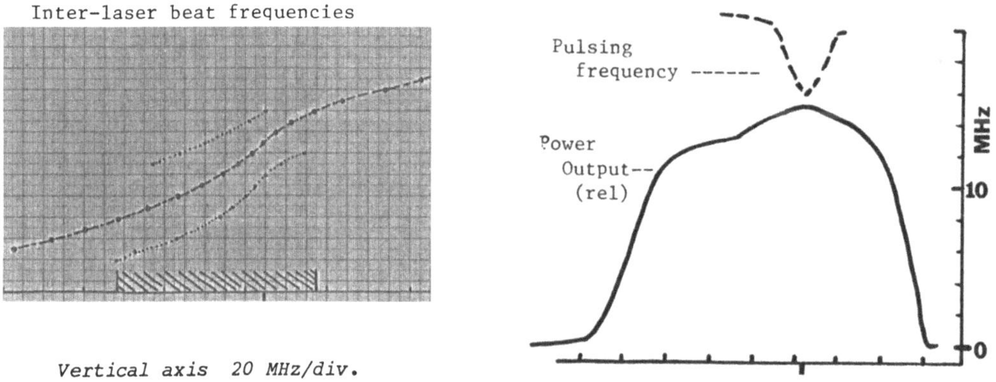

Vertical axis 20 MHz/div.

Horizontal axis in both cases is increasing cavity frequency, 80 MHz/div.

Figure 4: Power output, pulsing frequencies, and operating frequency versus cavity detuning for helium pressure of 1 Torr and discharge current of .5 mA. Peak power output at this current is roughly half that for the best current. Power output (solid right), relative scale; fundamental pulsing frequencies (dotted). Harmonics are not shown but were generally observed. t_c= .85 ns.

frequency moves to a minimum of 16 MHz and then rises to a value of 21 MHz. The minimum frequency occurs at the sharp kink in the frequency detuning plot. From our previous work [9], it appears that the pulsing frequency is closely linked to the intensity of the laser. This may be understood in part as due to power broadening of the holes burned in the gain profile. Note that the dip in pulsing frequency matches the peak output and does not coincide with the apparent dip in the observed asymmetric power output versus detuning.

Returning to the detuning curve for this case and Figure 3, the plotted dashes represent sidebands that were observed to appear in the spectrum of beat frequencies between the reference laser and the laser under study. Note that for the extreme detunings in the pulsing region, only one sideband was visible. The sidebands were more nearly equal in strength near line center corresponding to AM modulation. With detuning the residual pulsing again appeared to involve only one sideband. While theoretical and experimental evidence indicates that the line center instability represents a Hopf bifurcation [7,8], the single sideband evidence suggests the bifurcation may be different for the detuned case.

Independently Minden and Casperson [30] and Wesson [31] have attempted to model the detuning in the standing wave laser but the theory is quite complex. The initial work was for the so-called "no population pulsations" approximation which gives the correct qualitative behavior but fails to exactly incorporate the full nonlinearities of the medium. Recently Minden has completed an alternative analysis in the small signal regime which seems promising [32].

In Figure 5 we show similar data for detuning and pulsing frequencies at the highest gain obtained at this lowest helium pressure from the set shown in Figure 3. In this case we see considerably more interesting structure including the appearance of a first pulsing frequency at 14 MHz and its harmonics followed by the appearance of a second pulsing frequency at 20 MHz. These two incommensurate frequencies coexist over a narrow detuning range and at one point appear to lock at exactly a 3:2 ratio. With slightly greater detuning the higher frequency dominates showing a dip in frequency over a narrow range just after the lower frequency disappears. A strong dip in the pulsing frequency is then observed as the pulsations become quite chaotic, with the frequency shifting from a relatively high-Q pulsation at 20 MHz to a broadband spectrum with some broad peaks at 16 MHz. The more stable pulsing reappears and again terminates at a relatively large value of 24 MHz.

Figure 6 shows the pulsing frequencies and power output for scans of the gain in the region of the two incommensurate frequencies.

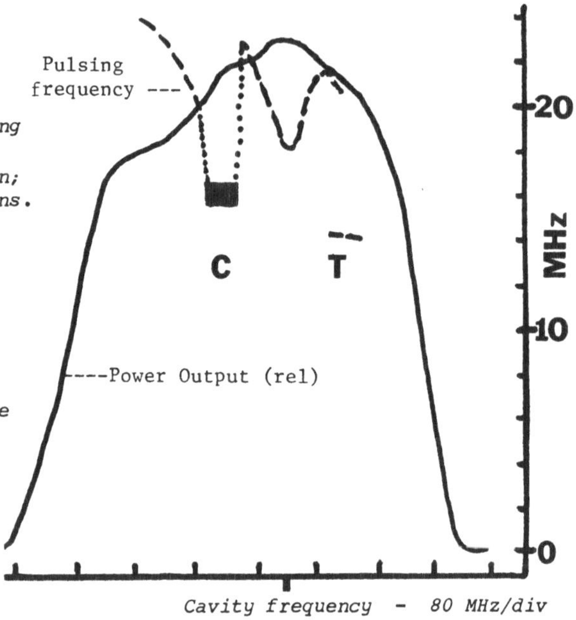

Figure 5: *Power output and pulsing frequencies versus detuning for optimum current at 1 Torr He, 175 mTorr xenon; current 1.4 mA.* $t_c = .85$ *ns.*

C - Chaotic pulsation region: period one chaos.

T - Coexistence region for two incommensurate frequencies. Values are close to a 1.5 ratio.

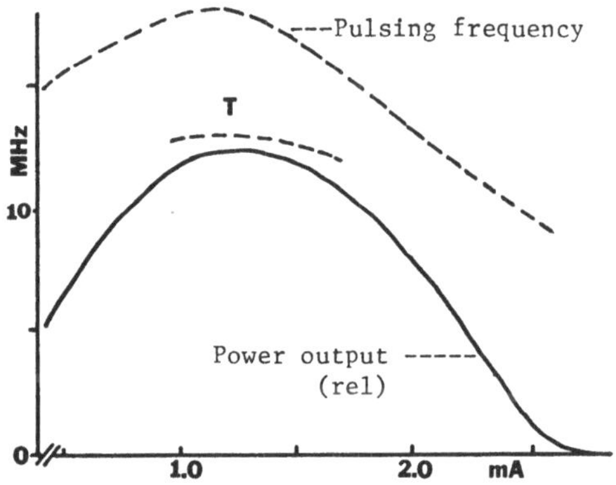

Figure 6: *Power output and pulsing frequencies versus discharge current for fixed detuning in region of two incommensurate frequencies from Fig. 5.*

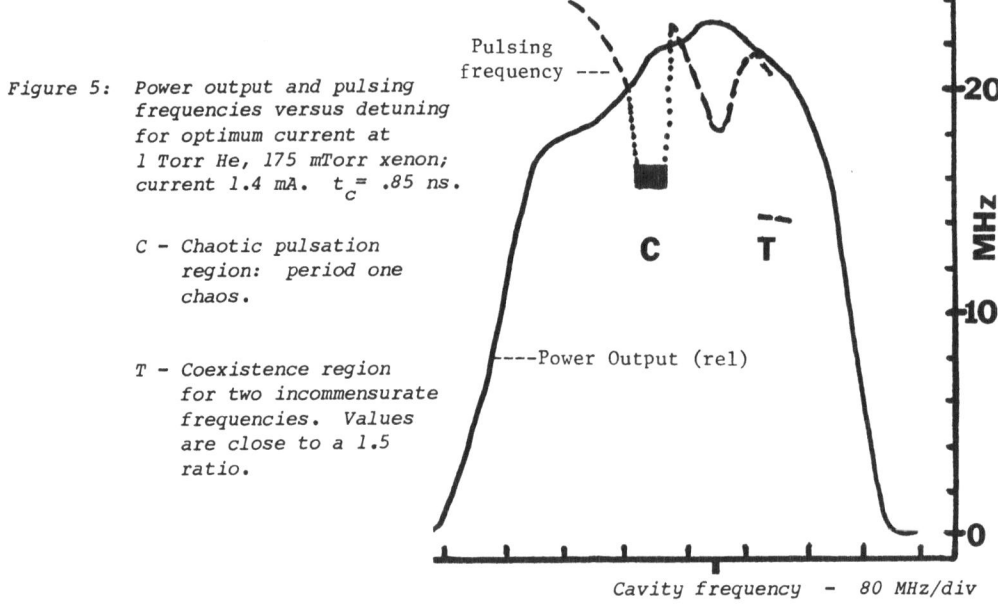

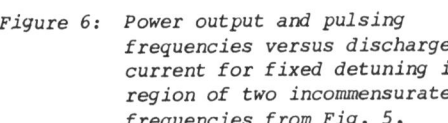

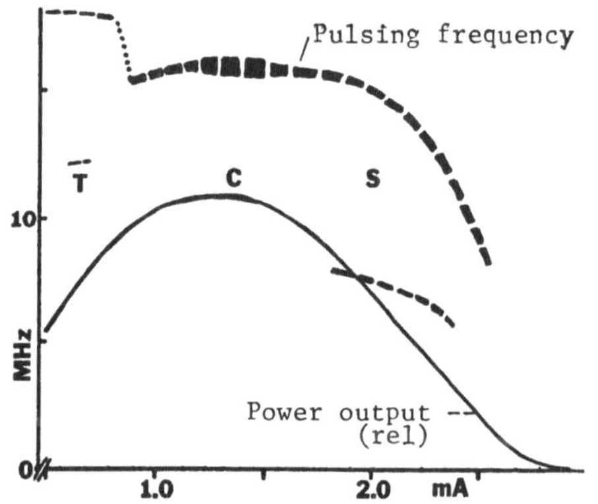

Figure 7: *Power output and pulsing frequencies versus discharge current for fixed detuning in region of chaotic spectra shown in Fig. 5.*

T -- two frequencies; c -- chaotic region; S -- subharmonics. Harmonics of pulsing frequencies are not shown. $t_c = .85$ *ns.*

In contrast to our earlier work, current studies show that maintaining a stable, low current discharge in the small capillary of the laser under study requires a relatively large ballast resistor, of order 500 kohms or more. With a large resistor, we can vary the discharge in the low current regime where the gain is roughly proportional to the current. However, the discharge spontaneously extinguishes for currents below some limit which prevented our work reported here from observing the laser threshold at low currents. We are also able to raise the current sufficiently to turn the medium into an absorber because the inelastic electronic collisions which excite the lasing levels reach the upper level by a single collision and the lower level by double collisions.

In this case we see that the pulsing exists over essentially the entire range of lasing action with pulsing frequencies observed as low as 9 MHz at very low gain settings. The termination of the pulsations at a threshold value of the gain just above lasing threshold indicates how unstable this laser system can be. This evidence that the minimum pulsation frequency is nonzero is consistent with the latest theoretical predictions [4,7,8] and indicates a weakness of the "no population pulsations" approximation [2] which predicted a zero frequency at threshold. Thus we find that a combination of cavity tuning and gain is required to observe the instability and not just absolute gain. The vanishing of pulsations at relatively large output power in the detuning curves may be interpreted as a requirement of a large gain at that particular detuning, but it does not necessarily indicate a requirement of a large gain everywhere. Of theoretical and practical interest is the lack of hysteresis in these curves, suggesting that the instabilities are relatively stable functions of the operating parameters. The pulsing frequency of 9 MHz when compared to the pressure-broadened linewidth of 23.2 MHz gives a value of .78 for the pulsing frequency normalized to the γ of the medium. Current theoretical models should shortly be able to predict this number given our cavity lifetime of about 1 ns.

Of some news to the theoretical considerations is the appearance of an asymmetry of the pulsing characteristics for two currents giving the same average power output. These two currents indicate different excitation of the two atomic levels involved in the transition and so such an asymmetry might be expected but it has not heretofore been investigated in the theoretical models.

The gain scan in Figure 6 also shows the region of two incommensurate frequencies without chaotic interaction. Figure 7 shows similar results for a gain scan through the narrow chaotic region. Here there is a broad chaotic power spectrum at the peak gain. At this setting there is a pulsing frequency and its harmonics superimposed on the chaotic background. This is sometimes referred to as "period one chaos". With detuning to the low current side the chaos reduces to 10 MHz wide peaks at 15 MHz and its harmonics and then separates into two competing frequencies at 18 MHz and 12 MHz before the 12 vanishes and only a narrow set of peaks for 18 MHz and its harmonics remains. On the high current side of the peak output the chaotic background is reduced and the fundamental broad pulsation peak at 16 MHz develops a subharmonic at 8 MHz. As the gain is decreased further with increasing current, the subharmonic vanishes and the still relatively broad pulsation frequency disappears almost at the laser threshold at a frequency of 8 MHz (.7 in normalized units).

Selections from the observed power spectra and real-time recording of the pulsings are shown in Figure 8.

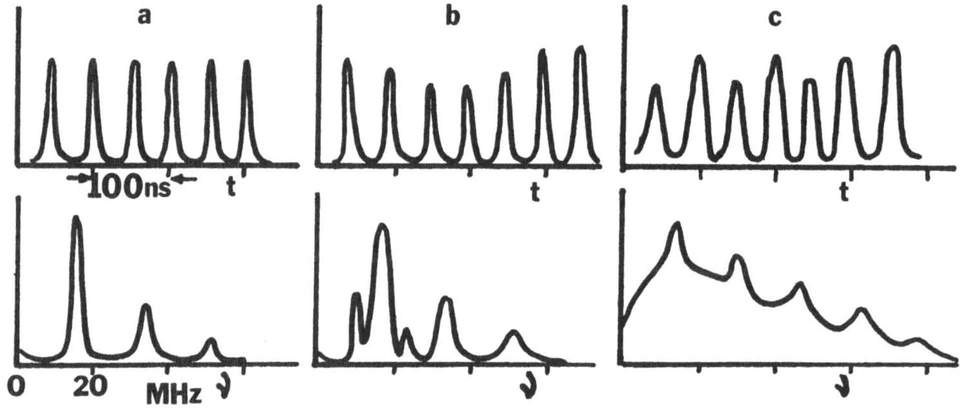

Figure 8: Real-time output and power spectra from regions of detuning in Figs. 6,7. Real-time displays (top); power spectra (log scale), bottom. Stable pulsing (a), two incommensurate frequencies (b), period one chaos (c).

At these helium pressures, the lasers typically showed some instabilities at the most extreme detunings observed. Features observed include pulsing frequencies and their harmonics and the appearance of subharmonics mixed with features similar to those observed near linecenter but both weaker and at lower pulsing frequencies. All of this occurred at about 50-70 MHz detuning from line center. In these particular cases, careful analysis of intra-laser and inter-laser beat frequencies established that some of these represented weak instabilities of a very weak transverse mode. However, some features disappeared when the longitudinal beat note vanished suggesting that there was an instability in the longitudinal mode in this tuning region which may properly be ascribed to the mode-splitting of Casperson and Yariv [16].

Selected results are shown in Figure 9 for earlier work at no helium pressure and higher cavity losses (t_c = .42 ns). Here for a small homogeneous linewidth, three regions of pulsing near line center are observed with definite gaps of no pulsing. This can be attributed to the complicated hole overlapping suppression in a slightly detuned Fabry-Perot laser and is satisfactorily explained by the theoretical analyses [30,31].

The pulsing observed away from line center is not linked to other transverse modes in this case. The mode-splitting instability [16] may explain the qualitative

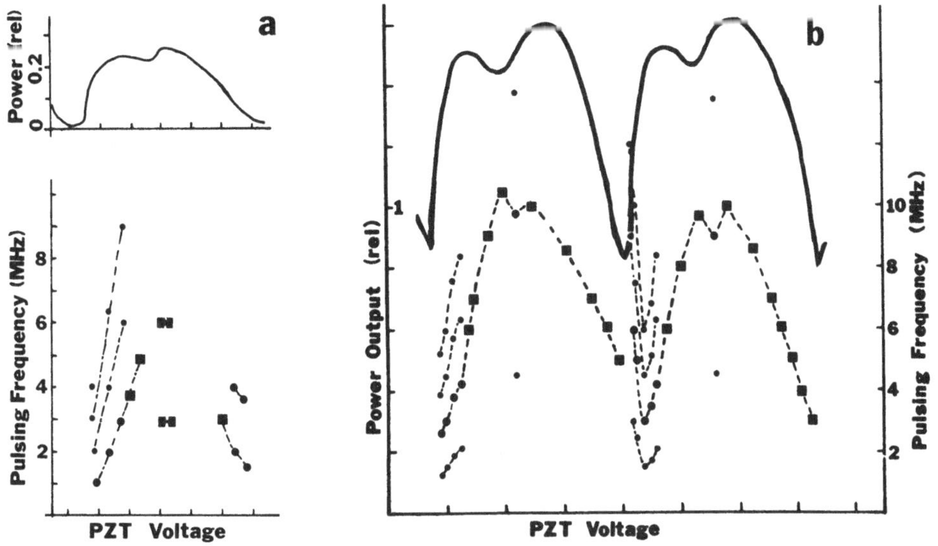

Figure 9: Power output and pulsing frequencies versus cavity detuning for 98 mTorr xenon and no helium at different discharge currents, after Ref. 5b. Large dot shows stable, high-Q pulsation frequency; small dot shows harmonics or subharmonics (b). Large square indicates broad spectral peak.

change from chaotic to stable pulsing in the wings and the appearance of subharmonics in Figure 9b.

Figure 10 shows a sequence of power spectra of the pulsations with increased detuning from 10a-10f. These show complicated pulsation frequencies away from line center including a high-Q pulsation frequency and a subharmonic and a broad asymmetric hump. Nearer to line center a simpler pattern of high-Q pulsations and an associated subharmonic is observed. Because this early data was taken before the current could be reduced to the highest gain region, no chaotic behavior was observed. In quite recent work at no helium, chaotic spectra near line center have routinely been found. A more detailed summary of these results will be published elsewhere [33].

In summary, the single-mode standing wave laser provides a wealth of evidence for the predicted laser instabilities. The appearance of such instabilities for relatively large homogeneous linewidths confirms that this effect does not rely solely on inhomogeneous broadening and that the instability is an enhanced version [4,7,8] of the self-pulsing predicted originally in models of homogeneously-broadened lasers [34-37].

There also seem to be important complications of the results for the detuned laser that may be properly attributed to the interaction of the two counter-propagating waves and the overlapping of the two holes burned in the gain

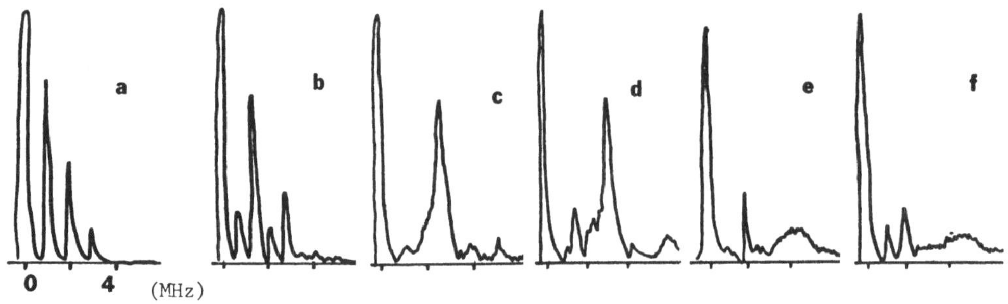

Figure 10: Power spectra with increased detuning (a-f) from line center for 140 mTorr of xenon and no helium. Vertical scale is logarithmic. t_c=1.2 ns.

profile. This suggests that detailed theoretical treatments will be difficult and even at line center the standing waves may lead to spatial variations that will be troublesome to deal with. Nevertheless this work has demonstrated that the instability can be observed for a wide range of laser parameters and suggests strongly that it may also appear in other suitable [4] high gain laser systems. (The experiments of Weiss et al [21] may indicate single-mode instabilities in a HeNe 3.39 laser, but at this time it seems more likely that theirs is a multimode effect.)

3. INSTABILITIES IN RING LASERS

One of the greatest complications of the standing-wave laser system (spatial inhomogeneities -- standing waves, spatial hole burning; and double spectral hole-burning by a single mode) can be eliminated if one goes to a travelling-wave ring laser. By convention the theoreticians can assume a unidirectional, single-mode laser to simplify their work and leave it to the experimentalists to realize it in practice.

Difficulties obviously arise because most optical systems are equally transmissive in the forward and backward directions. Thus in a ring laser, the two counter-propagating parts of the standing-wave laser become two modes which are in most cases degenerate in frequency. In an extremely Doppler broadened medium these modes will not significantly interact except in the vicinity of line-center tuning where the modes use the same atomic velocity classes.

Because we generally wish to study the mixed broadening cases and operation at line-center, it is important to ultimately suppress one of the two counter-rotating modes. The classic method for doing this is the use of two linear polarizers and two 45-degree Faraday rotators to accomplish a single-polarization, single direction operation. Unfortunately the process of designing and specifying the Faraday rotators for a wavelength lacking commercial applications has been long and tedious. Thus our preliminary work with ring lasers has accepted the complication of two modes. (It has been suggested that undirectional operation could be more simply achieved by a nonresonant reflection of one of the output beams, returning it into the laser to seed and thereby enhance the other beam and suppress the back-reflected beam. We have not been persuaded that this induced coupling between the two directions of operation would be desireable and have chosen to wait for the Faraday Rotators which are now available.)

Studies of instabilities and chaos in ring lasers proceeded with the two laser configurations shown in Figure 11.

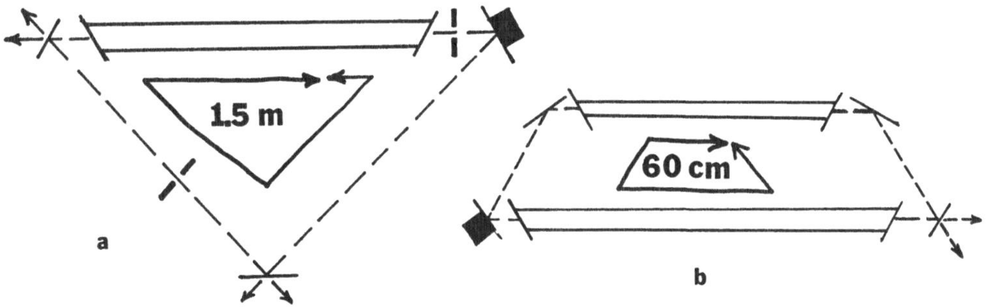

Figure 11: Designs for Xenon 3.51 micron ring lasers. a) 3-mirror ring with 2 90% mirrors (t_c=6.2 ns). b) 4-mirror ring with one 90% mirror (t_c=3.6 ns). Calculated t_c neglects gain and dispersion focussing and diffraction loss. Round-trip cavity lengths as shown.

The first laser had three mirrors and a gain medium in only one arm with output coupling through two 90% reflecting mirrors. The total path length of the ring was 1.5 m giving a free spectral range of 200 MHz. While some single-mode results were obtained, the second laser was designed to give higher gain and a greater free spectral range through the reduction of its path length to 60 cm.

Interesting results were obtained in comparing the two modes in simultaneous operation. Typically, for all detunings the pulsations were observed in both modes or not at all. For 140 microns of xenon-136, the puslation frequences were observed in the range of 4-6 MHz. When pulsations were observed, they appeared as either in-phase pulsations, suggesting an atomic-pulsation driving of the instability, or as out-of-phase pulsations suggesting a very weak coupling and a simple gain competition mechanism. The out-of-phase pulsations were observed closer to line center and for high helium pressures. In some cases the pulsations simply alternated strength between the two directions while occuring in-phase. In these initial studies the pulsations were not observed over a sufficiently wide range to investigate a fully uncoupled operation of the two modes. Some forms of mode switching are known to occur in two-mode lasers [38] so some degree of caution must be taken in interpreting this data.

The pressure dependence of the pulsing frequency for fixed ratio of helium to xenon pressures of 20:1 is summarized in Table 1. This data was taken for the long laser with t_c = 6.2 ns.

TABLE 1

Total pressure (Torr)	Pulsing frequencies	ω_p ($\times 10^6$ sec^{-1})	$\gamma=\pi(\Delta\nu_h)$ ($\times 10^6$ sec^{-1})
3.8 - 2.1	none		
1.9	weak pulsing at 10 MHz	62 $\pm$ 2	126 $\pm$ 10
1.6	weak pulsing at 10 MHz	62	108
1.2	moderate pulsing at 10 MHz	62	85
.9	moderate pulsing at 8 MHz	50	67
.7	strongest observed pulsing, 6 MHz	37	55
.4	moderate pulsing at 5 MHz	31	38
.25	weak pulsing at 5 MHz	31	29

Data for long ring laser with t_c = 6.2 ns (neglecting diffraction losses). The decline of pulsing strength at low pressures is due to overall weak laser output, while the weak pulsing at high pressures indicates that the pulsing is a small modulation of a strong laser output.

Figure 12 shows typical real-time, multiple-sweep oscilloscope traces of the two mode laser output.

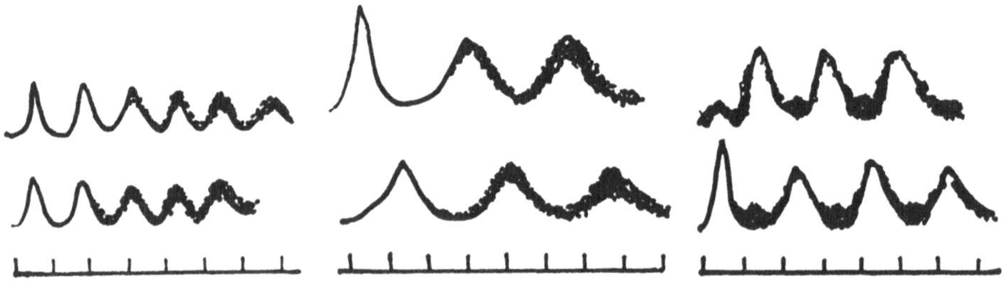

in-phase pulsing .1 s/div — *out-of-phase pulsing .05 s/div* — *in-phase, alternate size .2 s/div*

Figure 12: Multiple-trace, real-time oscilloscope traces of two counter-propagating signals in large ring laser showing in-phase and out-of-phase pulsations. Real-time signal ac-coupled. Here and elsewhere it is reasonable to claim that the signal went to approximately zero between pulses.

Figure 13 shows several single-trace pulse trains showing both a simple pulse height alternation (period doubling) and more complicated waveforms. In each case there seems to be a sort of relaxation oscillation in each pulse. In some cases there are waiting periods between pulses as observed in the early work by Casperson [1], in other cases the pulse trains occur steadily if chaotically in amplitude.

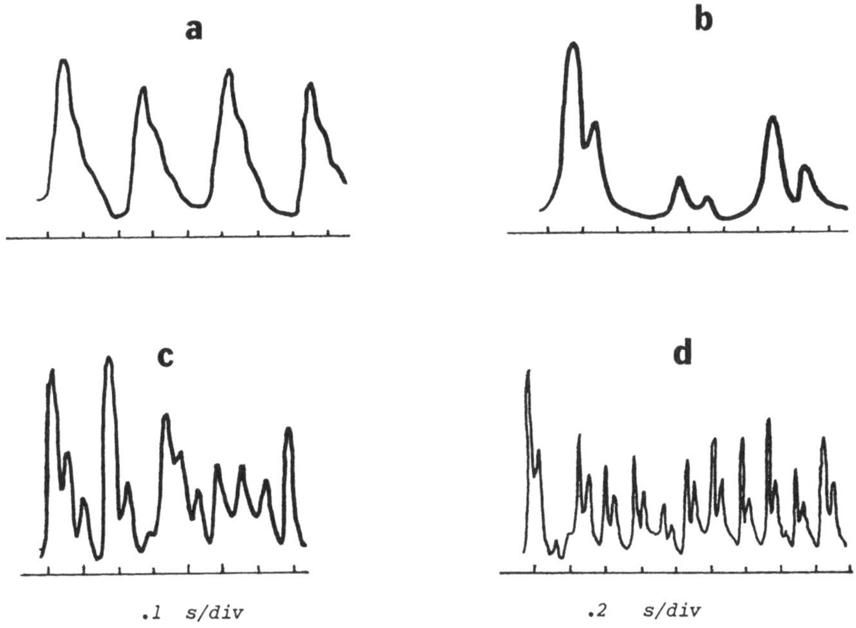

Figure 13: Real-time, single-sweep oscilloscope traces of large ring laser pulses.

To study variation of the pulsing phenomena in detail, we used the compact ring laser and found a clear sequence of transitions from stable to chaotic behavior. Figure 14 displays several stages of the laser operation with detuning.

Observations included simple pulsing, period doubling, and chaotic behavior represented by a substantial increase in broadband spectral noise. The chaos at times included some periodic behavior - period one, period two, period three, and period four were observed. Bursts of pulses that seemed to represent higher periodicities (period five, period six and period seven) were also observed at or near the threshold for chaotic behavior.

In some regions of fixed detuning, the laser seemed to spontaneously switch from chaotic behavior to stable pulsing behavior. It is not presently clear whether this was a form of chaotic intermittency ([39,40] which has been seen in numerical studies of lasers with modulated parameters [41] and lasers with injected signals

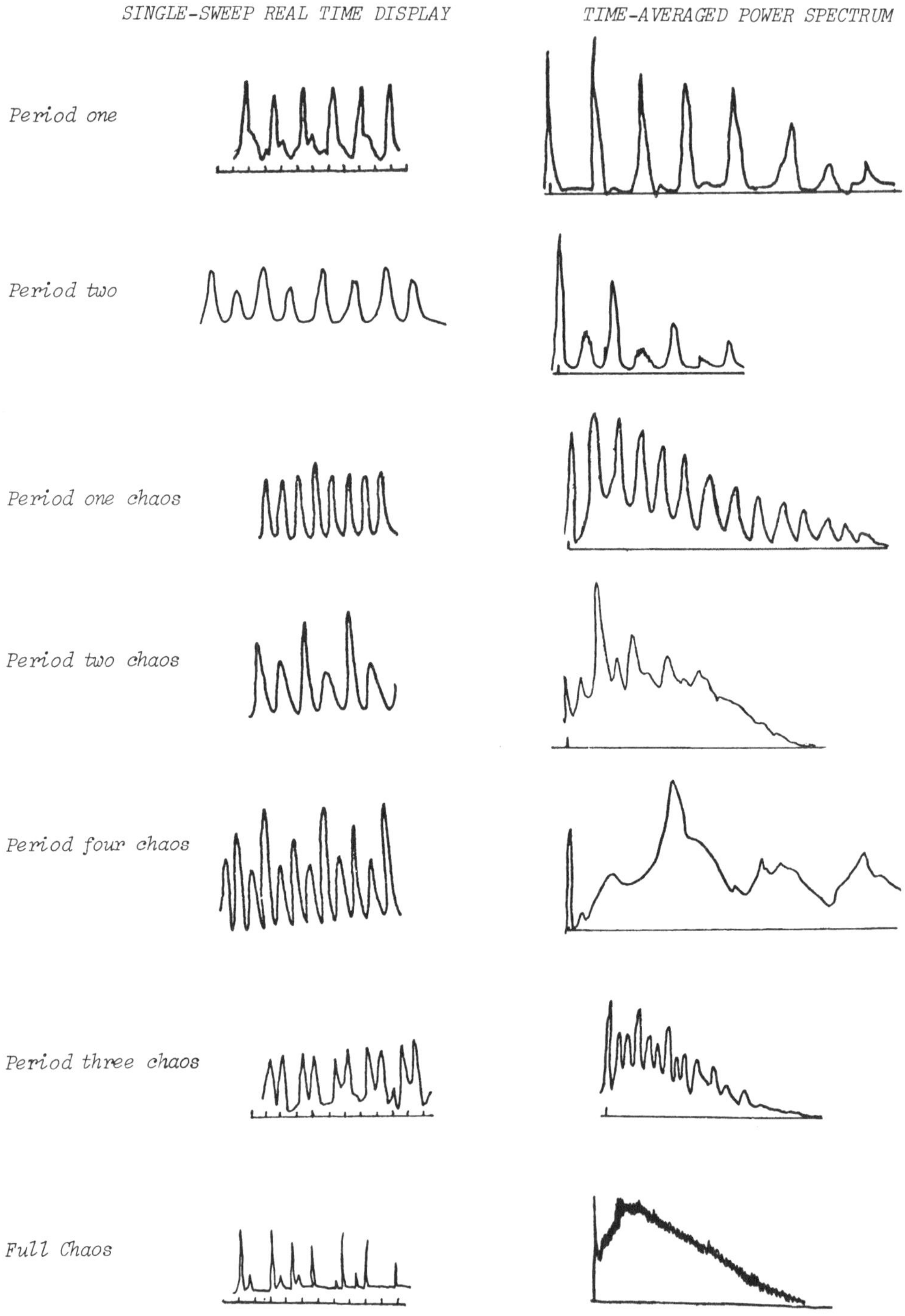

Figure 14a: *Real-time single-sweep oscilloscope traces and some associated power spectra from the output of the compact ring laser filled with 70 mTorr xenon and no helium.*

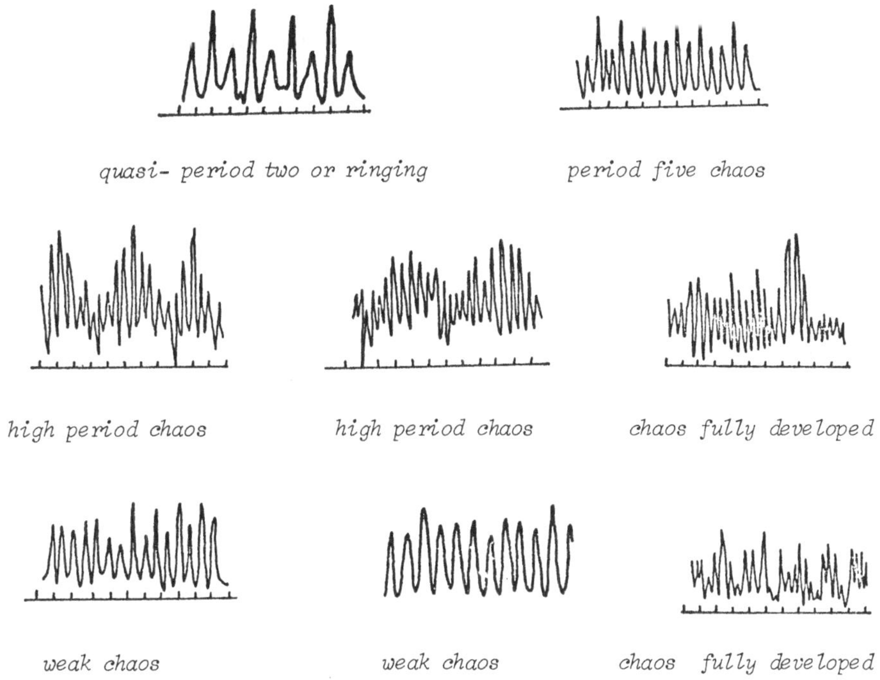

Figure 14b: *Real-time oscilloscope traces of various other characteristic features shown by the compact ring laser. From a brief time series such as those shown, it is often difficult to infer whether there is long-time regularity of high order or whether these represent part of even more complicated or chaotic behavior. The time-average power spectrum often provides the more definitive answers in these cases.*

[41,42]) or was a switching induced by environmental noise. Further isolation of the laser cavity and improved mechanical stability will be used to check these results.

The ring laser has given us the clearest indication that the basic laser equations may yield a simple sequence of periodic and chaotic states as various parameters are changed. This is consistent with observations by Haken (single mode) and Graham (multimode) [43] that the basic equations (in the homogeneous case) are analagous to the Lorenz equations which lead to instabilities and chaos in fluid dynamics. The additional laser boundary conditions may provide some limitations on direct comparisons, but the data suggests that periodic and chaotic behavior of the kinds now well documented for fluids will be observed. Obviously these lasers are the subject of intense current study.

4. CHAOTIC BEHAVIOR IN THE TRANSITION OF A THREE-MODE LASER TO A MODE-LOCKED CONDITION

One of the next most intriguing of the simple laser systems is the three-mode laser. In his now classic paper, Lamb [9] found that the nonlinearities in the medium would lead to the generation of combination tones. Under suitable conditions these combination tones would be at frequencies so close to the frequencies of the three oscillating modes that a locking would take place. Bennett's work [10] also considers these mode-mode couplings and these early studies led to our present understanding of mode repulsion, mode-pushing and pulling, and a wide variety of complex phenomena when modes in standing wave lasers were symmetrically tuned with respect to line center.

Interests in the operation of three-mode lasers came from a wide variety of sources. The basic locking mechanism which leads to equal spacing of the frequencies (mode-locking) results in very short, high power pulses which have been used to great advantage particularly for various kinetics and spectroscopy applications. Of course the very technical and practical applications require the locking of many modes to minimize the pulse length.

In the early laser instability analyses for homogeneously broadened lasers, Risken and Nummedal [36] and Graham and Haken [34] considered a breakdown of cw laser action that was distinct from the single mode instabilities discussed in the previous section. They discussed the onset of weak sideband frequencies as supplements to the initial operation of a strong single mode. These sidebands are at cavity modes that differ from that of the main mode and the instability can be viewed as the onset of mode-locked three-mode operation of the laser. Such instabilities have also been studied in optical bistability [12,44] and are conceptually simpler to understand than the single-mode instabilities discussed earlier.

In the laser case, the transition with increasing gain from free-running single-mode (or multimode) operation to mode-locked operation can be viewed as occurring at a "second laser threshold". Studies of mode-locked dye lasers have found such transitions [45,46]. The pulse to pulse irregularities in some mode-locked lasers [47] may be evidence of a kind of chaotic behavior above this mode-locked threshold which is linked to the decoupling of the modes [48] as predicted in the mode-locking theories [34,36,49].

With these two guiding interests, one would think that the three-mode case would have been well studied and would now be well understood. The literature of the three-mode lasers is quite extensive and several good reviews exist [49,50]. A variety of authors reported combination tones and low frequency noise which was variously attributed to phase fluctuations, mode-competition, or secondary beat frequencies generated by the nonlinear medium. However, most studies were completed before the current fascination with (and understanding of) chaotic behavior, so detailed investigations of the transition to mode-locing are not available.

When three independent modes coexist in a laser at three separate optical frequencies, the power spectrum of the output intensity will show only the three beat frequencies between the modes. A weak coupling of the modes will lead to locking of the relative phases without significantly altering the mode frequencies. If the phase-locked frequencies are not equally spaced, the nonlinearities of the medium can generate a low-frequency modulation of one or more of the modes at a frequency equal to the difference between the nearly equal beat frequencies.

Our interest in this problem begins with the appearance of such a "beat-beat" frequency [51] and follows the mode coupling as the gain or cavity tuning is adjusted to bring the modes to a frequency locked (equally spaced frequencies) condition as well.

In Figures 15 and 16 we show the power spectrum of the laser output for three uncoupled modes and for three phase-locked modes which show a low frequency "beat-beat" note and its harmonics. The three high frequency beat notes are the inter-mode beat frequencies that appear in the detected signal.

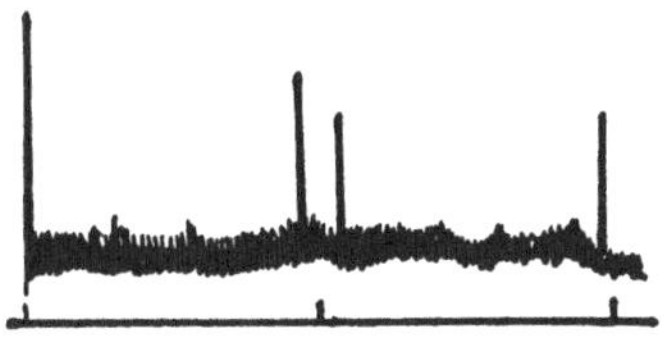

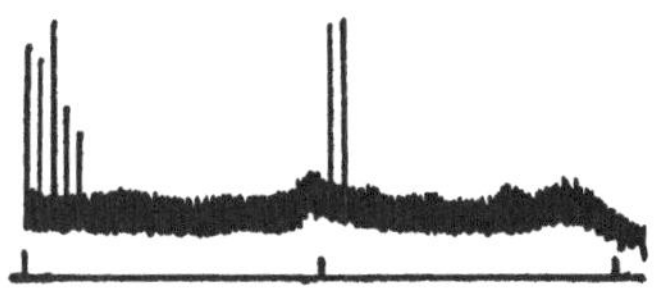

Frequency scale: 100 MHz/div

Figure 15: Beat notes observed for three-mode laser with uncoupled modes.

Figure 16: Beat notes observed for three-mode laser with phase-locked modes showing beat-beat notes.

In Figure 17 we show changes of the low frequency secondary beat note pattern as the laser is tuned toward the mode-locked condition.

As the central mode is tuned to line center, the beat-beat frequency goes to zero. For large values the spectrum is relatively stable but as the frequency approaches zero a form of critical fluctuations seems to appear as the mode-locking condition is neared. When the mode-locking occurs, the low frequency noise vanishes

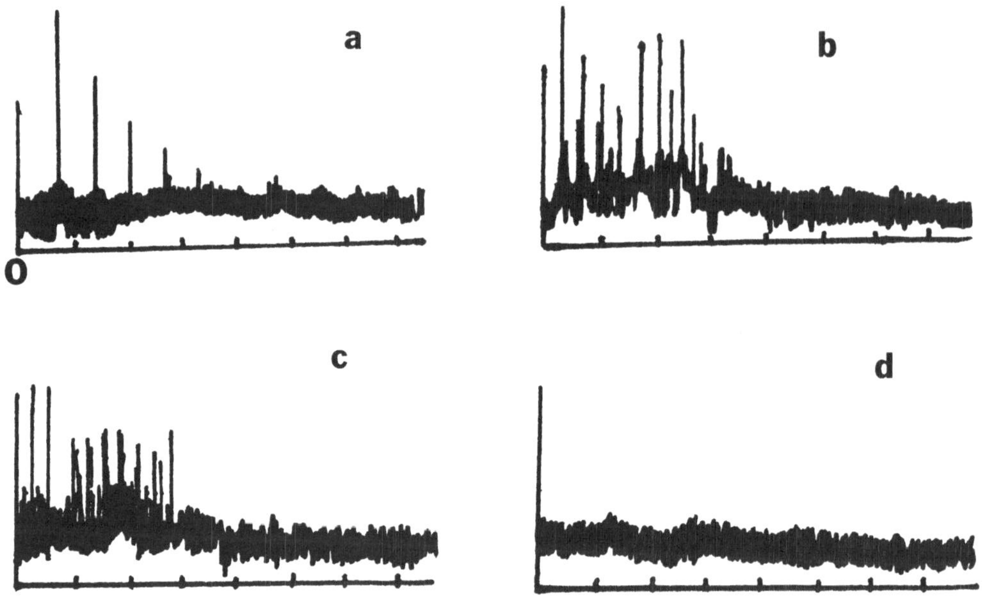

Figure 17: *Power spectra (low frequency region) of laser output as laser cavity length is tuned to achieve mode-locked operation. Vertical scale (logarithmic); horizontal scale 5.5 MHz/div.*

and the output is stable. The simple periodic pulsing at the dominant inter-mode beat frequency continues, but it is off scale in these figures.

In Figure 18 we show the behavior of the beat-beat frequency as the gain of

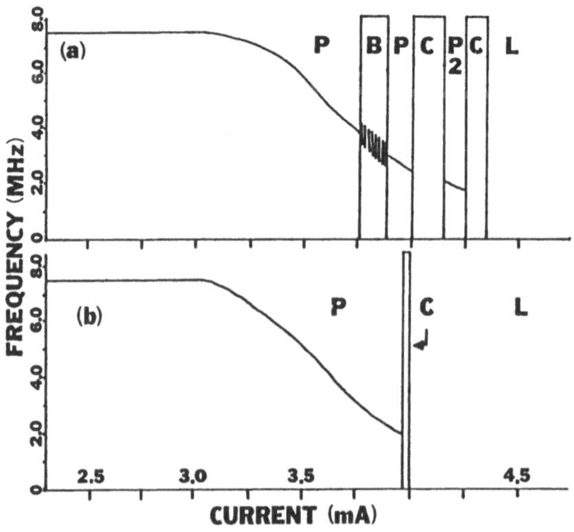

Figure 18: *Fundamental of secondary beat notes versus discharge current (approximately proportional to gain) after reference 36. P, periodic; B, broadened; C, chaotic; P2, period two; L, mode-locked.*

the laser was varied by changing the discharge current. Selected power spectra from various regions of distinctive behavior are shown in Figure 19.

The intermediate mode-coupled state of the laser passes from a stable pulsing regime to a much more unstable manner of pulsing evidenced by peaks in the power spectrum that were nearly 100 times broader than in the stable region. With further change in the gain, stable pulsing reemerged followed by a region of broad noise typical of chaotic behavior. Stability returned in a narrow window of period-doubled behavior followed by more chaos before the very quiet mode-locked region was reached.

On decreasing the gain, the mode-locked region persisted over a much wider range and only a remnant of the intermediate states was observed before the wide range of stable pulsing was obtained. Such hysteresis is to be expected in the locking and unlocking of harmonic systems.

Among the benefits of the understanding of these observations is their aid in interpreting recent results in both theories and experiments with other systems. Theoretical analyses of lasers with injected signal [42], and lasers with special phasing [52] show a form of "breathing", amplitude modulation of pulsing. A phase locking of two non-commensurate pulsing frequencies which are nearly identical would result in such a breathing effect.

Recent experiments on a "single-mode" HeNe laser as one mirror was misaligned [21] have shown a wealth of phenomena including a Grossmann-Feigenbaum-May type [40,53] period-doubling sequence into chaos and a Ruelle-Takens-Newhouse [40,54] route to chaos. It seems most likely from our own work that the tilted mirror caused coupling to transverse modes and led to the observed low frequency beats in a manner similar to the phase locking in our 3-mode laser system.

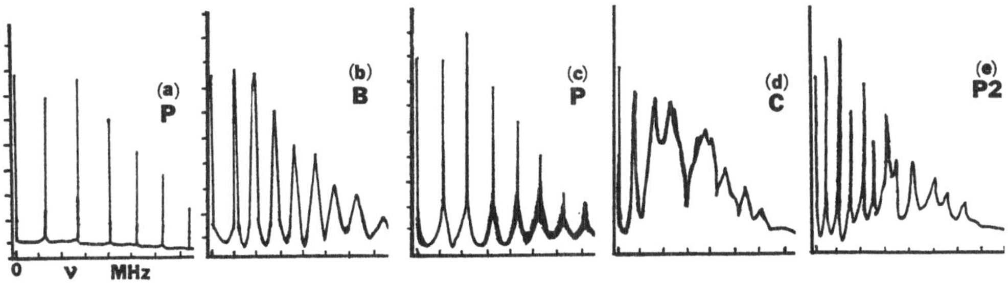

Figure 19: Power spectra of secondary beat frequency and harmonics for different discharge currents selected in scan of Fig. 18 (after reference 36). a) periodic (P); b) broadened (B); d) chaotic (C); e) period two (P2).

As a final general note, there are cases in these three-mode studies and in previous work using the 3.51 transition in xenon where we have observed a beat frequency characteristic of a difference frequency between a mode detuned from linecenter and a nearby hole burned by a second mode detuned to the other side of the center frequency. We have also observed instabilities when only a centrally tuned mode and a detuned mode were present. This is likely to be caused by a coupling of the detuned mode to a latent mode at a frequency close to the combination tone generated by the two oscillating modes.

Obviously, the nonlinearities in the laser medium will create many combination tones of various orders, and coupling between modes and combination tones should provide a fertile area for both experiments and theory. We should also note that such complexity observed in the simultaneous locking of three modes to nonlinearly generated injected signals may be modelled simply by the transition to slaved operation of a laser oscillator with an injected signal [42].

5. PROSPECTS FOR FUTURE STUDIES

Despite the previous work by others and our own study, there remain many exciting steps and areas of study. With Faraday Rotation Isolators in place we will be able to study the ring laser with unidirectional operation eliminating the two-mode complications in the present work and the standing wave complications of the Fabry-Perot laser work. We should have an experimental system that is finally so simple that the tests of theories and guidance of further theoretical work will both be possible.

The Faraday Isolators will also permit studies of the operation of lasers with an externally injected signal (variously referred to as "Laser with Injected Signal" or "Injection Locking"). Here too the theory is well developed and predicts a variety of intermediate pulsing phenomena and chaos enroute to the ultimate locking of the laser to the injected signal.

Coupling of several modes will also be a subject of continuing study. Variations on the mode structure of the modes being coupled will be studied and various forms of passive or active mode coupling may be tried to enhance the mode-locking regime and to perhaps (happily) complicate the transitions to mode-locking.

Finally, it is clear from recent work that several routes to chaos are observed when the systems have modulated parameters such as modulated gain of the laser or modulated frequency or amplitude of an injected signal [41]. Both from technical interests in mode-locking (with active modulators) or injection locking, and from the curiosity about the universality of routes to chaos, these predictions should spark broad interest and continuing research.

In summary, the results are both interesting and tantalizing and confirm the widely held view (among those of us with roots in laser physics) that the laser with its nonlinear field-atom interactions is a fertile testing ground and broadly applicable model for many proposed universal relationships in nonlinear, nonequilibrium dynamics.

6. ACKNOWLEDGEMENTS

This review would not have been possible without the extraordinary labors, patience and insight, of those who worked on these projects. The studies of the Fabry-Perot lasers still use the original lasers designed by Robert Tench and Robert MacDowall which were first studied by Janet Bentley. The detailed studies of Mari Maeda and Joel Wesson and the final careful work of Robert Gioggia have made the theoretician's nightmare a reasonably well-understood experimentalist's dream. The ring laser work has relied heavily on the proof of feasibility by Michael Coleman, the designs of Lois Hoffer and Alexander Rudolph, and the initial experiments by Lois Hoffer, Thomas Chyba and Christopher Chyba. The unidirectional designs of Su-Nin Liu and Lynn Urbach should shortly lead to results for unidirectional ring lasers and lasers with injected signal. The multimode laser phenomena were observed thanks to the careful notes of Mari Maeda, the patience of Su-Nin Liu in developing the HeNe system and the systematic studies by Nancy Halas.

Technical support from glassblower George Kusel and from machinists Bartholomew Hamilton, Harold Hartman, John Andrews, and Richard Willard was invaluable. Consistent help and advice from Steven Adams aided all of our recent experimental work.

Useful discussions with L. Narducci, L. Lugiato, and L. Casperson have greatly helped in the interpretation and understanding of these results.

This work was supported in part by a Cottrell College Science Grant and a Cottrell Research Grant from Research Corporation, by National Science Foundation Grants ENG-7823729 and ECS-8210263, and by an Alfred P. Sloan Research Fellowship.

7. REFERENCES

1. L. Casperson, IEEE J. Quant. Elec., QE-14, 756 (1978).
2. L. Casperson, Phys. Rev. A, 21, 911 (1980); 23, 248 (1981).
3. M. Minden and L. Casperson, IEEE J. Quant. Elec., QE-18, 1952 (1982).
4. L. Casperson, chapter in this volume.
5. J. Bentley and N.B. Abraham, Opt. Comm., 41, 52 (1982); M. Maeda and N.B. Abraham, Phys. Rev. A, 26, 3395 (1982).
6. S. Hendow and M. Sargent, III, Opt. Comm., 40, 385 (1982); 43, 59 (1982).
7. P. Mandel, Opt. Comm. (in press) and private communication.
8. L. Lugiato, L.M. Narducci, D.K. Bandy and N.B. Abraham, (manuscript in preparation).
9. W.E. Lamb, Jr., Phys. Rev., 134, A1429 (1964).
10. W.R. Bennett, Jr., Phys. Rev., 126, 580 (1962).
11. W.R. Bennett, Jr., The Physics of Gas Lasers, (Gordon and Breach: New York, 1977) page 128.
12. R. Bonifacio and L.A. Lugiato, Lett. Nuo. Cim., 21, 510 (1978); R. Bonifacio, M. Gronchi and L.A. Lugiato, Opt. Comm., 30, 129 (1979); L.A. Lugiato, L.M. Narducci, D.K. Bandy, and C.A. Pennise, Opt. Comm., 43, 281 (1982); J.Y. Gao, J.M. Yuan, L.M. Narducci, Opt. Comm., to be published.
13. R.R. Snapp, H.J. Carmichael, and W.C. Schieve, Phys. Rev. Lett., 40, 68 (1981). See also the chapters by H. Carmichael and P. Mandel in this volume.
14. K. Ikeda, Opt. Comm., 30, 257 (1979); K. Ikeda, H. Daido and O. Akimoto, Phys. Rev. Lett., 45, 709 (1980); K. Ikeda, K. Kondo and O. Akimoto, Phys. Rev. Lett., 49, 1467 (1982).
15. W.J. Firth, Opt. Comm., 39, 343 (1981); J. Chrostowski, C. Delisle, R. Vallee, D. Carrier, and L. Boulay, Can. J. Phys., 60, 1303 (1982); E. Abraham, W.J. Firth, and J. Carr, Phys. Lett., 91A, 47 (1982); C.M. Savage, H.J. Carmichael, and D.F. Walls, Opt. Comm., 42, 211 (1982).
16. L. Casperson and A. Yariv, Appl. Phys. Lett., 17, 259 (1970).
17. See for example: Hydrodynamic Instabilities and the Transition to Turbulence, Topics in Applied Physics: Vol. 45, ed. H.L. Swinney and J.P. Gollub (Springer-Verlag: NY, 1981).
18. P.S. Linsay, Phys. Rev. Lett., 47, 1349 (1981); I. Goldhirsch Y. Imry, and S. Fishman, Phys. Rev. Lett., 49, 1599 (1982).
19. F.A. Hopf, D.L. Kaplan, H.M. Gibbs, and R.L. Shoemaker, Phys. Rev. A, 25, 2172 (1982).
20. H. Nakatsuka, S. Asaka, H. Itoh, K. Ikeda, and M. Matsuoka, Phys. Rev. Lett., 49, 109 (1983).
21. C.O. Weiss and H. King, Opt. Comm., 44, (in press) and private
22. F.T. Arecchi, R. Meucci, G. Puccioni, and J. Tredicce, Phys. Rev. Lett., 49, 1217 (1982).
23. W.R. Bennett, S.F. Jacobs, J.T. LaTourette, and P. Rabinowitz, App. Phys. Lett., 5, 56 (1964).
24. H.S. Boyne, M.M. Birky and W.G. Schweitzer, App. Phys. Lett., 7, 62 (1965).
25. E.I. Tsetsegova, Sov. Phys.: Opt. and Spec., 32, 172 (1971).
26. P.W. Smith and P.J. Maloney, App. Phys. Lett., 22, 667 (1973).
27. L. Casperson and A. Yariv, App. Opt., 11, 462 (1972).
28. G. Stephan and M. Trumper, Phys. Rev. (submitted for publication); G. Stephan and H. Taleb, J. Physique, 42, 1623 (1981); G. Stephan, H. Taleb, C. Pesty and F. Legros, J. Physique, 43, 255 (1982).
29. D.H. Schwamb, Phys. Lett., 71A, 420 (1979).
30. M. Minden and L. Casperson, private communication.
31. J.C. Wesson, senior honors thesis, Swarthmore College 1982.
32. M. Minden, Ph.D. thesis, UCLA 1982.

33. R.S. Gioggia and N.B. Abraham, manuscript in preparation.
34. H. Haken, Z. Phys., 190, 327 (1966).**
35. H. Risken, C. Schmid, and W. Weidlich, Z. Phys., 194, 337 (1966).
36. H. Risken and K. Nummedal, J. Appl. Phys., 39, 4662 (1968).
33. R. Graham and H. Haken, Z. Phys., 213, 420 (1968).
37. S. Singh and L. Mandel, Phys. Rev. A, 20, 2459 (1979).
39. Y. Pomeau and P. Manneville, Commun. Math. Phys., 77, 189 (1980); Physica D, 1, 219 (1980).
40. J.-P. Eckmann, Rev. Mod. Phys., 53, 643 (1981).
41. T. Yamada and R. Graham, Phys. Rev. Lett., 45, 1322 (1980); H.J. Scholz, T. Yamada, H. Brand, and R. Graham, Phys. Lett., 82A, 321 (1981).
42. L.A. Lugiato, L.M. Narducci, D.K. Bandy and C.A. Pennise, private communication.
43. H. Haken, Phys. Lett., 53A, 77 (1975); R. Graham, Phys. Lett., 58A, 440 (1976).
44. R. Bonifacio and L.A. Lugiato, Lett. Nuo. Cim., 21, 505 (1978).
45. K. Kaufman and G. Marowsky, App. Phys., 11, 47 (1976).
46. R. Wilbrandt and H. Weber, IEEE J. Quant. Elec., QE-11, 186 (1975).
47. C. Kolmeder, W. Zinth and W. Kaiser, Opt. Comm., 30, 453 (1979).
48. W. Brunner and H. Paul, App. Phys., 28, 168 (1982).
49. P.W. Smith, Proc. IEEE 58, 1342 (1970); L. Allen and D.G.C. Jones, Progress in Optics, 9, 179 (1971); and P.W. Smith, M. DuGuay, and E. Ippen, Progress in Quantum Electronics, 3, 105 (1974).
50. N.J. Halas, S.-N. Liu, and N.B. Abraham, manuscript in preparation.
51. P. Mandel, Physica, 82C, 353 (1976).
52. M. Mayr, H. Risken and H.D. Vollmer, Opt. Comm., 36, 480 (1981).
53. F.J. Feigenbaum, J. Stat. Phys., 19, 25 (1978); S. Grossmann and S. Thomas, Z. Natur., 32A, 1353 (1967); R. May, Nature, 261, 459 (1976).
54. D. Ruelle and F. Takens, Commun. Math. Phys., 20, 167 (1971). S. Newhouse, D. Ruelle, and F. Takens, Commun. Math. Phys., 20, 167 (1971).

**For a thorough review of the very early literature on laser self-pulsing, see the chapter in this volume by L. Casperson.

SECOND HARMONIC GENERATION IN A RESONANT CAVITY.

Paul MANDEL

Service de Chimie-Physique II, Campus Plaine, C.P.231, Université Libre de Bruxelles, Bruxelles 1050, Belgium.

T. ERNEUX

Department of Engineering Sciences and Applied Mathematics,The Technological Institute, Northwestern University, Illinois 60201, U.S.A.

Introduction.

In these notes we consider the simplest case of harmonic generation namely the second harmonic generation. The fundamentals of SHG are found in the textbooks of Bloembergen[1] and Yariv[2] (see also the recent review articles[3,4]). Two different situations will be considered: in the first part we discuss SHG in a resonant cavity which causes only linear losses; in the second part we discuss SHG in an active laser cavity.

SHG in a passive resonant cavity is a subject which was recently revived by the Hamilton group[4,5,6,7] who showed that though the system is very simple it displays a rich variety of behaviors. In particular self-pulsing may arise from a destabilization of the field phases. In the first part of these notes we show how modern methods of bifurcation theory lead to an analytic description of self-pulsing. These results are supplemented by a numerical analysis describing a domain of bistability between two self-pulsing solutions.

SHG in an active laser cavity is a much more complex subject. We shall only describe the stationary solutions and indicate some of the results of the linear stability analysis in the second part.

SHG in a passive resonant cavity.

Consider a nonlinear medium which is the source of SHG and is placed in a resonant cavity. This nonlinear medium is pumped by an external coherent field whose frequency is doubled by the medium. To stick to essentials we consider the good cavity limit (also called adiabatic limit in [4]) in which the cavity is practically lossless for the driving field and has at least output coupling for the second harmonic; furthermore we assume perfect matching.

In a recent publication[8] we have shown that this situation is described by the equations:

$$dR(1)/dt=R'(1)=R^*(1)R(2)+E$$

$$R'(2)=-R(2)-R^2(1) \qquad (1)$$

where R(1) is the electric field of the fundamental mode, R(2) the electric field of the second harmonic and E the driving field real amplitude. Reduced variables have been introduced to simplify the analysis. In terms of the real functions

$$R(1)=X+iU \qquad R(2)=Y+iV \qquad (2)$$

eqs.(1) become

$$X'=XY+UV+E$$
$$Y'=-Y-X^2+U^2$$
$$U'=XV-UY$$
$$V'=-V-2UX \qquad (3)$$

The stationary solution is

$$U=0 \qquad V=0 \qquad X=E^{1/3} \qquad Y=-E^{2/3} \qquad (4)$$

The linear stability of this solution is easily tested and shows that perturbations will evolve according to an exponential law $\exp(\lambda t)$ where the four eigenvalues are:

$$\lambda(1),\lambda(2)=1/2\left[-1-X^2\pm(1+X^4-10X^2)^{1/2}\right]$$

$$\lambda(3),\lambda(4)=1/2\left[-1+X^2\pm(1+X^4-6X^2)^{1/2}\right]$$

The first two eigenvalues have always a negative real part. The last two eigenvalues can have a negative or a positive real part. A bifurcation occurs when the real part of these eigenvalues vanish i.e. when

$$X=1 \qquad E=1 \qquad \lambda=\pm i \tag{5}$$

This is a typical Hopf bifurcation[9]. It was discovered and analyzed numerically by Drummond et al.[5,6,7]. A first analytic study has already been presented[8]. In this lecture we follow an alternative and simpler procedure.

We characterize the vicinity of the bifurcation point by a small positive parameter ϵ such that:

$$E=1+\epsilon^2 E(2)+O(\epsilon^3) \tag{6}$$

and seek solutions to eqs.(3) in power series of ϵ :

$$\begin{aligned} X(\epsilon,T)&=1+\epsilon^2 X(2,T)+\ldots \\ Y(\epsilon,T)&=-1+\epsilon^2 Y(2,T)+\ldots \\ U(\epsilon,T)&=\epsilon U(1,T)+\epsilon^3 U(3,T)+\ldots \\ V(\epsilon,T)&=\epsilon V(1,T)+\epsilon^3 V(3,T)+\ldots \end{aligned} \tag{7}$$

where $T=\omega(\epsilon)t=(1+\epsilon^2\omega(2)+\ldots)t$ (8)

To first order in ϵ we have:

$$U'(1)=U(1)+V(1) \qquad V'(1)=-V(1)-2U(1) \tag{9}$$

whose general solutions are the time-periodic functions

$$U(1)=e^{iT}+e^{-iT}\ ,\quad V(1)=(i-1)e^{iT}-(i+1)e^{-iT} \tag{10}$$

To second order in ϵ we have:

$$X'(2)=-X(2)+Y(2)+U(1)V(1)+E(2) \tag{11}$$

$$Y'(2)=-Y(2)-2X(2)+U^2(1)$$

whose general solutions are:

$$X(2)=e^{-T}(c e^{iT2^{1/2}}+d e^{-iT2^{1/2}})+E(2)/3+(\frac{2+i}{1-4i}e^{2iT}+c.c.)$$

$$Y(2)=i2^{1/2}e^{-T}(ce^{iT2^{1/2}}-de^{-iT2^{1/2}})+2-2E(2)/3$$

$$+(\frac{3}{4i-1}e^{2iT}+c.c.) \qquad (12)$$

These solutions display damped oscillations which are negligible in the long time limit ($T \gg 1$) and undamped oscillations which remain in the final state.

To third order in ϵ we have:

$$U'(3)=U(3)+V(3)+P$$
$$V'(3)=-V(3)-2U(3)+Q \qquad (13)$$

where

$$P=P(3)e^{3iT}+P(1)e^{iT}+c.c.$$
$$Q=Q(3)e^{3iT}+Q(1)e^{iT}+c.c.$$

The difficulty with eqs.(13) is that P and Q have oscillations at the eigenfrequencies of the homogeneous equations (see 9 and 10).This induces in general secular terms which diverge in time. The requirement that these terms identically cancel will determine $E(2)$ and $\omega(2)$ via the solvability condition. To derive this condition it is better to write eqs.(13) in vector notation:

$$Z'=LZ+I \qquad Z=\mathrm{col}(U(3),V(3)) \qquad I=\mathrm{col}(P,Q) \qquad (14)$$

For 2π-periodic vectors $A(T)=A(T+2\pi)=\mathrm{col}(A(1),\ A(2))$ we define a scalar product:

$$(A,B)=1/(2\pi)\int_0^{2\pi}\left[A^*(1)B(1)+A^*(2)B(2)\right]dT \qquad (15)$$

Let the vector D be a solution of $D'+MD=0$ where M is the adjoint of L. Then for any Z we have $(Z,D'+MD)=0$ or$(-Z'+LZ,D)=0$. If Z is to be a solution of (14) then the last equality implies that D and I must be orthogonal. This is the solvability condition which is a necessary and sufficient condition for the absence of secular terms. This condition $(I,D)=0$ implies:

$$P(1)+(1/2-i/2)Q(1)=0$$

from which we obtain

$$\omega(2)=27/17=1.5882...$$

$$E(2)=99/34=2.9118... \qquad (16)$$

Hence we have constructed analytically a solution of eqs.(3) near the bifurcation point which becomes periodic in the long time limit. In addition the positivity of E(2) ensures, by Hopf's theorem[9], the stability of these periodic solutions near the bifurcation.

To investigate the solutions of eqs.(3) at a finite distance of E=1 we have made a numerical analysis. In the long time limit the solutions remain periodic over the range of E we studied. On fig.(1) we have plotted the minimum of X(T) for increasing E. This figure displays clearly an hysteresis domain with first-order transitions between two periodic solutions. Let us denote by J (K) the intensity of the modes on the upper (lower) branch. In the following four figures E=7. On fig.(2) we show the limit cycle in the (U, V) plane corresponding to the J solutions and fig.(3) displays the intensities of both modes. On fig.(4) we show the limit cycle in the (U,V) plane corresponding to the K solutions and fig.(5) displays the corresponding intensities.

Fig.1

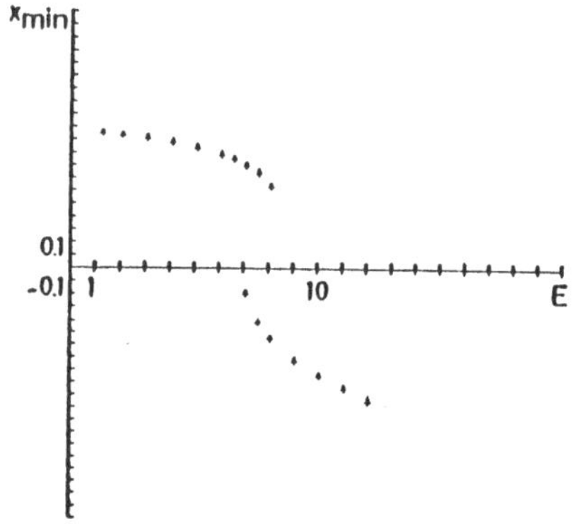

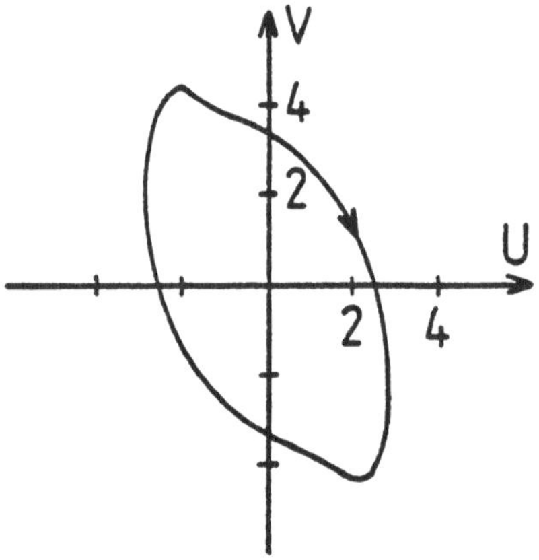

Fig.2

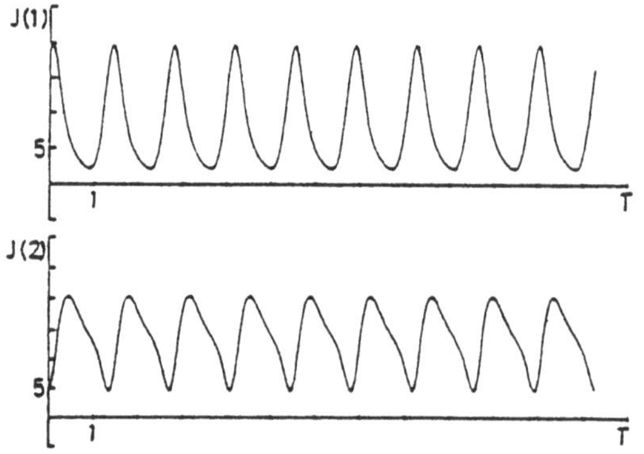

Fig.3

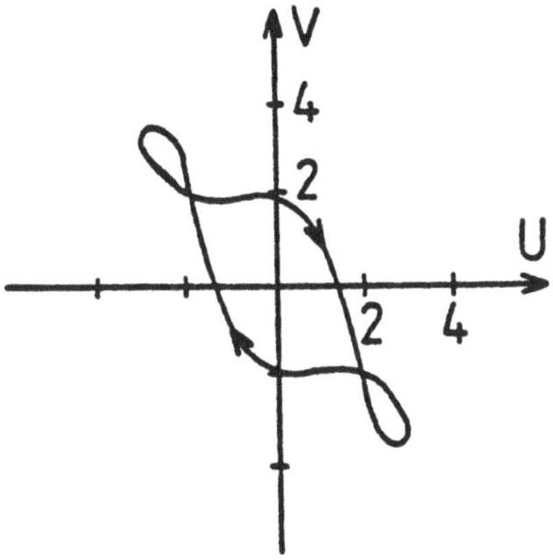

Fig.4

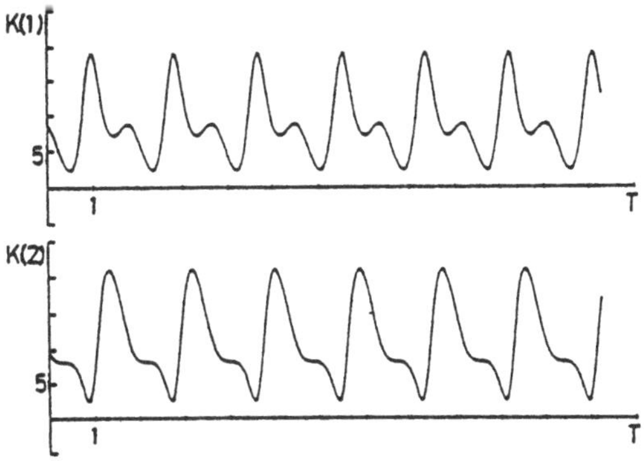

Fig.5

SHG in an active laser cavity.

In this paragraph we consider a nonlinear medium, which is the source of SHG, placed in a laser. The laser is pumped incoherently and produces a coherent field whose frequency is doubled inside the cavity. It is easy to couple eqs.(1) to the usual laser equations to obtain:

$$R'(1)=-kR(1)+R^*(1)R(2)+\bar{P}$$

$$R'(2)=-R(2)-R^2(1)$$

$$\bar{P}'=-k_\perp\bar{P}+DR(1)$$

$$D'=-k_{\parallel}(D-D_0)-g(\bar{P}R^*(1)+\bar{P}^*R(1)) \qquad (17)$$

In these equations k is the damping constant of mode 1, $\bar{P}$ is the atomic polarisation of the laser amplifying medium whose relaxation rate is $k_\perp$; D is the atomic inversion with relaxation rate $k_{\parallel}$ and initial inversion D_0; finally g measures the atom-field coupling and the time has been scaled by the damping constant of mode 2. To derive eqs. (17) we have neglected all dispersive contributions.

We introduce the polar decompositions:

$$R(j)=E(j)\exp(i\varphi(j)) \quad , \quad \bar{P}=P\exp(i\varphi(p)) \tag{18}$$

and the phase differences:

$$\Psi(2)=\varphi(2)-2\varphi(1), \Psi(p)=\varphi(p)-\varphi(1) \tag{19}$$

in terms of which eqs (17) become:

$$E'(1)=-kE(1)+E(1)E(2)\cos\Psi(2)+P\cos\Psi(p)$$

$$E'(2)=-E(2)-E^2(1)\cos\Psi(2)$$

$$P'=-k_{\perp}P+DE(1)\cos\Psi(p)$$

$$D'=-k_{\parallel}(D-D_0)-2gPE(1)\cos\Psi(p) \tag{20}$$

$$\Psi'(2)=(\ -2E(2)+E^2(1)/E(2)\ )\sin\Psi(2)-2P/E(1)\sin\Psi(p)$$

$$\Psi'(p)=-(\ (DE^2(1)+P^2)/E(1)P\ )\sin\Psi(p)-E(2)\sin\Psi(2) \tag{21}$$

Note that only the phase differences enter in the equations; the absolute phases are given by:

$$\varphi'(1)=E(2)\sin\Psi(2)+(P/E(1))\sin\Psi(p)$$

$$\varphi'(2)=(E^2(1)/E(2))\sin\Psi(2), \qquad \varphi'(p)=-(DE(1)/P)\sin\Psi(p) \tag{22}$$

Quite unexpectedly there are three stationary solutions to eqs.(20) and (21):

(1) First there is the trivial solution:

$$E(1)=0 \quad , \quad E(2)=0 \tag{23}$$

(2) Second there is the "normal" solution:

$$\Psi(2)=\pi \quad , \quad \Psi(p)=0 \quad , \quad E(2)=E^2(1)=I_+$$

$$I_+=(2S)^{-1}\left\{\ -1-kS+\left[(1-kS)^2+4AkS\right]^{1/2}\right\} \tag{24}$$

where $S=2g/k_{\parallel}k_{\perp}$ and $A=D_0/kk_{\perp}$ are the reduced saturation and pump parameters respectly.

(3) Third there is an "anomalous" solution:

$$\cos\Psi(2)=-(\ aE(2)\ )^{-1},\qquad \cos\Psi(p)=\left(1+\frac{a^2I(2)-1}{4k_\perp^2}\right)^{-1}$$

$$I(2)=b(A-A_0),\qquad I(1)=aI(2) \tag{25}$$

where we have introduced the notations:

$$I(j)=E^2(j),\quad a=(2k_\perp-1/(k+k_\perp),\qquad A_0=(2k+1)(2k_\perp+1)/(4kk_\perp)$$

$$b^{-1}=2g(1+ak)/(kk_\parallel k_\perp)-a(a-2)/(4kk_\perp)$$

It is also easy to determine the phases of the anomalous solutions:

$$\varphi(1,t)=\omega t+\varphi(1,0)$$

$$\varphi(2,t)=2\omega t+2\varphi(1,0)+\Psi(2)$$

$$\varphi(p,t)=\omega t+\varphi(1,0)+\Psi(p) \tag{26}$$

where $$2\omega=a(I(2)-a^{-2})^{1/2} \tag{27}$$

The concept of stationarity becomes somewhat ambiguous in the case of the anomalous solution. Indeed, if we had used the cartesian decomposition (2) for the fields and a similar decomposition for the atomic polarization, we would have obtained the stationary solutions (23) and (24) only. Then a linear stability analysis of (24) would have indicated a Hopf bifurcation leading to a time-periodic solution which is precisely (25). This is because the anomalous solution corresponds to the simplest example of a time-periodic function i.e. a single frequency of oscillation and a constant amplitude; on the contrary, the time-periodic solution constructed in the first part of these notes has amplitude modulation and therefore oscillations at the fundamental frequency and at all harmonics of the fundamental frequency.

The reason solution (25) is anomalous is that it does not oscillate at the same frequency as solutions (23) and (24). Indeed we have neglected dispersion in eqs.(17) and therefore the cavity has been described as purely absorptive; consequently all three phases vanish for "normal" solutions (23) and (24) (the unperturbed frequencies do not enter in eqs.(17) since we made a transformation to a rotating frame of reference). Hence the existence of anomalous solutions indicates that a purely absorptive cavity is able to act as a dispersive cavity under some conditions. Such a situation also occurs in the laser with a saturable absorber and in the dye laser[10]. The common feature of these models which explains the occurence of anomalous solutions and the dispersive behavior of an absorptive cavity is the interplay between nonlinear gain and nonlinear loss which leads to a nontrivial dispersion relation admitting new solutions.

The linear stability analysis of the trivial solution (23) indicates that it is stable for $A \leq 1$.

The linear stability analysis of the normal solution (24) is somewhat more elaborate. As in the case of SHG in a passive cavity, there is a factorization of the equations governing the stability of the phases and of the amplitudes. The salient property is that the normal solution is always destabilized, at least by the phases; in addition other bifurcations may arise from the amplitudes in some domains of the parameter space. In these notes we shall discuss the phase instabilities only. From eqs.(21) we easily derive the characteristic equation:

$$\lambda^2+\lambda(1+k+k_\perp-I_+)+k+k_\perp+(1-2k_\perp)I_+=0 \qquad (28)$$

We define four domains in the parameter space $(k,k_\perp)$:

D(1): $0<k_\perp<2^{-1/2}$

D(2): $2^{-1/2}<k_\perp<1$ and $2k>(2k^2-1)/(1-k_\perp)$

D(3): $2^{-1/2}<k_\perp<1$ and $2k<(2k^2-1)/(1-k_\perp)$

D(4): $k_\perp>1$

In the domains D(3) and D(4) the phases are destabilized when $k+k_\perp+(1-2k_\perp)I_+=0$. At this bifurcation point $I_+=I(1)$ i.e. the anomalous solution begins to exist. In the domains D(1) and D(2) the phases are distabilized when $I_+=1+k+k_\perp$; at this point a Hopf bifurcation occurs which leads to a time-periodic solution as in the first part of these notes.

Acknowledgments.

This research has been supported by the National Fund for Scientific Research (Belgium) and by the Association Euratom-Etat Belge.

References.

1. N. Bloembergen Nonlinear Optics (Benjamin, NY, 1965).
2. A. Yariv Introduction to Optical Electronics 2nd ed. (Holt, Rinehart and Winston, NY, 1976).
3. D.S. Chemla, Rep. Prog. Phys. 43 (1980) 1191.
4. D.F. Walls, P.D. Drummond and K.J. McNeil in Optical Bistability ed. by C.M. Bowden, M. Ciftan and M.R. Robl (Plenum, NY, 1981).
5. P.D. Drummond, K.J. McNeil and D.F. Walls, Optica Acta 27 (1980) 321; 28 (1981) 211.
6. K.J. McNeil, P.D. Drummond and D.F. Walls,Optics Comm. 27 (1978) 292.
7. P.D. Drummond, K.J. McNeil and D.F. Walls,Optics Comm. 28 (1979) 255.
8. P. Mandel and T. Erneux, Optica Acta 29 (1982).
9. G. Iooss and D.D. Joseph Elementary stability and bifurcation theory (Springer Verlag, Heidelberg, 1980).
10. P. Mandel, Phys. Letters A83 (1981) 207.

ON THE GENERATION OF TUNABLE ULTRASHORT LIGHT PULSES

A. Seilmeier and W. Kaiser
Physik Department der Technischen Universität München
München, Germany

During the past decade light pulses in the picosecond range have received increasing attention. Numerous applications were found in optoelectronics, molecular physics, chemistry, biology and solid-state physics /1/. It is not surprising, therefore, that considerable effort has been spent to develop new and more flexible laser systems for the generation of the appropriate light pulses.

A light pulse is characterized by a series of parameters: Pulse duration and temporal pulse shape, emission spectrum, beam divergence, and pulse energy. The generator and the following tuning system add more parameters such as repetition rate, tuning range, and reproducibility in frequency and time (jitter).

At the present time there exists a series of experimental arrangements which generate pulses with widely differing properties. In this short summary we concentrate on five systems (Fig. 1), the performance of which has been studied in various laboratories.

I. The Flashlamp-Pumped Passively Mode-Locked Nd:Glass Laser

The first high-power picosecond laser system dates back to 1966, when the passive mode-locking of a flashlamp-pumped Nd:glass laser was reported /2/. Such systems have been investigated in great detail /3/ and were applied in numerous experiments. It is now well established that the pulse properties (e.g., pulse duration and frequency bandwidth) vary substantially during the pulse train which is generated for each pumping flash. Cutting electronically a single pulse from the leading part of the pulse train gives very favorable and reproducible pulses. Typically, one obtains pulses at 1.06 μm of $t_p = 5 \pm 1$ picosecond duration, which are of Gaussian temporal shape and are bandwidth limited /4/, i.e., the spectral width is $\Delta\nu = 0.44\ t_p$. The peak power of the single pulse extracted from the oscillator may readily be amplified in subsequent Nd:glass rods to approximately 100 mJ or 3×10^{10} W, i.e. 10^{18}

	pulse duration	repetition rate	energy / pulse
Nd: glass laser flashlamp pumped	5 ps	< 1 Hz	50mJ
Nd: YAG laser flashlamp pumped	25 ps	1 - 50 Hz	30mJ
Ar^+, Kr^+ laser cw mode-locked	150 ps	80MHz	10 nJ
Nd: YAG laser cw mode-locked	80 ps	100MHz	100 nJ
cw Ar^+ laser and mode-locked dye laser	1 ps	80MHz	10 nJ

Fig. 1 Five laser systems which generate picosecond light pulses (see text for details).

photons per pulse. (Large glass-laser systems for fusion research, where several 10^2 or even 10^3 Joules are generated, will not be discussed here.)

The high output power of the mode-locked Nd:glass laser suggests the use of nonlinear optical systems for the generation of pulses of variable frequencies. In fact, the parametric three-photon process proves to be a highly versatile method to produce picosecond pulses over a wide frequency range from the infrared to the ultraviolet /5/. Various nonlinear crystals have been studied /5-8/, e.g., $LiNbO_3$, $LiJO_3$, KDP, ADP, and Ag_3AsS_3. Detailed data are available of the angle and temperature tuning of $LiNbO_3$ crystals /5/. Using the fundamental laser frequency for pumping the parametric process one obtains tuning between 2700 and 8300 cm^{-1}; with the second harmonic frequency as pump the tuning range is extended to 17,000 cm^{-1} and frequency doubling of the parametrically generated pulses leads to light pulses up to 32,000 cm^{-1} (see Fig. 2). On the lower side of the frequency scale, very recently the travelling wave parametric process was extended down to the medium infrared range /8/ at 1600 cm^{-1} (see Fig. 3). Proustite crystals of large optical nonlinearity and favorable IR-transmission were used in these investigations.

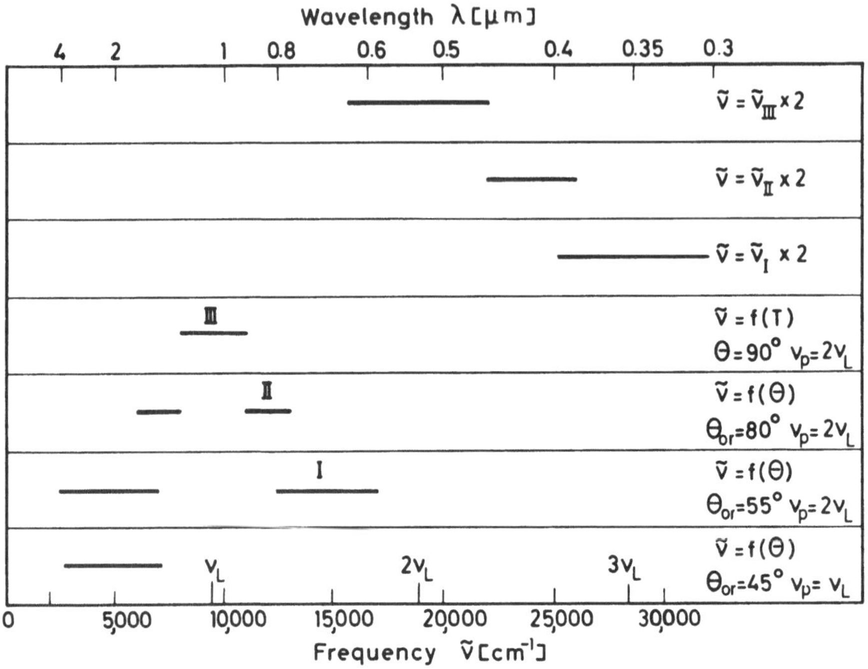

Fig. 2 Frequency regions covered by parametric amplification in $LiNbO_3$ crystals. Θ_{or} is the crystal's orientation angle. (Ref. 5)

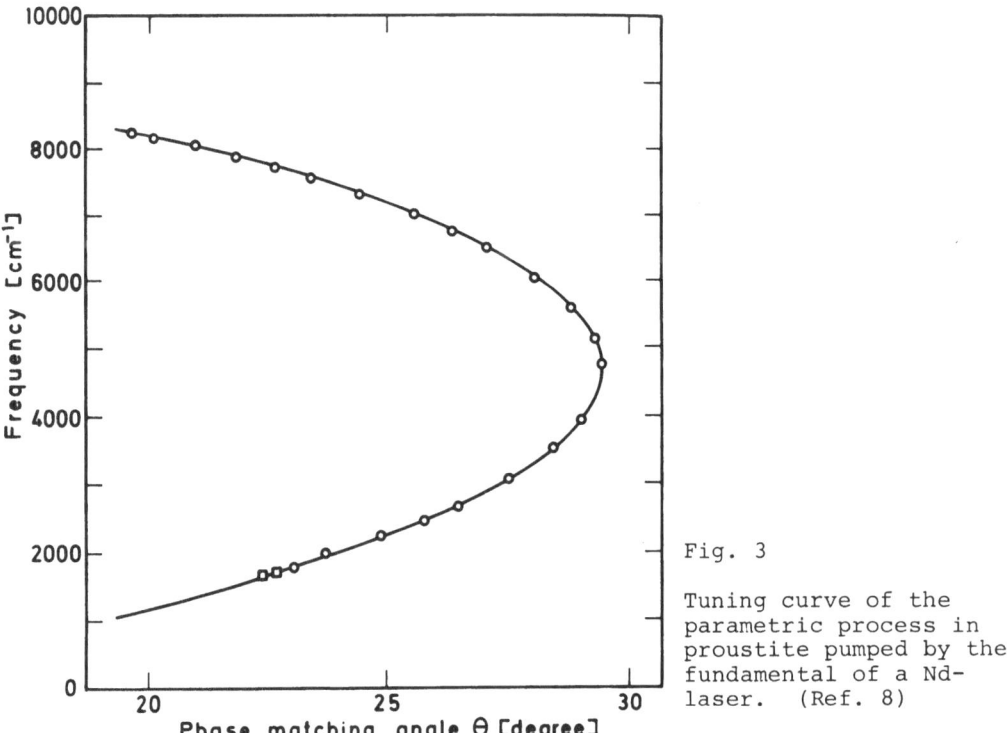

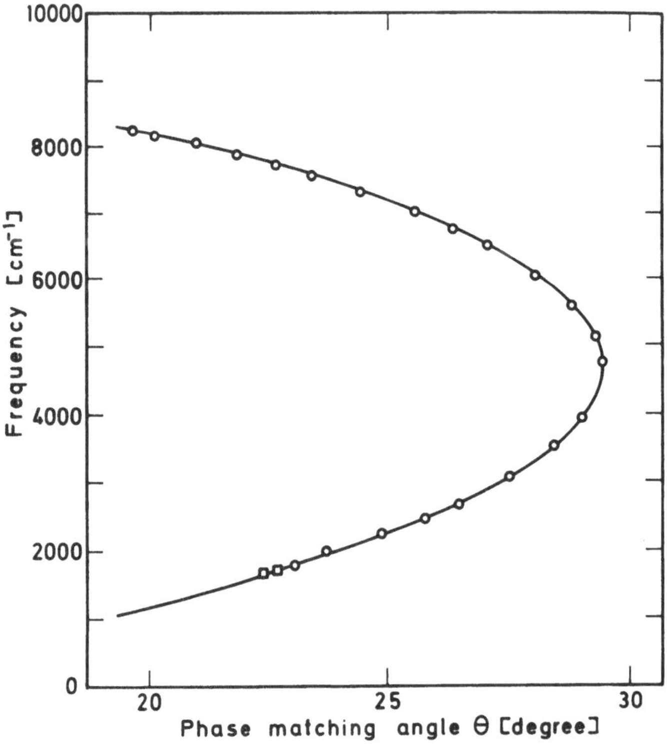

Fig. 3

Tuning curve of the parametric process in proustite pumped by the fundamental of a Nd-laser. (Ref. 8)

Two points in favor of the parametric process should be mentioned here. (i) The bandwidth of the generated pulses may be predicted from the dispersion of the medium and from other system parameters (e.g., the pump divergence). In a wide frequency range, pulses (of several picoseconds duration) are observed with bandwidths of approximately 10 cm^{-1}. (ii) The parametric process, electronical in nature, is very fast. As a result, the jitter between two parametric systems operating in parallel was found to be less than one picosecond.

The main disadvantage of a laser with glass as the active material is the small repetition rate of operation. The low thermal conductivity of glass requires dead times of seconds depending upon the diameter of the glass rod. Thus, the acquisition of large amounts of data is limited with glass-laser systems. To overcome this problem, solid-state lasers with crystals of favorable thermal conductivity have been devised. We discuss a frequently used system in the next section.

II. The Flashlamp-Pumped Passively Mode-Locked Nd:YAG Laser

This laser system is able to operate - with a small Nd:YAG crystal - at a rate of several kilohertz.With larger amplifier crystals the repetition rate goes down to one to ten Hz. The pulse trains are shorter in the Nd:YAG system consisting of approximately ten individual pulses of varying intensity. Cutting out the center pulse of the oscillator train for further amplification is a common procedure in many systems. The main disadvantage of the Nd:YAG laser is the longer pulse duration of 25 ps at 1.064 μm. The sharper emission lines of the Nd-ion in the crystalline host allow a smaller number of modes to be locked together and thus give rise to longer laser pulses.

With sufficient amplification the pulses of the Nd:YAG laser are of high enough peak intensity to operate a parametric single path system as discussed in the previous section. Tunable picosecond pulses have been generated in this way over a wide wavelength range from the UV to the IR around 3 μm /9/. The parametric process has the additional advantage of shortening the laser pulses. In fact, parametrically generated pulses of 5 ps /9/ duration have been generated for pump pulses of 22 ps.

There exists a second method to generate tunable picosecond pulses starting from a laser system of fixed frequency. In this case, a dye laser is synchronously pumped by the whole pulse train of the primary laser or by the second harmonic thereof. Various laser dyes have been used successfully, which absorb at the fundamental /10/, 1.06 μm, or at the second harmonic /11/, 0.53 μm; typical tuning ranges are 1.15 to 1.24 μm and 550 to 700 nm, respectively.

III. The CW Mode-Locked Ar^+ or Kr^+ Laser For Synchronous Pumping

In recent years, substantial effort has been spent to make the cw mode-locked ion laser a reliable device /12-20/ (see Fig.4). It has been recognized by people working with these systems, that the quality of the acousto-optic mode-locking device is essential for the stability of the pulse train. A precise match is necessary between the frequency of the

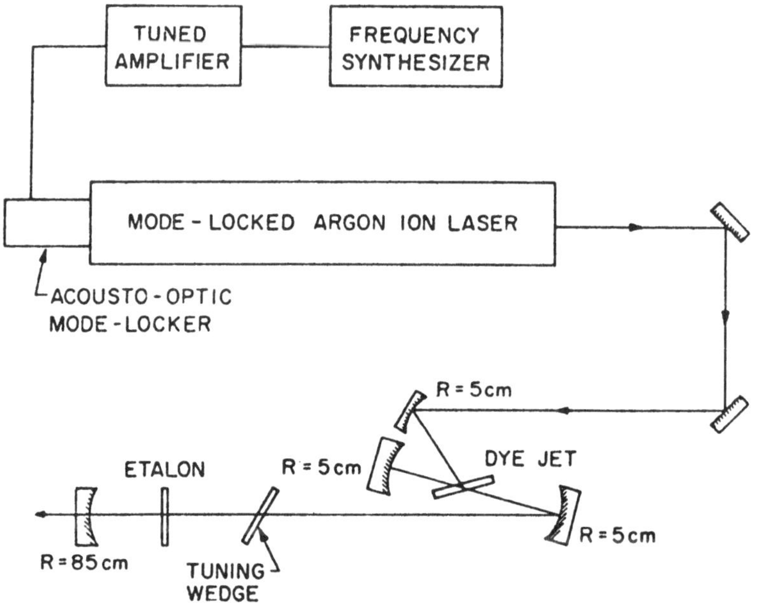

Fig. 4 Experimental system of a synchronously pumped dye laser (see Ref. 14).

driving power and one of the resonances of the acoustic modulator and with the frequency difference of the neighboring cavity modes. High quality frequency synthesizers are mandatory. Of special importance is the observation that noisy pump trains seriously affect the operation of a synchronously pumped dye laser. Thus, the performance of the primary pump laser is transferred to the quality of the subsequent tunable pulse source /14/.

The characteristics of synchronously mode-locked cw dye lasers have been studied in great detail /12-20/, both experimentally and theoretically. Under carefully controlled conditions it is possible to obtain highly stabilized pulse trains. Then the background-free autocorrelation curves - most frequently used for the control of the laser performance - show one sharp maximum without broad wings or sockets.

Of interest for spectroscopic applications is the generation of synchronized trains of picosecond pulses at two independently tunable wavelengths /13/. Two cw mode-locked dye lasers were synchronously pumped by one mode-locked argon-ion laser. Cross-correlation measurements indicate - for optimum performance of the whole system - a jitter of a few picoseconds between the two pulse trains.

In a cw mode-locked system, the pulse sequence is quite large with approximately 10^9 pulses per second. For an average output of 0.1 W one obtains small peak powers of the order of watts. To boost the peak power of the picosecond pulses multi-stage dye laser amplifiers have been designed which were pumped by a Nd:YAG laser operated at 10 Hz repetition rate /20/. Dye laser pulses of a few picoseconds duration with peak powers of the order of megawatts were observed.

Synchronous pumping of semiconductor platelets turns out to be an interesting source of tunable picosecond pulses /21/. Various semiconductor crystals, e.g. CdS, CdSe or InGaAsP were pumped by mode-locked ion lasers (of $\sim$ 150 ps) to give tunable pulses of 4 to 10 ps duration. Cooling of the samples is required in most cases. It is believed that lasers of this type have the capability for single-frequency operation tunable throughout most of the visible and near IR.

IV. The CW Mode-Locked Nd:YAG Laser For Synchronous Pumping

A cw mode-locked Nd:YAG laser provides certain advantages over the conventional argon and krypton pumping systems. The shorter pumping pulses of approximately 80 ps support shorter pulses in the synchronously pumped dye lasers and the higher component lifetime promisses reduced maintainance cost.

Using a frequency doubled cw mode-locked Nd:YAG system of high stability synchronous pumping of a Rhodamine 6G dye-laser has been investigated /22,23/. Red dye laser pulses shorter than 0.5 ps were observed. In a subsequent two-stage Rhodamine 6G dye amplifier a gain of 2×10^6 was obtained with a repetition rate of 2 Hz.

Very recently, the direct synchronous pumping of an IR-dye laser was achieved using a cw mode-locked Nd:YAG laser /24/. Infrared laser dyes of high photochemical stability have recently been developed which made this laser system possible. The tuning from 1.20 to 1.32 μm (see Fig.5) covers the interesting wavelength range around 1.3 μm, where the absorption of optical fibers has its minimum. Pulse durations of several picoseconds are observed at a laser wavelength of 1.3 μm. It should be noted that this laser system consists of commercially available components and operates at room temperature.

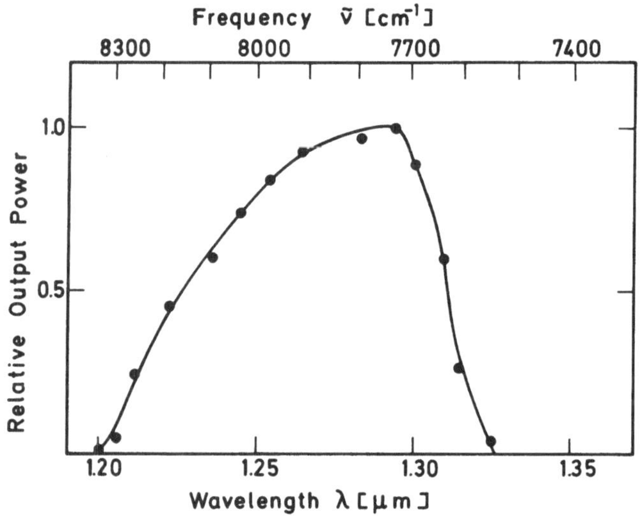

Fig. 5 Tuning curve of a synchronously pumped IR dye laser (Ref. 24)

Besides laser dyes, alkali halide crystals containing various color centers were shown to be useful active materials for near IR lasers /25/. Mode-locking by synchronous pumping produced pulses as short as several picoseconds. The 1.064 μm wavelength of a Nd:YAG laser was used as the pump source. Most of the crystals require high energy electron beams or x-ray irradiation for the generation of the color centers. The laser has to be operated at low temperature (∿ 77K) making the experimental system more elaborate.

V. The Passive Mode-Locking of a CW Dye Laser

Using a cw Ar^+ laser as a pump source passive mode-locking of a Rhodamine 6G laser is possible /26/. The multiple-folded dye cavity contains an amplifier cell (or jet) and a saturable absorber (see Fig. 6). Pulses with a duration of a few picoseconds were achieved and a tuning range of 590 to 610 nm was observed.

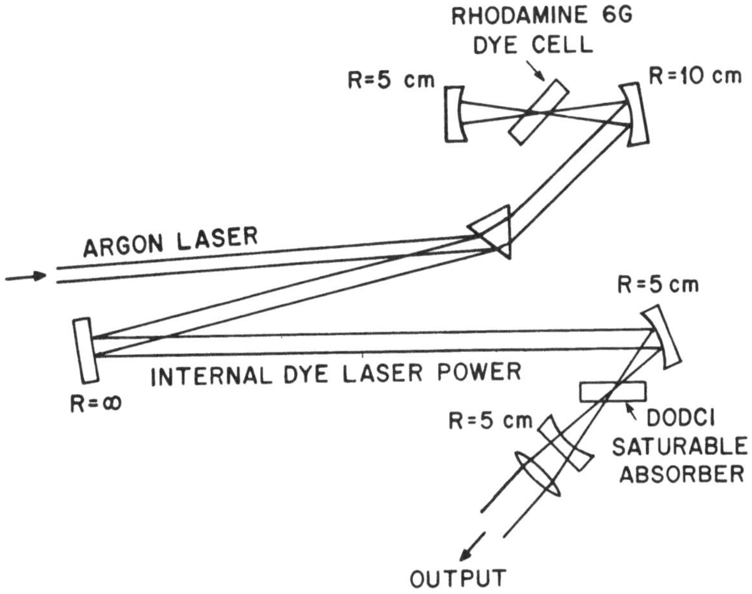

Fig. 6 Experimental system of a passively mode-locked cw dye laser (Ref. 26)

The passive mode-locking of a dye laser was investigated recently in a ring laser system /27/. The interaction of two oppositely directed pulses in a thin saturable absorber gave continuous pulse trains with pulses as short as 90 femtoseconds in a Rhodamine 6G dye laser system. The counter-propagating pulses create a transient grating in the population of the absorber molecules which synchronizes and shortens the pulses in both pulse trains. This process is called colliding pulse mode-locking (CPM).

Amplifier systems for femtosecond pulses have recently been developed /28,29/. The four dye-laser stages are pumped with a frequency doubled Nd:YAG laser (10 Hz). During amplification an incident pulse of 90 femtoseconds duration is broadened to 400 fs. Using a grating compressor the pulse width can be restored to 30 fs with a peak intensity at the gigawatt level /28/. These systems operate best at a wavelength of 618 nm.

Concluding Remarks

In this short summary we have not yet mentioned two systems which produce pulses in the picosecond regime: Flashlamp pumped dye lasers /30/ and distributed feed-back dye lasers pumped by a low-pressure N_2 laser /31/. The latter generates - close to the threshold - pulses of several picoseconds duration. With the laser dye Rhodamine 6G one obtains an output wavelength of 590 nm. The tuning range is quite limited.

The quest for coherent ultrashort pulses in the UV or even XUV has lead to the development of very complex laser systems consisting of numerous amplifiers and frequency doubling and/or tripling devices. The reader is referred to the literature /32,33/.

In summary we wish to say that a variety of systems is now available for the generation of the desired pulse properties. The strong activity in the field of picosecond phenomena will continue to stimulate the development of new, pulse generating, devices.

References

1. Picosecond Phenomena III, Springer Series in Chemical Physics 23, eds. K.B. Eisenthal, R.M. Hochstrasser, W. Kaiser, A. Laubereau, Springer-Verlag,Heidelberg 1982.
2. A.J. DeMaria, D.A. Stetser and H. Heyman, Appl. Phys. Lett. 8 (1966) 174.
3. D. von der Linde, IEEE J. Quant. Electron. QE-8 (1972) 328.
4. W. Zinth, A. Laubereau and W. Kaiser, Optics Commun. 22 (1977) 161.
5. A. Seilmeier and W. Kaiser, Appl. Phys. 23 (1980) 113.
6. G. Dikchyus, R. Danelyus, V. Kabelka, A. Piskarskas, T. Tomkyavichyus and A. Stabinis, Sov. J. Quant. Electron. 6 (1976) 425.
7. Y. Tanaka, T. Kushida and S. Shionoya, Optics Commun. 25 (1978) 273.
8. T. Elsässer, A. Seilmeier and W. Kaiser, to be published in Optics Commun. (1983).
9. W. Kranitzky, K. Ding, A. Seilmeier and W. Kaiser, Optics Commun. 34 (1980) 483.
10. W. Kranitzky, B. Kopainsky, W. Kaiser, K.H. Drexhage and G.A. Reynolds, Optics Commun. 36 (1981) 149.
11. L.S. Goldberg and C.A. Moore, Appl. Phys. Lett. 27 (1975) 217.
12. C.P. Ausschnitt and R.K. Jain, Appl. Phys. Lett. 32 (1978) 727.

13. R.K. Jain and J.P. Heritage, Appl. Phys. Lett. 31 (1978) 41; J.P. Heritage and R.K. Jain, Appl. Phys. Lett. 32 (1978) 101.
14. A.I. Ferguson, J.N. Eckstein and T.W. Haensch, J. Appl. Phys. 49 (1978) 3389.
15. C.P. Ausschnitt, R.K. Jain and J.P. Heritage, IEEE J. Quant. Electron. QE-15 (1979) 912.
16. D.M. Kim, J. Kuhl, R. Lambrich and D. von der Linde, Optics Commun. 27 (1978) 123.
17. N.J. Frigo, T. Daly and H. Mahr, IEEE J. Quant. Electron. QE-13 (1977) 101.
18. J. Kuhl, H. Klingenberg and D. von der Linde, Appl. Phys. 18 (1979) 279.
19. H. Klann, J. Kuhl and D. von der Linde, Optics Commun. 38 (1981) 390.
20. S.R. Rotman, C. Roxlo, D. Bebelaar, T.K. Yee and M.M. Salour, Appl. Phys. B26 (1982) 319.
21. M.M. Salour, in Ref. 1, p. 53; T.C. Damen, M.A. Duguay, J.M. Wiesenfeld J. Stone and C.A. Burrus, in "Picosecond Phenomena II", Springer Series in Chemical Physics 14, eds. R.M. Hochstrasser, W. Kaiser, C.V.Shank, Springer-Verlag, Heidelberg 1980, p. 38.
22. T. Sizer II, G. Mourou and R.R. Rice, Optics Commun. 37 (1981) 207.
23. T. Sizer II, D. Kafka, A Krisiloff and G. Mourou, Optics Commun. 39 (1981) 259.
24. A. Seilmeier, W. Kaiser, B. Sens and K.H. Drexhage, to be published
25. L.F. Mollenauer and D.M. Bloom, Optics Lett. 4 (1979) 247; L.F. Mollenauer, N.D. Vierira and L. Szabo, Optics Lett. 7 (1982) 414.
26. E.P. Ippen, C.V. Shank and A. Dienes, Appl. Phys. Lett. 21 (1972) 348.
27. R.L. Fork, B.I. Greene and C.V. Shank, Appl. Phys. Lett. 38 (1981) 671.
28. C.V. Shank, R.L. Fork and R.T. Yen, in Ref. 1, p. 2.
29. A. Migus, J.L. Martin, R. Astier, A. Antonetti and A. Orszag, in Ref. 1, p. 6.
30. D.J. Bradley, B. Liddy, A.G. Raddie, W. Sibbett and W.E. Sleat, Optics Commun. 3 (1971) 426; W. Sibbett and J.R. Taylor, Optics Commun. 43 (1982) 50.
31. Z. Bor, A. Müller, B. Râcz and F.P. Schäfer, Appl. Phys. B27 (1982) 9; ibid, Appl. Phys. B27 (1982) 77; Z. Bor, B. Râcz, G. Szabo and A. Müller, in Ref. 1, p. 62.
32. P.H. Bucksbaum, J. Boker, R.H. Storz and J.C. White, Optics Lett. 7 (1982) 399.
33. T. Srinivasan, K. Boyer, H. Egger, T.S. Luk, D.F. Muller, H. Pummer and C.K. Rhodes, in Ref. 1, p. 19.

RAMAN SPECTROSCOPY WITH ULTRASHORT COHERENT EXCITATION. NARROWING OF SPECTRAL LINES BEYOND THE DEPHASING LINEWIDTH.

W. Zinth and W. Kaiser
Physik Department der Technischen Universität München
München, Germany

Spectroscopists are constantly faced with the task of improved spectral resolution. Two points are of major interest: (i) The precise frequency of the quantized transition and (ii) the detection of new neighboring transitions. Besides experimental factors the ultimate spectral resolution is determined by the inherent linewidth of the transition. Optical spectroscopists have to deal with different line-broadening processes; for instance with the Doppler effect or with collision broadening in gases, with dephasing processes in condensed systems and with the population relaxation which results in the natural linewidth.

In recent years, different novel techniques have been devised which provide spectral resolution beyond the transition linewidth. For instance, Doppler broadening can be eliminated by saturation spectroscopy or by two counter-propagating beams for two-photon transitions/1/. Even measurements beyond the natural linewidth have been performed taking biased signals from the fluorescent decay /2-5/. Techniques have been proposed where the difference between the decay rates of the two states rather than their sum determines the linewidth /6,7/, and narrowing of the natural linewidth by decaying-pulse excitation has been discussed /8/.

Very recently, we have demonstrated substantial line narrowing of Raman type transitions in condensed phases /9-11/. The lines were broadened by vibrational dephasing. New information was obtained in congested spectral regions.

Theory

In a transient experiment the spectral resolution is not limited by the lifetime of the investigated levels, but is determined by the specific experiment. Under favorable conditions the observed line may become substantially narrower than the spontaneous width measured in a steady-state experiment. We have treated an ensemble of two-level systems of frequency difference ω_o using the density matrix formalism. Of importance are the time constants of the system. The population of the upper level decays with the damping constant T_1 and the off-diagonal elements relax with the dephasing time T_2. The latter determines the linewidth of homogeneously broadened transitions in stationary experiments, $\Delta\nu_{spont}=1/\pi T_2$. Quite generally, we may write $1/T_2=1/2T_1+1/T_{ph}$, where T_{ph} is related to pure phase disturbing processes.

The spectral resolution may be improved beyond the limit imposed by the spontaneous linewidth $\Delta\nu_{spont}$ by coherent transient interaction between the electromagnetic fields and the atomic system. We introduce an observable quantity, the expectation value of the transition operator $\langle r\rangle$. It has been shown that $\langle r\rangle$ obeys the equation of a damped harmonic oscillator with driving force $\tilde{A}(t)$ /12,13/.

$$\langle\ddot{r}\rangle + \frac{2}{T_2}\langle\dot{r}\rangle + \omega_o^2 \langle r\rangle \propto \tilde{A}(t) \tag{1}$$

Introducing plane waves for A(t) and $\langle r\rangle$ with slowly varying amplitudes propagating in the x-direction we write:

$$\tilde{A} = 1/2\ A(t)\ \exp(-i\nu t + ik_A x) + c.c. \tag{2}$$

$$\langle r\rangle = 1/2\ R(t)\ \exp(-i\omega t + ik_R x) + c.c. \tag{3}$$

and obtain

$$\frac{\partial R}{\partial t} + \left[i(\omega_o-\omega) + \frac{1}{T_2}\right]R = \kappa A(t) \tag{4}$$

κ stands for a proportionality constant, ω is the momentary frequency of the transition amplitude. During the excitation process we have $\omega=\nu$, i.e. the system is driven off resonance by $\Delta\omega = \omega_o-\omega$. Eq.(4) is readily integrated to give

$$R(t,\Delta\omega) = \kappa e^{-t/T_2} \int_{-\infty}^{t} e^{(i\Delta\omega+1/T_2)t'}\ A(t')dt' \tag{5}$$

At this point we wish to specify the investigation we have in mind. We deal with vibrational transitions in molecular liquids. In this case, the transition amplitude $\langle r\rangle$ corresponds to the expectation value of the operator q of the vibrational coordinate. The molecular system is first coherently excited by the stimulated Raman process and the coherent vibrational excitation is subsequently monitored by a properly delayed long probe pulse. The Hamiltonian for the molecular system may be written in the form /13/:

$$H = H^{o} - \frac{1}{2} q \sum_{h,i} \left(\frac{\partial\alpha}{\partial q}\right)_{hi} E_h E_i \qquad (6)$$

where $(\frac{\partial\alpha}{\partial q})$ is the Raman susceptibility which couples the vibration with the electric field E. The subscripts h,i refer to a coordinate system which is fixed to the symmetry axes of the individual molecules. The force exerted by the electromagnetic field E on the vibrating molecules is:

$$A = \frac{1}{2} \sum_{h,i} \left(\frac{\partial\alpha}{\partial q}\right)_{h,i} E_h E_i \qquad (7)$$

In our experiments we excite our molecular system by two light pulses. The electric fields of the laser pulse, E_L, and the Stokes shifted pulses, E_S, have the frequency difference $\nu = \nu_L - \nu_S$, the frequency of the driving force.

We recall that the propagation of the light pulses and the interaction with the vibrating molecules are described by Maxwell's equation which leads to the nonlinear wave equation

$$\Delta E - \frac{1}{c^2}\frac{\partial^2}{\partial t^2}(\mu^2 E) = \frac{4\pi}{c^2}\frac{\partial^2}{\partial t^2} P^{NL} \qquad (8)$$

μ denotes the refractive index of the medium and the nonlinear polarisation P^{NL} couples the light fields and the vibrational mode. Under simple conditions we have for the Raman process:

$$P^{NL} = N\left(\frac{\partial\alpha}{\partial q}\right) \langle r\rangle\, E \qquad (9)$$

where N stands for the number density of molecules. In the probing process the vibrational material excitation, R(t), and the electric field of the probe pulse, E_p, generate a scattered Stokes (and anti-Stokes) wave E_{S2}. It has been shown that Eqs.(8) and (9) give Eq.(10) /13/:

$$\frac{\partial E_{S2}}{\partial x} \propto R(t)\, E_p(t) \qquad (10)$$

Experimentally we observe the time-integrated scattered intensity I_{S2} as a function of time delay T_D between the exciting and the probing pulse.

Very recently we have demonstrated that coherent probe scattering may lead to sub-linewidth resolution of Raman transitions /9-11/. Short excitation and prolonged interrogation (SEPI) was used. The method is shown schematically in Fig.1. A short driving force A(t) at a frequency ν (i.e., two pulses E_L and E_S with $\nu = \nu_L - \nu_S$) near the resonance ω_o of the

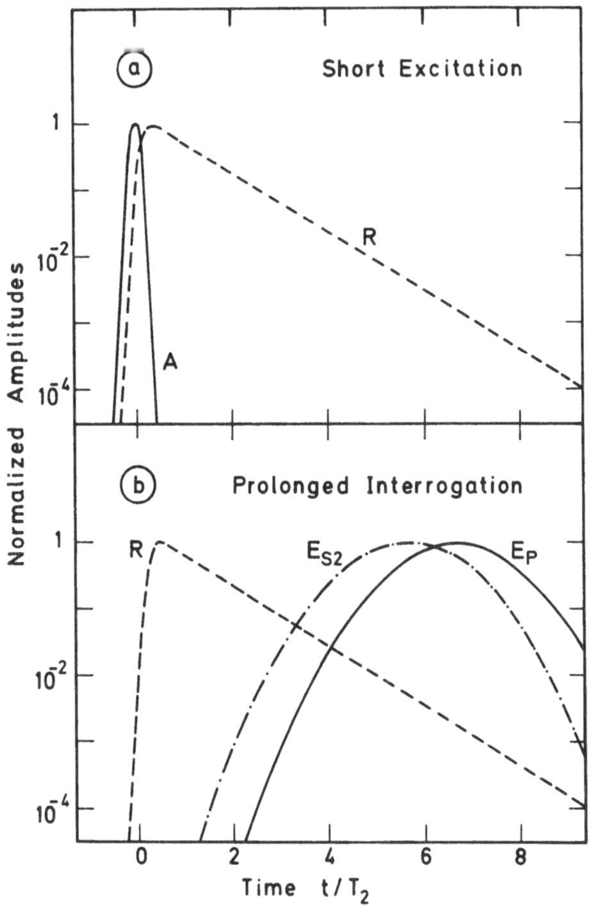

Fig. 1 The short excitation and prolonged interrogation (SEPI) technique. (a) A short driving pulse, A(t), excites the exponentially decaying transition amplitude R(t). (b) The transition amplitude R(t) is interrogated by a long pulse E_p giving rise to the scattered pulse $E_S(t)$.

quantum system generates a transition amplitude R(t) (see Fig.1a). The driving force has a short duration or is switched off rapidly at t=0. After the excitation the exponentially decaying transition amplitude R(t), oscillating freely at ω_0, is investigated by a prolonged probing pulse at frequency ω_p. A pulse with suitably shaped amplitude $E_p(t)$ generates the scattered signal $E_{S2}(t)$ with frequency $\omega_{S2} = \omega_p \pm \omega$. The frequency spectrum of the scattered intensity is observed. $I_{S2}(\omega_{S2})$ has the form:

$$I_{S2}(\omega_{S2}) \propto \left| \int_{-\infty}^{+\infty} dt\; E_{S2}(t)\; e^{i\omega_{S2}t} \right|^2 \propto$$

$$\propto \left| \int_{-\infty}^{+\infty} e^{i\omega_{S2}t}\; E_p(t) R(t)\; e^{-i(\omega_p \pm \omega)t} dt \right|^2 \qquad (11)$$

Introducing $\Delta\omega = \omega_{S2}-\omega_p\pm\omega_o$ and $t_e = t_p^2/(T_2\times 4\ln 2)$ and using a Gaussian shape $E_p(t) = E_{p_o}\exp(-((t-T_D)/t_p)^2 2\ln 2)$ for the delayed probing pulse we obtain at late delay time T_D a Gaussian shaped spectrum centered at the frequency $\omega_p\pm\omega_o$:

$$I_{S2}(T_D,\Delta\omega) \propto e^{-2T_D/T_2 - (t_e/t_p)^2 4\ln 2} \times$$

$$\times \left| \int_{-T_D}^{\infty} e^{i\Delta\omega t} e^{-[(t-t_e/t_p)]^2 2\ln 2} dt \right|^2 \qquad (12)$$

For a long delay time T_D the width of the observed spectrum is only determined by the duration of the probing pulse. For a sufficiently long pulse, $t_p > 1.4\ T_2$, the SEPI lines are narrower than the spontanous width.

The probing with longer pulses and at later times T_D leads to a loss of scattered signal (see Eq.(12)). We have calculated the peak intensity of the scattered signal as a function of spectral narrowing. We find a signal reduction of approximately 10^6 for a narrowing of four. These values are experimentally feasible.

Experimental System

The short dephasing times in molecular liquids require picosecond pulses in order to measure SEPI spectra. We use exciting pulses with a band width of $\simeq 10\ cm^{-1}$ tuned in steps over a larger frequency range. For each excitation band the coherent spectrum was recorded by a spectrograph with sufficient resolution. Fig.2 shows the schematic of our experimental system. At the top, l.h.s.,a single frequency doubled pulse from a mode-locked Nd-glass laser system enters the figure. This pulse of frequency $\nu_1 = 18{,}990\ cm^{-1}$ is split in three parts by two consecutive beam splitters. The pulse in beam 1 passes through the polarizer P1 and the sample, but is blocked by the polarizer P2 in its straight path. In the center beam 2 of Fig.2 a new frequency ν_S is produced in the generator by a stimulated Raman process. Changing the medium of the generator one readily obtains pulses of different frequencies ν_S. These pulses are blocked by the polarizer P2 and are spectrally monitored by spectrograph SP2. On account of the transient generation process the pulses of frequency ν_S are shorter in duration than the incident pulses ν_L by a factor of approximately three /14/. The two pulses of the beams 1 and 2 enter simultaneously the sample coherently exciting molecular

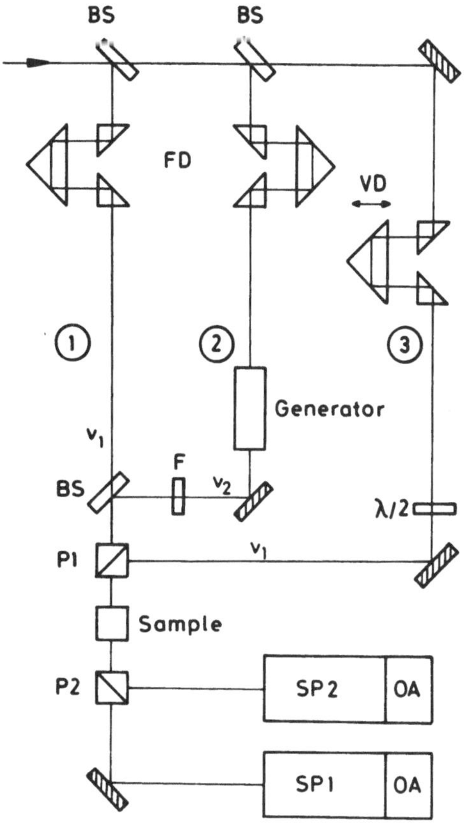

Fig. 2 Schematic of the experimental system used for the study of SEPI spectra of liquids. Beam splitters BS, variable and fixed delay VD and FD, polarizers P1 and P2, blocking filter F, spectrographs SP1 and SP2, used in conjunction with optical spectrum analysers OA.

vibrations via transient stimulated Raman scattering at frequency $\nu = \nu_L - \nu_S$. In the optical path 3 a delayed pulse with polarisation perpendicular to the pulses of 1 and 2 is produced. This third pulse interacts with the coherently excited molecules of the sample producing a Raman shifted signal pulse. Using Stokes scattering the three pulses travel collinearly through the sample. When anti-Stokes scattering is used, the probing pulse crosses the beam direction of the exciting pulses at the phase matching angle. The spectrum of the coherently scattered light is studied by a 2 m spectrograph SP1 and a cooled optical spectrum analyser OA. The experimental system has a resolution of 0.2 cm^{-1} per channel and an absolute accuracy of the frequency scale of 0.4 cm^{-1}.

Experimental Results

We have performed SEPI measurments using a number of organic and anorganic liquids. First we present results on liquid CH_3CCl_3, where a single Lorentzian shaped Raman line exists at 2939 cm^{-1}. Second, we compare the spontaneous Raman spectrum of CCl_4 at 460 cm^{-1} with a SEPI spectrum. Third, we show results on liquid C_6H_{12}, where broad and overlapping lines occur between 2850 and 2940 cm^{-1}, in the common spontaneous Raman spectrum.

A. Liquid 1.1.1. Trichloroethane

In Fig.3 we show the Raman band of a CH_3-stretching mode at 2939 cm^{-1} of liquid CH_3CCl_3 measured with a standard laser Raman spectrometer. The

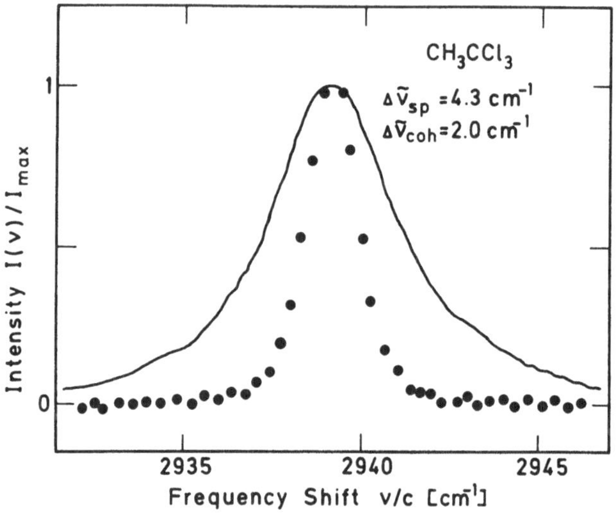

Fig. 3 Spontaneous Raman spectrum (solid curve) and coherent probing (SEPI) spectrum taken at t_D=18.5 ps. The CH_3-stretching mode of CH_3CCl_3 is investigated. Note the narrow SEPI spectrum.

Lorentzian shaped line (solid curve) has a bandwidth of $\Delta\nu_{spont} = 4.3$ cm^{-1}. Quite different is the bandwidth of the coherent probing (SEPI) spectrum taken at a delay time of $t_D = 18.5$ ps. Now we find a bandwidth of $\Delta\nu_{coh} = 2.0$ cm^{-1}; i.e., we have a spectral narrowing of a factor of two. The frequency of the Stokes shift of the SEPI spectrum is 2938.2 cm^{-1} in agreement with the spontaneous value of 2939 cm^{-1}. In this experiment the pulse duration of the exciting and the probing pulse is 5 ps and 10 ps, respectively.

B. Carbon Tetrachloride

The natural abundance of chlorine, $^{35}Cl:^{37}Cl$ = 75.5:25.5, lead to five components of CCl_4: 32.5% $C^{35}Cl_4$, 42.2% $C^{35}Cl_3{}^{37}Cl$, 20.5% $C^{35}Cl_2$, 4.4% $C^{35}Cl^{37}Cl_3$, and 0.4% $C^{37}Cl_4$. As a result, the symmetric vibrational mode $\nu_1(a_1)$ at 460 cm^{-1} consists of four major components /15/. The spontaneous Raman spectrum of Fig.4 shows, indeed, four peaks, the intensity of which is in good agreement with the distribution of the four major

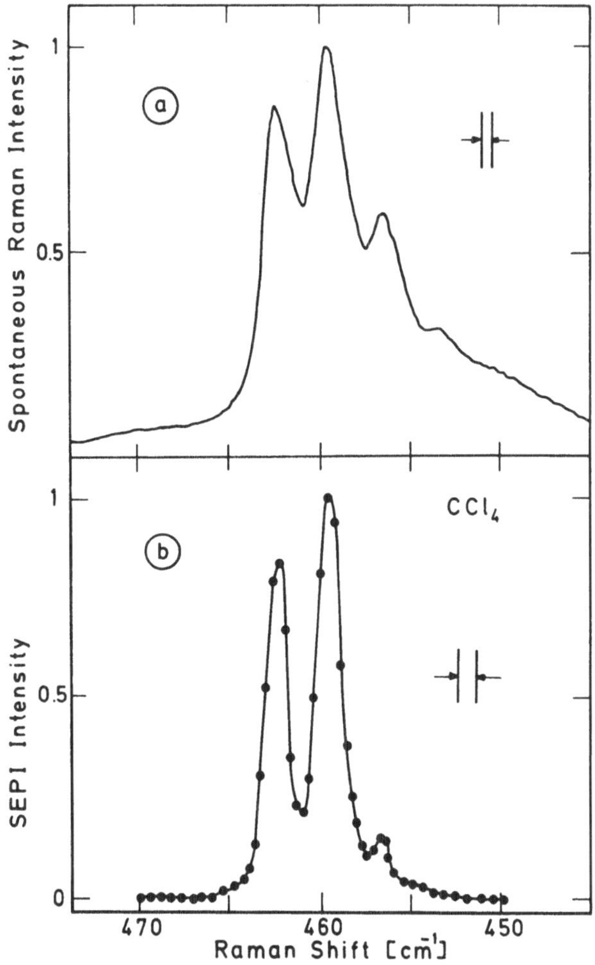

Fig. 4 Raman spectra of the ν_1 mode of liquid CCl_4. (a) Polarized spontaneous Raman spectrum, instrumental resolution 0.5 cm^{-1}. (b) SEPI spectrum of CCl_4. Instrumental resolution 1 cm^{-1}.

molecular species. The different Raman lines in Fig.4a overlap strongly inspite of the high resolution of the Raman spectrometer of 0.5 cm^{-1}. The individual Raman lines are broadened by the dephasing time $T_2 = 6.0$ ps, which gives rise to the observed width $\Delta\nu_{spont} = 1/\pi c T_2 = 1.8\ cm^{-1}$ /16/.

The SEPI spectrum of the same vibrational mode of CCl_4 is depicted in Fig.4b. In this case the sample was excited for approximately 7 ps by a laser pulse and a second Stokes shifted pulse ($\nu_L - \nu_S \simeq 458\ cm^{-1}$). Under these conditions the two neighboring major molecular components are strongly excited and give rise to the observed strong scattering spectrum. The data of Fig.4b were obtained with long probing pulses of $t_p = 20$ ps and at a delay time of 40 ps. We point to the sharp lines in the SEPI spectrum which are narrower than $\Delta\nu_{spont}$ in the spontaneous spectrum of Fig. 4a. The depicted linewidths in Fig.4b are determined by the limited spectral resolution of the spectrometer.

C. Liquid Cyclohexane

As another example for the short excitation and prolonged interrogation (SEPI) technique we present Raman data of cyclohexane in the small frequency range between 2850 cm^{-1} and 2940 cm^{-1}. In Fig.5b the polarized spontaneous Raman spectrum is depicted. This spectrum was taken with an Ar^+ laser and a Raman spectrometer with a resolution better than 1 cm^{-1}. The three strong Raman bands correspond to CH-stretching modes and the diffuse spectrum between 2860 cm^{-1} and 2920 cm^{-1} is considered to be due to overlapping overtones and combination modes which are enhanced by Fermi resonance with the fundamentals /17,18/.

In Fig.5c we show three SEPI spectra on an expanded scale (factor 3.7). Each spectrum was obtained by a single laser shot. On the r.h.s. we present the sharp SEPI band corresponding to the CH-stretching mode at 2923 cm^{-1}. The small linewidth of 2.3 cm^{-1} allows to determine the peak position accurately to $2922.0 \pm 0.7\ cm^{-1}$. We note that the SEPI band is considerably smaller than the corresponding band in the spontaneous Raman spectrum of Fig.5b, the latter being asymmetric on account of other smaller Raman transitions. The Raman transition at 2923 cm^{-1} was excited using ethylene glycol, $(CH_2OH)_2$, in the generator cell of Fig.2.

The SEPI spectrum of Fig.5c, middle, shows four Raman transitions. Lines as close as 2.5 cm^{-1} are clearly resolved. The four transitions are hidden under the wing of the strong Raman band at 2923 cm^{-1}; they cannot be detected in the conventional Raman spectrum of Fig.5b. The SEPI spectrum is obtained by using an exciting pulse ν_S with a frequency

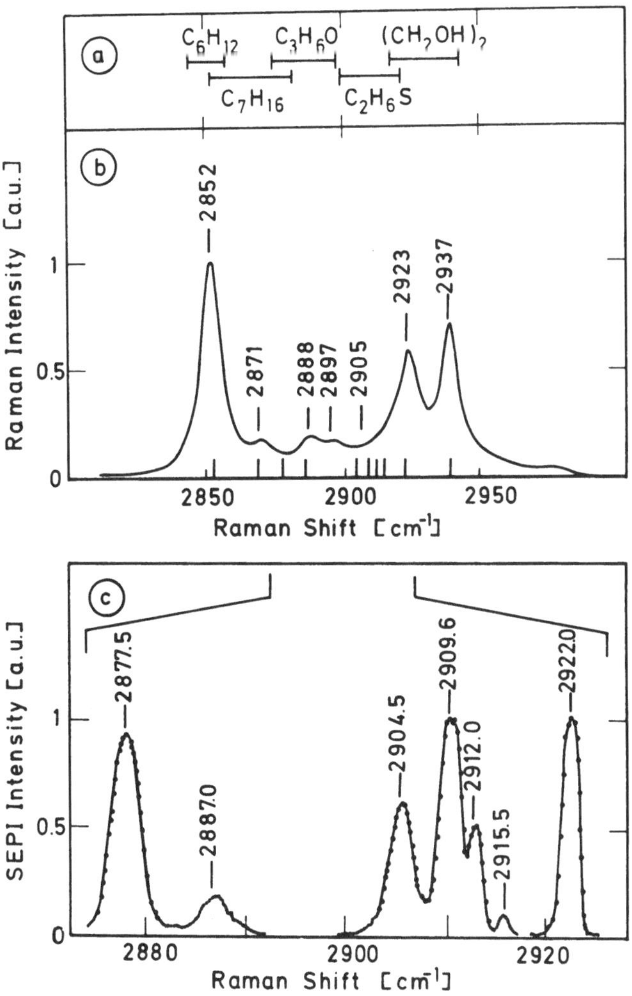

Fig. 5 Experimental results of short excitation and prolonged interrogation (SEPI) spectroscopy of C_6H_{12}. (a) Frequency range of the various generator liquids used in the experiment. (b) Polarized spontaneous Raman spectrum of C_6H_{12} recorded with a resolution of 1 cm^{-1}. The frequency positions of the resonances found in SEPI spectra are marked by vertical lines. (c) Three SEPI spectra taken with different generator liquids. New Raman lines are detected and the spectral resolution is improved. (Note, the frequency scale of c is 3.7 times larger than the one of b).

band extending from 2900 cm^{-1} to 2920 cm^{-1} (dimethyl sulfide, C_2H_6S, in the generator).

In Fig.5c, l.h.s., we depict a SEPI spectrum obtained after excitation by a ν_S pulse by a spectral band width extending from 2875 cm^{-1} to 2890 cm^{-1} (propylene oxide, C_3H_6O). We find two distinct Raman bands at

2877.5 cm^{-1} and 2887 cm^{-1}. The band at 2877.5 cm^{-1} has never been reported previously. It is buried in the diffuse part of the conventional Raman spectrum (see Fig.5b).

A final assignment of the new Raman lines between 2870 cm^{-1} and 2920 cm^{-1} has not yet been made. Inspection of the lower fundamental modes suggest overtones and combination modes in this frequency range. Of special interest is the new Raman line at 2912 cm^{-1} which coincides precisely with an infrared active mode of the molecule. It appears that we observe here a Raman forbidden mode.

Additional Observations and Comments

The following points are relevant for the application of the SEPI technique: (i) The frequency positions of the observed Raman lines are independent of the excitation conditions since we observe freely relaxing molecules. We have tested this notion by exciting our sample with narrow or with broad pulses of similar central frequency ν_S. This experiment is readily performed using different media in the generator cell. The advantage of a broad frequency spectrum of the incident pulse is to provide initial conditions for several Raman transitions in congested frequency regions. One can observe several Raman lines with one shot (see Fig.5c, middle). (ii) In SEPI experiments the exciting and interrogating pulses should not overlap temporarily in order to avoid the generation of a coherent signal via the nonresonant four-photon parametric process. For this reason, the delay time of the third probing pulse has to be sufficiently large. One roughly estimates delay times of $t_D = 20$ to 25 ps for dephasing times $T_2 \simeq 1$ ps and Gaussian probing pulses of 8 ps duration. The SEPI spectra are observed with good accuracy, approximately five orders of magnitude below the peak value at $t_D = 0$. (iii) The maxima of the SEPI spectra are not proportional to the Raman scattering cross-section, since the initial conditions of the exciting pulses and the T_2 times are important parameters for the observed magnitude of the generated signal. SEPI spectra taken for different delay times allow an estimate of the dephasing times T_2 . (iv) The frequency precision of the generated Stokes spectrum depends upon the frequency stability of the interrogating pulse. For highest accuracy the frequency ν_L has to be measured simultaneously with the SEPI spectrum. Interrogating pulses with a chirped frequency spectrum give unwanted shifts of the SEPI spectra and should thus be avoided. (v) The scattering process may also be performed on the anti-Stokes side of the spectrum. The disturbing interference found in stationary CARS spectroscopy does not occur for the delayed probing used with the SEPI spectroscopy /9,19/. (VI) A simultaneous measurement of the coherent Stokes and anti-Stokes SEPI spectra allows to eliminate the effect of a chirped probing pulse. In this way, the absolute frequency position is obtained with high accuracy /9/.

Concluding Remarks

The data presented in this article convincingly show the usefulness of the short excitation and prolonged probing technique. It is possible to obtain molecular information which are not found by other existing spectroscopic techniques.

References

1. For a review see: H. Walther, in "Laser Spectroscopy of Atoms and Molecules", Topics in Applied Physics, Vol.2, Ed. H. Walther, Springer, Heidelberg 1976, p.1.
2. G. Copley, B.P. Kibble and G.W. Series, J. Phys. B1 (1968) 724.
3. H. Figger and H. Walther, Z. Physik 267 (1974) 1.
4. H. Metcalf and W. Phillips, Opt. Lett. 5 (1980) 540.
5. F. Shimizu, K. Umezu and H. Takuma, Phys. Rev. Lett. 47 (1981) 825.
6. P. Meystre, M.O. Scully and H. Walther, Opt. Commun. 33 (1980) 153.
7. H.-W. Lee, P. Meystre and M.O. Scully, Phys. Rev. A24 (1981) 1914.
8. P.E. Coleman, D. Kagan and P.L. Knight, Opt. Commun. 36 (1981) 127.
9. W. Zinth, Opt. Commun. 34 (1980) 479.
10. W. Zinth, M.C. Nuss and W. Kaiser, Chem. Phys. Lett. 88 (1982) 257.
11. W. Zinth, M.C. Nuss and W. Kaiser, to be published in Opt. Commun.
12. J.A. Giordmaine and W. Kaiser, Phys. Rev. 144 (1966) 676.
13. A. Laubereau and W. Kaiser, Rev. Mod. Phys. 50 (1978) 607. A. Penzkofer, A. Laubereau and W. Kaiser, Progr. Quant. Electron. 6 (1979) 55.
14. R.L. Carman, F. Shimizu, C.S. Wang and N. Bloembergen, Phys. Rev. A2 (1970) 60.
15. J. Brandmüller, K. Buchardi, H. Hacker and H.W. Schrötter, Z. Angew. Phys. 22 (1967) 177.
16. W. Zinth, H.J. Polland, A. Laubereau and W. Kaiser, Appl. Phys. B26 (1981) 77.
17. K.W.F. Kohlrausch and W. Wittek, Z. Phys. Chem. B48 (1941) 177.
18. K.B. Wiberg and A. Shrake, Spectrochim. Acta 27A (1971) 1139.
19. W. Zinth, A. Laubereau and W. Kaiser, Opt. Commun. 26 (1978) 457.

AN AUTOCORRELATOR FOR THE MEASUREMENT OF CW ULTRASHORT OPTICAL PULSES HAVING FREQUENCY VARIATIONS

Y. Cho, T. Kurobori, Y. Matsuo

The Institute of Scientific and Industrial Research

Osaka University

1. INTRODUCTION

Since the speed realized in ultrashort optical pulses obtainable today from mode-locked lasers exceeds far off the response of any presently available high-speed opto-electronic devices (photodetectors or streak camera), measurements of pulse widths of those ultrashort optical pulses must depend on a rather indirect method - second-order autocorrelation technique, which was first reported by Maier, Kaiser, and Giordmaine [1], and by Armstrong [2]. Although the autocorrelation technique includes ambiguity in determining the pulsewidth, it essentially contains more information than normally utilized in the pulsewidth measurement. In particular, for ultrashort optical pulses obtainable from cw mode-locked lasers, precise scan in an interferometric configuration necessary for obtaining the autocorrelation function can provide the resolution - in space in place of in time - less than the wavelengths of optical pulses to be measured. Hitherto, however, for the pulsewidth measurement, only rather coarse intensity autocorrelation component of the pulse has normally been measured whereas the included phase component has been discarded. This ignorance of the phase component in the conventional autocorrelation measurement is due to its fast scan giving an average action in its interferometric measuring process. Retaining of the phase component in the interferometric measurement can be attained by a slow enough scan ensuring the response of a detection system and a corresponding stability needed for the interferometric driving mechanism.

For various applications of the ultrashort optical pulses, a so-called Transform-Limited Pulse (TLP), i.e. a pulse having no phase (frequency) variation within its pulse duration, is often desired. This causes the necessity of checking the phase (frequency) characteristic of an ultrashort optical pulse. An autocorrelator capable of retaining the phase component was already reported by Diels employing a gas pressure scan [3]. In this autocorrelator, however, the conventional intensity correlation information is masked by the phase correlation component, and hence for the pulse containing any frequency variation such as chirping, as being normally the case, this correlator is not suited for pulsewidth measurements.

Here, we describe an autocorrelator capable of providing both the intensity correlation and an autocorrelation including phase components within a single measurement run. In this autocorrelator, the scanning in its interferometer is alternatively switched over two different speeds of fast and slow, thereby the above-mentioned two different autocorrelations are recorded in a single run. This autocorrelator also has an operation mode as a real-time autocorrelation monitor, which is useful in adjusting stages of a source mode-locked laser and the interferometer itself. With the use of this autocorrelator, not only the behaviour of ultrashort optical pulses can be monitored, but also the phase correlation width can be measured, from which degrees of frequency variations contained in those pulses can be estimated. Measurements of this kind are usually done by a combination of an intensity autocorrelator and a spectrometer.

For obtaining the above-mentioned two different scanning speeds alternatively in a single run, we employed mechanical scanning by a piezoelectric translator element. For the monitor mode operation, a speaker was used. The stability of the interferometer obtained by the above scheme was found quite satisfactory for our autocorrelator purpose.

2. AUTOCORRELATION INCLUDING PHASE COMPONENTS

Setup of the constructed autocorrelator is shown in Fig.1. The nonlinear interference between mutually delayed pulses, which is necessary for obtaining the second-order autocorrelation, is yielded with a combination of a Michelson-type interferometer section and an SHG crystal (shown as KDP), and is detected by a slow response detector (shown as PMT). A recorded curve $S(\tau)$ obtained through a slow response detector can be expressed as follows [3],[4]:

$$S(\tau) \propto \overline{|E(t)|^4} + \overline{|E(t-\tau)|^4} + \overline{4|E(t)|^2.|E(t-\tau)|^2}$$

$$+ \overline{4\{|E(t)|^2 + |E(t-\tau)^2\}|E(t)|.|E(t-\tau)|.\cos\{\omega\tau+\phi(t)-\phi(t-\tau)\}}$$

$$+ \overline{2|E(t)|^2.|E(t-\tau)|^2.\cos 2\{\omega\tau+\phi(t)-\phi(t-\tau)\}}$$

where the electric field $\varepsilon(t)$ of an optical pulse is assumed as

$$\varepsilon(t) = E(t)\exp i\omega t + c.c. \tag{1}$$

here

$$E(t) = |E(t)|\exp i\phi(t) \tag{2}$$

and ω is the optical frequency, τ is the delay of the autocorrelation, which is varied by scanning one of two arms of the interferometer, and $\phi(t)$ represents the possible phase variation contained in the optical pulse.

In (1), the first and the second terms correspond to the background, the third term gives the intensity autocorrelation function, and the fourth and the fifth terms give the phase correlation information. These last two terms are not observed by a conventional fast scan because they include rapidly varying $\omega\tau$ and $2\omega\tau$ under their cosine terms, but a stable slow scan recovers these terms. In our autocorrelator fast and slow scans are repeated alternatively, then the intensity correlation and the correlation including the phase component i.e. $S(\tau)$ are obtained separately in a single recording. Hereupon it will be helpful to mention that the contrast ratio (peak (τ=0) to background ($\tau=\infty$) ratio) expected for the complete mode-locked pulses (those pulses having zero-E(t) periods in (2)), becomes 3:1 for the intensity correlation comprising up to the third term and becomes 8:1 for the whole of $S(\tau)$.

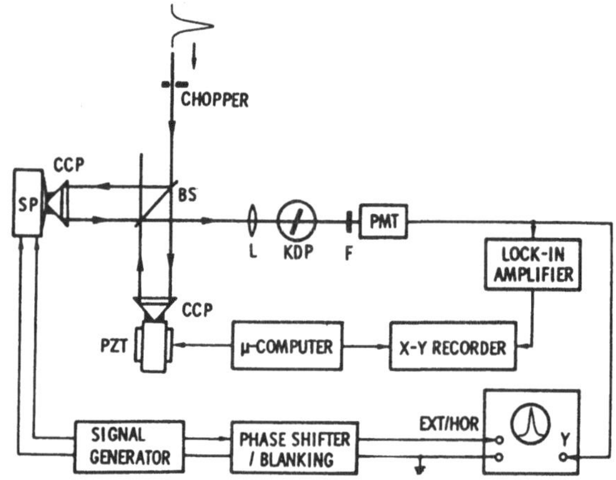

Figure 1: Set up of the autocorrelator

CCP : corner cube prism
SP : speaker
PZT : piezoelectric translator
PM : photomultiplier tube
BS : beam splitter
F : filter
L : lens

3. COMPUTER SIMULATIONS

In the following, the correlation including only up to the third term in (1) is called the intensity correlation and the correlation including up to the fifth term, i.e. the whole of $S(\tau)$ the phase correlation.

Computer simulations were tried to see the effects on the intensity and the phase correlations when variations of the optical frequency are included in an optical pulse. The simplest case of the optical frequency variation may be expressed by the phase variation $\phi(t)$ as follows:

$$\phi(t) = 4A(t/\Delta t_p)^2 \tag{3}$$

This means the instantaneous frequency $d\phi(t)/dt$ contains the following linear frequency variation (chirp) with respect to time:

$$d\phi/dt = 8At/(\Delta t_p)^2 \tag{4}$$

where Δt_p is the pulsewidth (intensity FWHM) and A indicates the chirp factor. Value of A corresponds to the phase difference at the intensity half-maximum point ($t=\Delta t_p/2$) with respect to the pulse center (t=0).

Ippen et.al. [5] reported the existence of a positive chirp in the ultrashort optical pulse from a cw mode-locked dye laser by their dynamic spectroscopy. Also they showed that pulses can be compressed by compensating accompanying chirps by using a pair of diffraction gratings having an opposite sign of dispersion with respect to the chirp and they concluded that the chirp contained in the pulse from their cw mode-locked dye laser was almost linear. Considering such experimental evidence, the simplest phase variation $\phi(t)$ expressed by (3) was assumed. For the pulse waveform $|E(t)|$ in (2), the sech-shape was used with considering theory [6] and experiment [7] as follows:

$$|E(t)| = \mathrm{sech}(1.76t/\Delta t_p) \qquad (5)$$

Computed results are shown in Fig.2, where Fig.(a) shows amplitude, real part, and imaginary part respectively of the envelope component of the optical field, where the time scale is normalized by Δt_p, Fig.(b) shows the intensity (white dotted curve) and the phase (group of vertical lines) autocorrelations of the optical pulse, where the mutual delay is normalized by the FWHM of the intensity autocorrelation, $\Delta\tau_i$, and Fig.(c) shows corresponding power spectra, where the frequency is normalized by the FWHM of the power spectrum for A=0. When A=0 (chirpless), that is, for the case of TLP (Transform-Limited-Pulse) a product of the pulsewidth Δt_p and the power spectrum width $\Delta\nu$ (FWHM) becomes $\Delta t_p \Delta\nu = 0.32$ [8]. For A=0.3, $\Delta t_p \Delta\nu = 0.65$, double of that for TLP, and for A = 0.8, $\Delta t_p \Delta\nu = 1.23$, about four times of that for TLP. The relation between A and $\Delta t_p \Delta\nu$ is shown later in Fig.7. The reported experimental values on $\Delta t_p \Delta\nu$ obtained from passively mode-locked dye lasers are mostly in a range of 0.4 - 0.8 [7],[9]. If these spreadings are caused by the simple chirp of (3), corresponding A values are ranging over 0.1 - 0.5. From Fig.(b), it is worthwhile to notice that the intensity correlation (white dotted curve) is independent from A, but the phase correlation (group of vertical lines) is quite sensitive to A.

4. CONFIGURATION OF THE SYSTEM

The configuration of the constructed autocorrelator was already shown in Fig.1. In the construction of the system, particular attention was payed to make the system stable. The whole structure of the interferometer section is built with rigid element-holders installed on a massive table and retained in an isolation container for the elimination of air-borne vibration and air turbulence. In order to avoid the instability of a source laser due to an unnecessary feedback of the pulse into the laser, a double-path type in Fig.1 is used for the interferometer section. In this configuration, as for the end reflectors of the interferometer, the retro-reflection type is necessary. We used corner-cube prisms CCP because of its easiness of installation. The corner cubes are mounted on holders with a small leaning angle so that the light paths reflected at their surfaces are separated from their incoming light paths. The beam splitter we used has a reflective coating of 50% on one side and AR coating on the other side, and has a small wedge angle between those two planes. Mutually delayed pulses prepared in the interferometer section are focussed by a lens

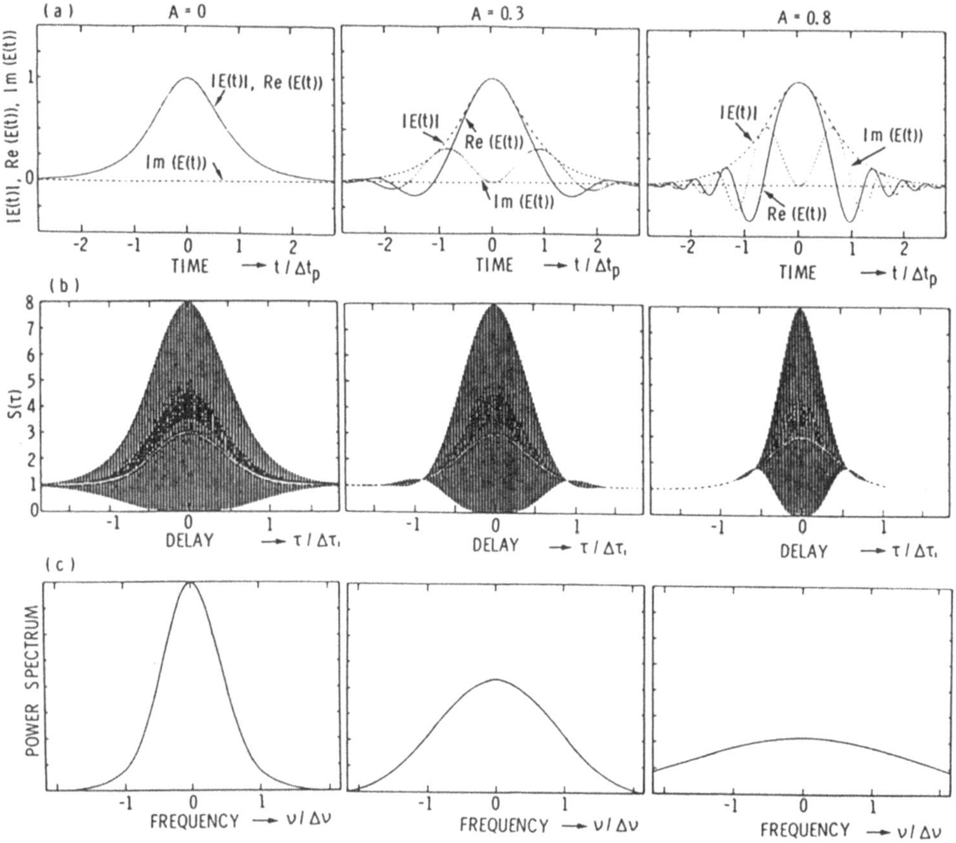

Figure 2: Computer simulation of optical pulses (a) field, (b) autocorrelation $S(\tau)$ and (c) power spectrum. In (b), bell-shaped broken white curves show the intensity autocorrelation having 3:1 contrast ratio, and groups of vertical lines show the phase autocorrelation having 8:1 contrast ratio.

L of 5cm focal length into an SHG crystal KDP. The SHG crystal we used is a 45°z-60° y'-cut KDP of 170μm thick. Second harmonic pulses generated in the KDP are, after filtering out by a fundamental-cut filter F, sent to a photomultiplier tube PMT.

In this system the following two modes of operation are possible:

(a) Real-time intensity autocorrelation monitor mode,

(b) Intensity/phase autocorrelation measurement mode.

For accomplishing these two operation modes in a single system, on one arm of the interferometer a two-speed (fast and slow) scanner using a precision piezoelectric translator PZT is installed, and on the other arm a repetitive fast scanner utilizing the piston motion of a speaker SP is installed. For mode (a), the repetitive fast scanner of SP is used, and for mode (b) the two-speed scanner of PZT is used. The difficulty of this system is the compatibility of these two modes, because, for the phase autocorrelation measurement a high degree of stability is needed while the speaker is sensitive to the external disturbances. The countermeasure for it is described in 5. The photodetected current by PMT giving S(τ) of (1) is monitored with a CRT in mode (a), or recorded on an xy-recorder after amplified with a lock-in amplifier in mode (b).

Real-time monitor using speaker

In general, for the generation of ultrashort optical pulses from a cw mode-locked laser, precisely controlled setting of various parameters of the laser is necessary in its adjusting stage. Also the adjustment of the autocorrelator itself is a time-consuming task. For these adjustments and the monitoring purposes, a real-time monitor-type autocorrelator is very useful. Already various types of these real-time monitors were reported [10],[11],[12]. Here we employed the type using a speaker because of its easiness of installation. Checking points of a speaker when it is used for these purposes are (a) linearity of movement, (b) response linearity of distance versus applied voltage, (c) stability against external disturbances. Since we found most commercially available audio speakers were not satisfactory for the above points, a reinforced type based on a small-size audio speaker (FW-100, Fostex) was constructed as shown in Fig.3. After removing a cone, its voice-coil cylinder was extended and was fixed with two reinforced dampers to its frame for ensuring its stability and linear movement. The linear movement of the voice-coil was checked by the movement of a laser beam spot on the screen at a distance reflected from a mirror attached on the top of the voice-coil. This reflected laser beam spot mostly shows a small Lissajous motion with conventional audio speakers, but no Lissajous motion was observed with the reinforced type. For the operation mode (b), this speaker must be fixed stably. This is accomplished by short-circuiting the voice-coil. The stability of this fixing is discussed in 5. The scanning sensitivity of this speaker is 2 mm/V at 60 Hz. This corresponds to a delay sensitivity of 13 ps/V, which provides enough coverage for the delay up to ∿ 10ps. The repetition frequency was 60Hz. One of the corner cube prisms is attached on the top of the extended voice-coil cylinder.

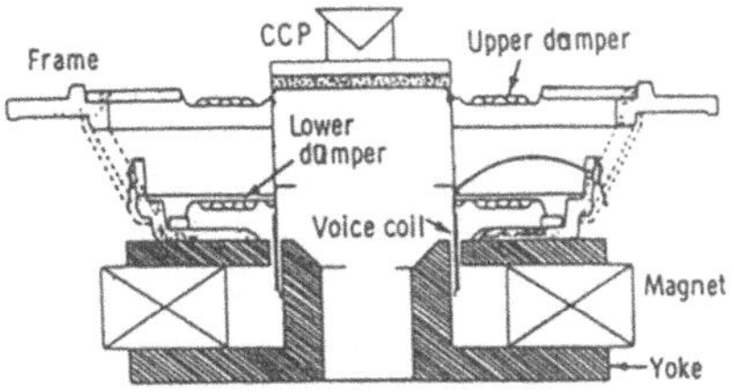

Figure 3: Cross sectional view of the speaker.

Piezoelectric translator

The piezoelectric translator we used is Burleigh's Inchworm System (PZ-551). A corner cube prism is attached on the piezoelectric translator. The scanning speed is switched over alternatively from the slow scan of 0.03μm/s to the fast scan of 30μm/s by the instruction of a microcomputer. When the time constant of the whole system including an xy-recorder is set to 300ms, with scanning speeds more than 15μm/s, the phase component disappears and only the intensity correlation is observed. The speed of the x-axis advance of the xy-recorder is also changed in proportion with the above scanning speeds.

5. EXPERIMENTS

The optical pulses for the experiments were obtained from an argon-ion laser pumped Rhodamine-6G/DODCI cw passive-mode-locked dye laser. This laser has a single-plate etalon in its cavity, lengthening the pulsewidth to around 1.5ps. The oscillation wavelength was around 605nm and the average output was about 20mW.

Fig.4 shows an example of the CRT-trace of the real-time intensity autocorrelation monitor in the operation mode (a). The contrast ratio of this trace is about 3:1, indicating the complete mode-locked pulses. The FWHM of the trace, i.e. $\Delta\tau_i$, is 2.4ps which corresponds to a pulsewidth, Δt_p, of 1.6ps with assuming the sech-pulse shape.

In Fig.5, an example of the interference trace obtained with the interferometer system with the slow scan at around the delay $\tau = 0$ in the operation mode (b) is shown. In the left-half part, the voice-coil of SP was kept opened. No regular interference pattern is seen due to the fluctuating motion of the open-circuited voice-coil. In the right-half part, the voice-coil was short-circuited, where, owing to the current

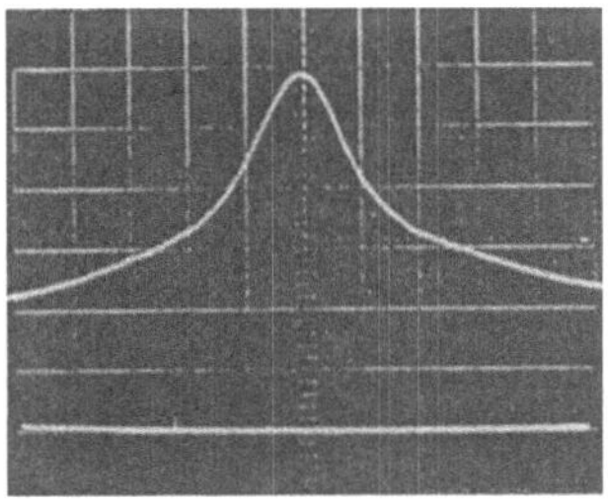

Figure 4:
CRT-trace of the real-time intensity autocorrelation monitor. Horizontal: 1ps/div. 3:1 contrast ratio is seen. Pulsewidth of 1.6ps is estimated by assuming a sech-pulse.

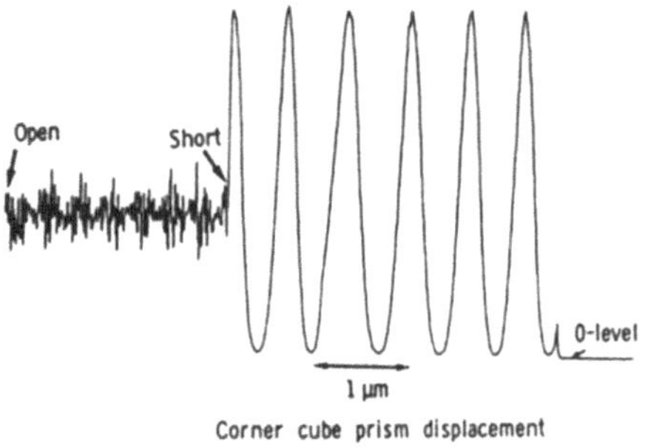

Figure 5:
Typical interference trace obtained with the slow scan of PZT. Irregular trace was obtained while the voice-coil was opened.

feed-back action of the voice-coil to its fluctuating motions, those fluctuating movements are well compensated and hence the stabilized interference pattern is recovered. The compatibility of the operation modes (a) and (b) mentioned in 4 is thus accomplished by short-circuiting the voice-coil of SP during the operation mode (b). The standard deviation of included periods in this interference pattern is less than 6%. The visibility obtained is almost 1, indicating the optical quality and the achieved adjustment of the interferometer section.

Fig.6(a) shows an example of the intensity/phase autocorrelation trace of the operation mode (b), and Fig.6(b) shows its corresponding power spectrum. The bell-shaped trace seen in Fig.(a) is the intensity autocorrelation curve, which was obtained by the fast scan periods, and the group of vertical lines is the phase auto-correlation trace, which was obtained by the slow scan periods. They show respectively 3:1 and 8:1 contrast ratios which are expected values for the complete mode-locked pulses and can serve as a check of the performance of this autocorrelator. From the intensity autocorrelation trace, the intensity correlation width $\Delta\tau_i$ is 2.0ps, while from the amplitude variation of the phase correlation vertical lines, the phase correlation width $\Delta\tau_\Phi$, defined as a FWHM of the amplitude variation of those correlation bars, is 1.4ps. Assuming the pulse shape of sech, the pulsewidth Δt_p is deduced to be $\Delta t_p = \Delta\tau_i/1.55 = 1.3$ps [8]. From Fig.6(b), the spectral width of this pulse is 11Å, which corresponds to a power spectrum width $\Delta\nu$ of 0.9THz at 605nm. Thus, the observed pulsewidth-spectral width product $\Delta t_p \Delta\nu$ becomes 1.2.

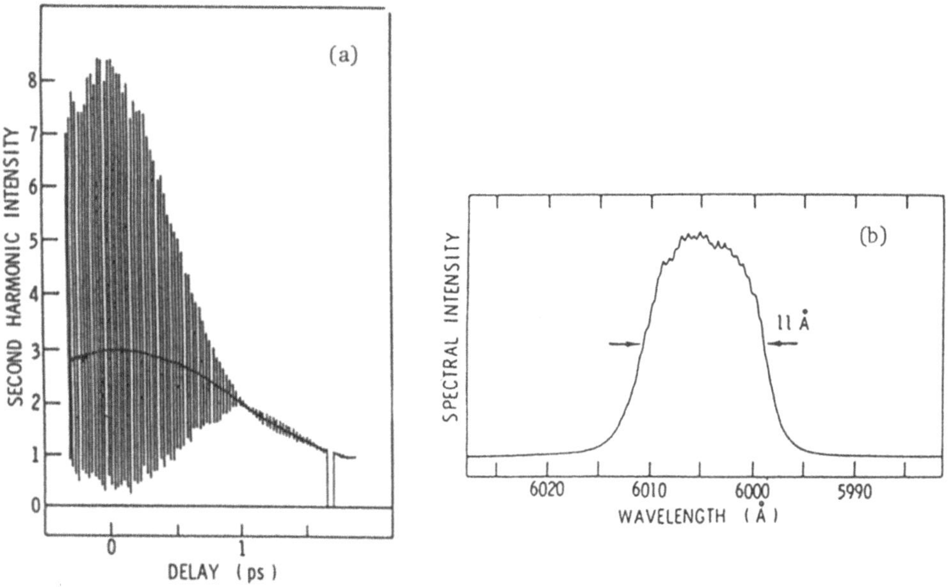

Figure 6: Observed autocorrelation trace (a) and its corresponding power spectrum (b). In (a), the bell-shaped curve corresponds to the intensity autocorrelation trace, which has a 3:1 contrast ratio, and the group of vertical lines corresponds to the phase autocorrelation trace, which shows an 8:1 contrast ratio.

6. ESTIMATION OF FREQUENCY VARIATION

From the result of the computer simulation shown in Fig.1, it is expected that any inclusion of simple frequency chirp in an optical pulse should sensitively reflect on the observed phase autocorrelation trace while it has no influence on the intensity autocorrelation trace. Therefore, a measure of the frequency variation on the observed intensity/phase autocorrelation trace may be defined as

$$f_c = \Delta\tau_\phi/\Delta\tau_i \tag{6}$$

Calculated f_c vs. chirp factor A for the sech-pulse is shown in Fig.7. Along with the f_c-value of Fig.7 the pulsewidth-spectral width product $\Delta t_p\Delta\nu$ vs. A is also shown.

The observed f_c-value in Fig.6 is $\Delta\tau_\phi/\Delta\tau_i = 1.4/2.0 = 0.7$. If we assume this observed reduction of f_c-value is caused by a linear frequency chirp, the chirp factor A of this observed pulse can be read from the f_c vs. A curve of Fig.7 to be 0.8. Also from $\Delta t_p\Delta\nu$ vs. A curve, $\Delta t_p\Delta\nu$ is read to be 1.2. This product value obtained from the intensity/phase autocorrelation measurement agrees with the previously mentioned value of 1.2 obtained from the measurements of its spectrum and intensity autocorrelation. This result partly warrants the measurement by this autocorrelator.

Since the chirp factor A of the observed intensity/phase autocorrelation trace of Fig.6 is estimated to be 0.8 as mentioned above, this trace should be compared with the computer simulation for A = 0.8 in Fig.2(b). A close resemblance between the whole shapes of observed and simulated traces is evidently seen, justifying the assumption of sech-pulse shape and linear chirp in this pulse. Particularly, it should be noted that, in either trace, the small dip appears in the amplitude

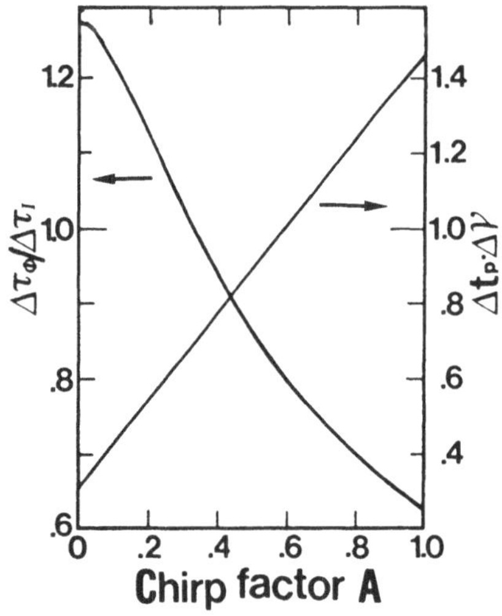

Figure 7: $\Delta\tau_\phi/\Delta\tau_i$ vs. A and $\Delta t_p\Delta\nu$ vs. A curves for the sech-pulse.

variation of the vertical lines at the outskirt part. The appearing positions of these dips in the observed and in the simulated traces also show an agreement, $\tau = \Delta t_p/2$, in this case. The dip of this kind can be caused when a monotonically varying frequency sweep is present due to the cumulative interference between the fourth and the fifth terms in (1) at a large value of $|\tau|$, but it cannot be caused by any randomly fluctuating phase variation. Therefore, the observation of the dip serves as a direct evidence supporting the presence of the chirp in the optical pulse. This is a superior point of the phase autocorrelation measurement compared with the spectral measurement.

7. CONCLUSION

It has been shown that, with using the constructed autocorrelator, which can provide both the usual optical pulse intensity autocorrelation and the autocorrelation including the phase component, the frequency characteristic included in the optical pulse could be estimated. As an example, the deviation from the transform-limited relation for the optical pulse from a cw mode-locked dye laser could be explained with the inclusion of a linear frequency chirp. This autocorrelator also has a monitor mode in addition to the above measurement mode, bringing more easiness to the ultra-short optical pulse measurement.

REFERENCES

1. M. Maier, W. Kaiser and J.A. Giordmaine, Phys.Rev.Lett., 17, 1275, (1966).
2. J.A. Armstrong, Appl.Phys.Lett., 10, 16, (1967).
3. J.-C. Diels, E.W. Van Stryland and D. Gold, Picosecond Phenomena, C.V. Shank, E.P. Ippen and S.L. Shapiro, Eds., Berlin: Springer-Verlag, 117, (1978).
4. T. Kurobori, Y. Cho and Y. Matsuo, Opt.Comm., 40, 156, (1981).
5. E.P. Ippen and C.V. Shank, Appl.Phys.Lett., 27, 488, (1975).
6. H.A. Haus, IEEE Jour.Quant.Electron., QE-11, 736, (1975).
7. I.S. Ruddock and D.J. Bradley, Appl.Phys.Lett., 29, 296, (1976).
8. E.P. Ippen and C.V. Shank, Ultrashort Light Pulses, Ed. S.L. Shapiro, Berlin: Springer-Verlag, 83, (1977).
9. Z.A. Yasa, A. Dienes and J.R. Whinnery, Appl.Phys.Lett., 30, 24, (1977).
10. Y. Ishida, T. Yajima and Y. Tanaka, Japan Jour.Appl.Phys., 19, L289, (1980).
11. K.L. Sala, G.A. Kenney-Wallace and G.Y. Hall, IEEE Jour.Quant.Electron., QE-16, 990, (1980).
12. H. Harde and H. Burggraf, Opt.Comm., 38, 211, (1981).

MULTIPHOTON IONIZATION OF ATOMS

- Cross Sections and Photoelectron Angular Distributions*) -

G. Leuchs
Sektion Physik der Universität München
D-8046 Garching, FRG

I. Introduction

The investigation of multiphoton ionization is important for applications where high power lasers interact with matter as e.g. for the study of laser induced plasmas. It can also be a loss mechanism for any application making use of other nonlinear processes. In addition, multiphoton ionization enhanced by resonant intermediate states has been used to design an efficient single atom detector [1]. From a more fundamental point of view theoreticians are challenged since the cross sections and angular distribution of photoelectrons as well are the result of the interplay between saturation, light shifts, effects of photon statistics and temporal and spatial intensity distributions [2-8].

If an atom is irradiated by photons having an energy less than the ionization potential of the atom, the atom cannot be ionized in a single photon process. It is, however, possible to ionize the atom by the simultaneous absorption of several photons. In general the cross section for such a nonlinear process is very low and high power lasers are required for the observation of multiphoton ionization. The cross section may increase by several orders of magnitude if the photons are tuned into resonance with one or more intermediate states of the atom. The study of resonant multiphoton ionization provides matrix elements for bound free transitions and information about excited states. In this context the measurement of the angular distribution of photoelectrons is interesting as it can be highly anisotropic and as its shape critically depends on the quantum numbers of the excited state. As a result angular distributions can serve as "fingerprints" of the resonant intermediate states making the assignment easier. Photoelectron angular distributions also provide information about the phases of bound-free matrix elements, not available in total cross section measurements.

The following chapters are not intended to be a review of the field. Instead, the purpose of the paper is to focus on a few selected experiments performed recently *). In the first part ionization probabilities i.e. total, angle integrated cross sections are discussed both for nonresonant multiphoton ionization and for ionization through resonant intermediate states, where the resonances are transitions to the successive members of a series of Rydberg-states i.e. highly excited states. The second part deals with angular distributions of photoelectrons in multiphoton ionization, again emphasizing stepwise ionization through resonant intermediate states. A brief derivation of the general formula describing the angular distributions of electrons in multiphoton ionization is given.

II. Ionization Probabilities

Nonresonant ionization

A first approach to calculate the probability for nonresonant multiphoton ionization is to use perturbation theory in the lowest nonvanishing order [3]. The absorption of at least N photons is required to ionize the atom if the ratio of the atomic ionization potential and the photon energy is between N-1 and N. In N-th order perturbation theory the probability for multiphoton ionization of the atom is then given by

$$P_N = \sigma_N \; n \int_V dV \, \frac{1}{T} \int_T dt \; I^N(r,t) \qquad (1)$$

n is the density of atoms in the interaction volume V. The N-th power of the laser intensity I is averaged in time and space. If pulsed lasers are used T is the duration of the laser pulse. σ_N is called the generalized cross section and can be written as the absolute square of an infinite sum over all possible intermediate states of the atom. As a result several excitation and ionization channels interfere with each other and the phases of transition matrix elements play an important role [3].

If the laser intensity is measured in photons/(cm^2·s) σ_N has the dimension cm^{2N}s^{N-1}. A rough estimate for the magnitude of σ_N can be obtained in the following way [9]. An atom in an excited state has a finite life time determined by the rate of decay to lower lying states. If the frequency of the exciting laser is detuned by $\Delta\upsilon$, however, the atom can exist at this energy only for times Δt shorter than $(\Delta\upsilon)^{-1}$ as determined by Heisenberg's uncertainly principle. This is sometimes referred to as a virtual excitation of the atom. The cross section for an electric dipole transition between bound states in the optical region is $\sigma \approx 10^{-16}$ cm^2 and a typical value for Δt is 10^{-14}s. Since the rate for the first virtual excitation is $\sigma \cdot I$, and each of the following steps of the multiphoton ionization process has to occur within the time Δt, the probability for each of the following steps is $\sigma \cdot \Delta t \cdot I$ and the rate for N-photon ionization is $\sigma \cdot I\ (\sigma \cdot \Delta t \cdot I)^{N-1}$. The typical generalized cross section for nonresonant five photon ionization is therefore

$$\sigma_N \approx \sigma^N \cdot \Delta t^{N-1} \approx 10^{-136} [cm^{10} s^4]$$

Using this rough number for the generalized five photon cross section one can estimate the laser intensity necessary to ionize every single atom within the duration $T \approx 10^{-8}$s of a typical high power laser pulse.

$$P_N \cdot T \approx \sigma_N \cdot I^N \cdot T \approx 1$$

$$\curvearrowright I \approx 10^{29} \text{ photons}/(cm^2 s)$$

For 1eV photons this corresponds to 10^{10}W/cm^2. Experimental results for multiphoton ionization cross sections of several alkali atoms by 1.2eV Nd:YAG laser radiation and its second harmonic is listed in Table 1.

Table 1

Atom	K	Na	K	Na	References
Photon energy [eV]	2.36	2.36	1.18	1.18	
N	2	3	4	5	
σ_N[cm^{2N}s^{N-1}]	$10^{-48\pm0.8}$	$10^{-79.6\pm1.1}$	$10^{-107.3\pm1.7}$		[11]
				$10^{-136.9\pm0.5}$	[10]
				$10^{-141\pm3.0}$	[40]

The above estimate of the required intensity is very crude since the average of the fifth power of the intensity in equ.(1) was replaced by the fifth power of the average intensity, which in general is not correct. On the contrary, the ionization probability critically depends on the temporal and spatial distribution of the laser intensity [7]. The ionization probability is then determined by the higher order correlation functions of the field. For an N-photon process induced by chaotic light the ionization probability is N!-times larger than in the case where coherent light of the same average intensity is used [3,12].

Equ. (1) also does not consider the case where intermediate states may be shifted into resonance with the laser frequency, due to the high intensities necessary for multiphoton ionization [6,4,13]. In this case the signal will not increase with the N-th power of the intensity as suggested by equ.(1).

Resonant ionization

If the laser frequency is tuned into resonance with an atomic transition, population will be created in the intermediate state and the saturation of this transition has to be considered. This complication can be avoided by using a sufficiently low laser intensity. Often this is not practical since, due to low ionization cross sections out of the resonant intermediate state, a signal can be observed only after the laser intensity has been increased beyond the onset of saturation and light shifts.

If the interest is not focussed on these intensity effects but on the spectroscopy of the intermediate state and on the bound-free matrix element it is of advantage to use several lasers differing in frequency and intensity. The intensities of the lasers inducing transitions between bound states will be kept low to avoid high intensity effects. In the case of pulsed lasers the ionizing laser pulse can be time delayed with respect to the laser pulses exciting the intermediate resonance, so that the excitation is not perturbed by the intensity of the ionizing laser pulse. The ionization probability is then determined by the one photon ionization out of the aligned or polarized intermediate state.

In the following, experiments on sodium and barium are discussed where the intermediate states are members of a Rydberg-series. In the case of sodium the ionization cross section decreases monotonically with increasing principal quantum number of the Rydberg-state and is well suited to check simple series formulae. The barium Rydberg-states, however, are perturbed by doubly excited states and might be considered more interesting from a spectroscopic point of view.

Photoionization of highly excited states

The cross section for photoionization out of excited atomic states depends on the frequency of the ionizing light. In many cases the cross section decreases monotonically with frequency but it may also show minima [14]. The cross section also depends on the quantum numbers of the excited state. Highly excited states have been referred to as Rydberg states since their binding energy is described by Rydberg's famous formula

$$E_n = - R/n^{*2},$$

where n^* is the effective principal quantum number which is obtained from the principal quantum number n by subtracting the quantum defect δ, $n^*=n-\delta$. δ describes the deviation from the hydrogen spectrum and depends on the angular momentum quantum numbers ℓ and J and to a good approximation accounts for the interaction between the highly excited electron and the inner electron shells, one of these interactions being the polarization of the ionic core. Many other properties of Rydberg-states are also described by scaling laws. The radius of the charge distribution of the outer electron increases as n^2 and the radiative life time of the states is proportional to n^3. Regarding the bound free transitions, the cross section for radiative recombination of an electron at a given energy into states of different principal quantum numbers n scales as n^{-3} and the inverse process, photoionization out of Rydberg states decreases proportional to n^{-5} [15]. The difference between the cross sections for radiative recombination and photoionization is due to the $2n^2$-fold degeneracy of the level n. Radiative recombination leaves the atom in any of these states whereas photoionization starts from one well prepared substate. All the scaling laws discussed hold for hydrogen and to a good approximation for other one electron systems.

Ionization of sodium Rydberg-states

Relative photoionization cross sections can be obtained from photoionization spectra. In a thermal beam experiment [16] sodium atoms were excited to high lying d-states using two pulsed dye lasers pumped by the second and third harmonic of a Nd:YAG laser (Fig. 1). The dye laser pulse duration was 5 ns. The output pulse of the Q-switched Nd:YAG laser at 1.06 μm was delayed by 20 ns and used to ionize the excited atoms.

Fig.1: Sketch of the experimental set up. The copper plates ensure a field free region at the interaction volume. The half wave plate was used in the experiments of section III.

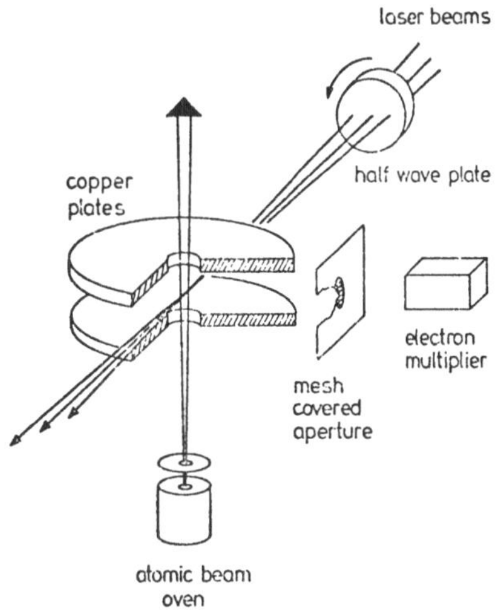

The intersection of the atomic beam and the laser beams was in a field free region. Electrons emitted in a direction mutually perpendicular to the atomic beam and the laser beams penetrated through a wire mesh covered aperture and were accelerated towards the electron multiplier capable of detecting single electrons. The multiplier signal was plotted on a chart recorder as a function of the wavelength of the second dye laser inducing transitions from the 3^2P-state of sodium to n^2D-states. Fig. 2 shows a recorded spectrum. The intensities of the two dye lasers were high enough so that the population of the n^2D state was saturated i.e. independent of n. Therefore the relative intensities of the spectral lines yield the relative photoionization cross sections. In Fig. 3 the line intensities are plotted versus n on a log-log scale. The solid line corresponds to a slope of five. It is clearly shown that the n^{-5} scaling law is in fairly good agreement with the data.

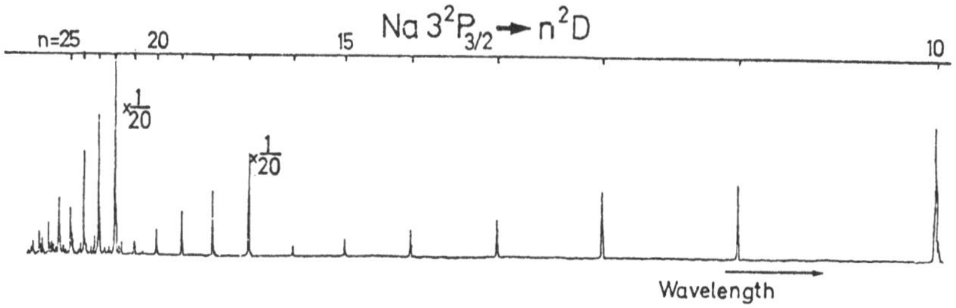

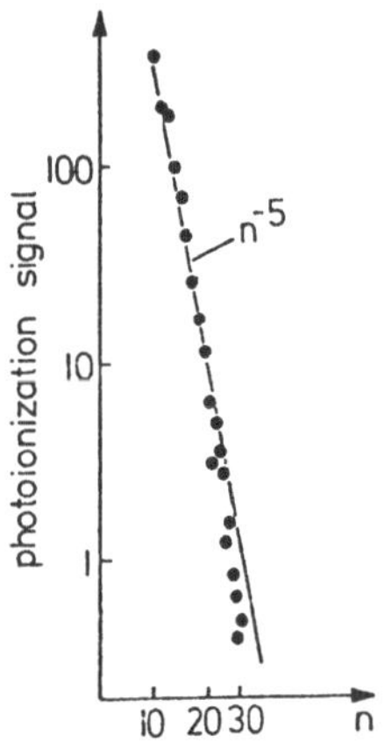

Fig.2: Photoionization spectrum via n^2D states of sodium. n^2S resonances do not appear due to the low ionization cross section [16].

Fig.3: Relative photoionization cross sections of sodium n^2D states versus n [16].

The measurement of absolute cross sections requires the absolute measurement of the laser intensity and the atomic number density. The atomic density, however, is not required if the saturation of the ionization signal with laser intensity is measured [9]. This technique has been used to measure absolute cross sections of the 5s and 4d state of sodium [17].

Ionization of barium Rydberg-states

For atoms with two valence electrons bound Rydberg-states correspond to the excitation of only one of the valence electrons. Such a Rydberg series may, however, be perturbed by configuration interaction with doubly excited states. These are states where both valence electrons are excited and the corresponding energy levels lie close to highly excited

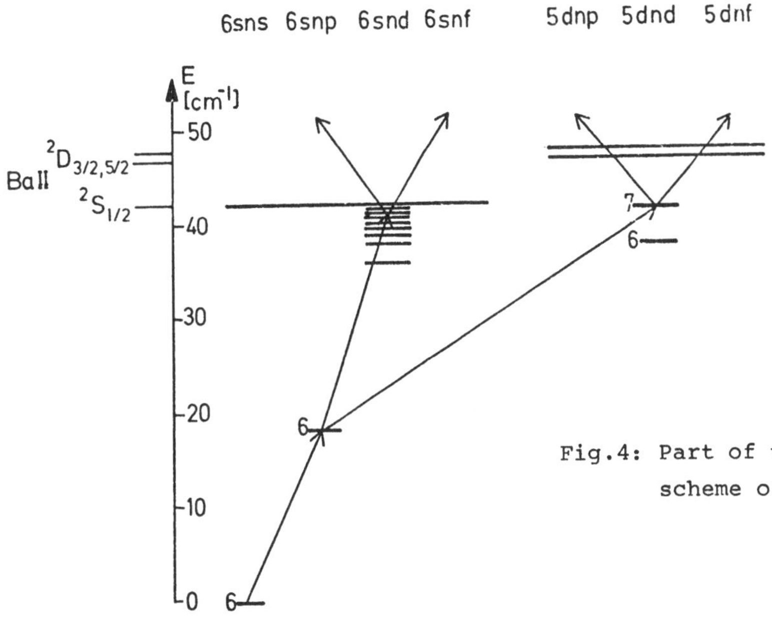

Fig.4: Part of the level scheme of barium

members of the Rydberg series. This is the case in barium where the 6snd series is perturbed e.g. by the doubly excited 5d 7d $^1D_2$ state which lies near the 6s 27d $^1D_2$ state (Fig. 4). An experimental arrangement similar to the one used for sodium was used to take photoionization spectra in the region of the perturbing doubly excited state 18]. As can be seen from the energy level diagram of barium (Fig. 4) a further complication arises as the first excited states of the singly charged ion are the 5d $^2D_{3/2}$ and 5d $^2D_{5/2}$ states lying about 0.6 eV above the 6s $^2S_{1/2}$ groundstate of the barium ion.

Using 1.2 eV photons to ionize a highly excited state of the neutral barium atom the remaining ion might be left in either the 6s or the 5d state, corresponding to velocities of the emitted electron differing by a factor of two. The atoms are ionized in a field free region. Since in addition pulsed lasers are used for ionization the energy of the emitted electrons can be analyzed using the time of flight method. The photoionization spectrum has been taken by recording either the slow, the fast or the total electron signal. In the photoionization of barium in a 6s nd state the most likely process is to emit the nd electron leaving the ion in the 6s state resulting in a fast outgoing electron. If the 5d 7d state is photoionized the ion will most likely be left in the 5d state emitting a slow electron. As a result the admixture of the 5d 7d state into the

6s nd series by configuration interaction is correlated with the probability of measuring slow photoelectrons and anticorrelated with the probability of measuring fast photoelectrons (Fig. 5). In addition the total photoionization signal along the Rydberg series has a maximum near the perturber since the cross section for the photoionization of the 5d 7d state is larger than the one for high lying 6s nd states. Similar perturbations of the Rydberg series can be observed in the line positions [19,20], the lifetime [21-23], the g-factor [24], the hyperfine splitting [25], and the isotope shift [26]. Photoionization is an alternative scheme for the investigation of excited states and also yields information about the bound-free matrix elements especially in connection with angle resolved measurements of photoelectrons as will be seen in the following.

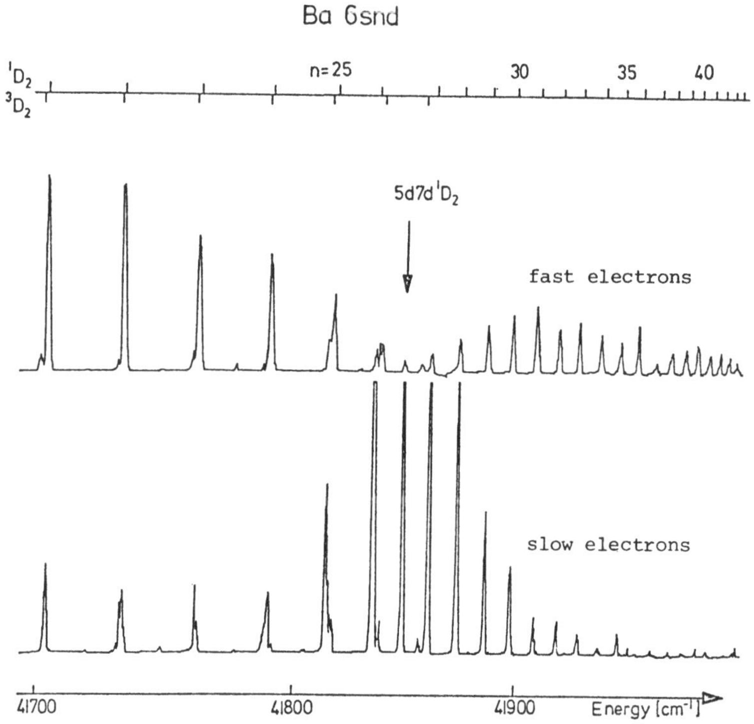

Fig.5: Photoionization spectra of Ba [18]

III. Angular Distribution of Photoelectrons

Differential photoionization cross sections have first been observed in photoionization of ground state molecules with x-rays [27]. For such high energy photons the photon momentum is not negligible with respect to the momentum of the bound electron. This results in the famous distortion of the photoelectron angular distribution due to momentum transfer. The first measurements of the angular distribution of photoelectrons in single photon ionization of ground state atoms and molecules with photon energies less than 10 eV have been performed by Berkowitz and Ehrhardt [28]. At these photon energies the photon momentum transfer can be neglected. The angular distribution of photoelectrons in one photon ionization of ground state atoms is given by the differential cross section for photoionization and can be written as [29]

$$d\sigma^{(1)}/d\Omega = \frac{\sigma_{Tot}}{4\pi}\left(1 + \beta_2 P_2(\cos\theta)\right) \qquad (2)$$

θ is the angle between the direction of the outgoing electron and the quantization axis which is most conveniently taken to be parallel to the direction of polarization of the ionizing light. P_2 is the second Legendre polynomial and σ_{Tot} the angle integrated (total) photoionization cross section. This equation for the angular distribution reflects the dipole character of the interaction between the photon and the atom. In experiments the anisotropy parameter β_2 is measured e.g. as a function of the wavelength of the ionizing light.

Equation (2) is correct only if all m-substates of the initial state of the atom are equally populated. It therefore does not apply to resonant or nonresonant ionization with two or more photons. In this case higher order Legendre polynomials have to be included.

The angular distribution of photoelectrons in multiphoton ionization was first observed by Edelstein et al. [30] when ionizing titanium atoms in a resonant two photon process. In subsequent experiments on sodium, cesium, strontium, and neon [31-38] the influence of resonant intermediate states on the angular distribution was studied. First experiments on angular distributions of electrons in nonresonant multiphoton ionization were performed recently [39-41].

In the case of resonant multiphoton ionization the resonant intermediate state in general has spatial anisotropy i.e. not all magnetic sublevels are populated. Therefore, the calculation of the electron angular distribution in photoionization out of a state with a well defined magnetic quantum number m will be sketched briefly. The result can also be used to give the general formula for the angular distribution of photoelectrons in nonresonant multiphoton ionization.

Calculation of photoelectron angular distributions

The differential photoionization cross section is obtained by evaluating the electric dipole matrix element for the transition from the initial bound state to the continuum. The effects of spin orbit and hyperfine coupling have been discussed in several theoretical papers [42-44] but for the sake of simplicity will be ignored here. The wave function of the bound state is characterized by the principal, angular momentum and magnetic quantum numbers. For a potential with point symmetry it factorizes into a radial and an angular part:

$$|n\ell m\rangle = R_{n\ell}(r)\, Y_{\ell m}(\theta,\phi) \tag{3}$$

The continuum state is taken as a superposition of an incoming spherical and an outgoing plane wave and can be expanded in Legendre polynomials [3,45] which in turn can be expanded in products of spherical harmonics:

$$|\vec{k}\rangle = \sum_{\ell'=0}^{\infty} i^{\ell'} e^{i\delta_{\ell'}}\, 4\pi G_{k\ell'}(r) \sum_{m=-\ell'}^{\ell'} Y_{\ell' m}(\Theta,\Phi)\, Y_{\ell' m}(\theta,\phi) \tag{4}$$

The arguments of the spherical harmonics contain the angles Θ,Φ and θ,ϕ, which describe the direction of the wave vector $\vec{k}$ and the radius vector $\vec{r}$, respectively. The angular distribution of the photoelectrons is then proportional to the absolute square of the electric dipole matrix element

$$d\sigma/d\Omega \approx |\langle\vec{k}|\vec{\varepsilon}\cdot\vec{r}|n\ell m\rangle|^2 \tag{5}$$

where $\vec{\varepsilon}$ is the polarization of the ionizing light. For light polarized linearly in the z-direction we have $\vec{\varepsilon}=(0,0,1)$. The corresponding selection rules are $\Delta\ell=\mp1$. With these selection rules only two terms in the infinite sum (4) describing the continuum state have to be considered.

$d\sigma/d\Omega$ is then the absolute square of the amplitudes of two outgoing partial waves $\ell'=\ell+1$ and $\ell'=\ell-1$. The absolute square contains an interference term which depends on the difference of the corresponding scattering phases $\delta_{\ell+1}-\delta_{\ell-1}$. The information about the scattering phase is available only in the angular distribution [3]. When measuring the total cross section σ_{tot} the angular distribution has to be integrated over Θ and Φ. In this case the interference term vanishes due to the orthogonality of the spherical harmonics.

For the purpose of deriving equ. (2) for one-photon ionization and the corresponding formula for N-photon ionization the problem is simplified here by neglecting the $(\ell-1)$-partial wave. Of course for almost any special case this simplification would yield wrong numbers. It is, however, useful for the purpose of determining the highest order of anisotropy to be expected for the angular distribution. Using this simplification the angular distribution is

$$d\sigma/d\Omega \sim \left|Y_{\ell+1,m}(\Theta,\Phi)\right|^2 \frac{(\ell+1)^2-m^2}{(2\ell+1)(2\ell+3)} A_{\ell+1}^2 \tag{6}$$

Here the electric dipole operator $\vec{\varepsilon}\cdot\vec{r}$ is written as $r\cdot\cos\Theta$ for linearly polarized light. $A_{\ell+1}$ represents the radial integral

$$A_{\ell+1} = 4\pi\int_0^\infty dr\cdot r\cdot G_{k,\ell+1}(r)\cdot R_{n\ell}(r) .$$

The angular part of the electric dipole matrix element has already been evaluated in equ. (6):

$$\int_0^\pi d\theta \sin\theta \int_0^{2\pi} d\phi\, Y^*_{\ell+1,m}(\theta,\phi) \cos\theta\, Y_{\ell,m}(\theta,\phi) = \sqrt{\frac{(\ell+1)^2-m^2}{(2\ell+1)(2\ell+3)}} \tag{7}$$

In the case where all magnetic sublevels of the bound state are equally populated $d\sigma/d\Omega$ of equ. (6) has to be summed over m. Using the following relations

$$\sum_{m=-\ell}^{\ell} \left|Y_{\ell,m}(\Theta,\Phi)\right|^2 = (2\ell+1)/4\pi$$

and more general

$$\sum_{m=-\ell}^{\ell} m^{2M} |Y_{\ell,m}(\Theta,\Phi)|^2 = \sum_{j=0}^{M} a_j \sin^{2j}(\Theta) \qquad (8)$$

one obtains for the m-averaged angular distribution a constant plus a $\sin^2\theta$ term and thus the same general form as equ. (2) [45].

The physical interpretation is that the original system consists of the spatially isotropic atom and the incoming photon. Since the interaction of the photon and the atom has dipole character the total system has dipole anisotropy which is also displayed in the photoelectron angular distribution. Considering the case of N-photon absorption the single dipole interaction leads to an angular distribution containing even powers of m up to 2N. The angular distribution is again averaged over the m sublevels of the initial state. The result is the general formula for N-photon ionization

$$d\sigma^{(N)}/d\Omega = \frac{\sigma_{Tot}}{4\pi} \sum_{j=0}^{N} \beta_{2j} P_{2j}(\cos\Theta) \qquad (9)$$

Again the physical interpretation is that in each absorption step the spatial anisotropy of the initially isotropic atoms can be increased due to the dipole character of the photon-atom interaction. Since the physical interpretation is not affected by the coupling scheme equ. (9) holds also in the case of spin-orbit and hyperfine coupling.

The coefficient β_{2N} of the most anisotropic Legendre polynomial in equ. (9) is zero if the anisotropy of the atom is not increased in one of the absorption steps. This can take place only in resonant multiphoton ionization whenever two successively excited states have the same number of m-sublevels and the laser light is linearly polarized.

Another factor complicating the calculation of the photoelectron angular distribution is that the remaining ion may carry some anisotropy. This, however, does not apply to the ionization of alkali atoms with photons having an energy of a few eV. The ground state of singly ionized alkali atoms is a $^1S_0$ state and the excited states of the ions lie so high that they are not accessible by photoionization with visible laser light. If several lasers are used which do not all have the same direction of linear polarization the angular distribution also depends on Φ. The maximum possible anisotropy, however, is still determined by the number of photons absorbed.

Nonresonant multiphoton ionization

Many experiments have been devoted to the study of the dependence of the total ionization rate on the laser intensity and laser light statistics [2,46] and to the investigation of N+M-photon ionization in the presence of N-photon ionization [47]. First measurements of the photoelectron angular distribution have been performed recently [39-41]. In one of them [40] sodium atoms of a thermal beam were nonresonantly ionized by the absorption of five photons of a Nd:YAG laser beam having a wavelength of 1.06 μ. The angular distribution of photoelectrons was measured in a plane perpendicular to the direction of propagation. The

Fig.6: Polar diagram of the measured (x) photoelectron angular distribution. The double arrow indicates $\Theta = 0$, the direction of linear polarization of the laser beam. The full curve is the result of a least squares fit [40].

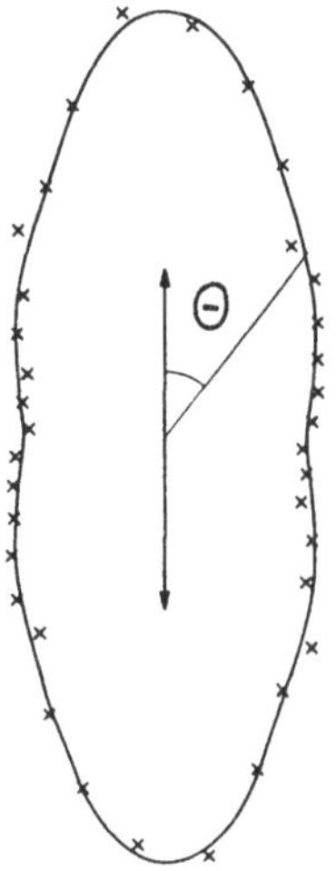

resulting angular distribution shown in a polar diagram in Fig. 6 is highly anisotropic and a least squares fit of equ. (9) (solid line) to the data (crosses) yields significant coefficients β_{2j} up to j=5 as expected. The value of the coefficients β_{2j}, however, still have to be interpreted theoretically [40]. The photoelectron angular distribution offers information in addition to what can be learned from total cross section measurements and it can be hoped that it will help to interpret nonresonant multiphoton ionization processes.

Resonant ionization of sodium

As suggested by equ. (6) the angular distribution of photoelectrons critically depends on the ℓ and m quantum numbers of the intermediate state and thus provides information about this intermediate state. Using a time delayed step-wise two photon ionization of sodium e.g. quantum beats have been observed in the angular distribution yielding the hyperfine splitting of the $3^2P_{3/2}$-state [33]. In the case of resonant multiphoton ionization with several lasers different m-substates of intermediate states can be populated by choosing appropriate combinations of laser polarizations. This effect has been demonstrated in resonant three photon ionization of sodium [34,36,48,49]. As already discussed in the paragraph about the photoionization spectra of sodium ^2D states (Fig. 1) two dye lasers and a Nd:YAG laser were used to resonantly ionize sodium atoms of a thermal beam through the 3^2P and 20^2D intermediate states [48]. The three laser beams were propagating nearly collinearly, the maximum angle between the photon wave vectors being less than 2 degrees. Polarizers were used to define the direction of polarization of the lasers and the angular distributions were recorded in a plane perpendicular to the photon wave vectors by inserting one achromatic half wave plate in the two dye laser beams and one low order half wave plate in the Nd:YAG laser beam and rotating them simultaneously with stepping motors. Fig. 7 shows polar diagrams of photoelectron angular distributions recorded for various combinations of polarizations and for two different first intermediate states $3^2P_{1/2}$ and $3^2P_{3/2}$ respectively. On the top of each column in Fig. 7 the three arrows indicate the polarizations of the lasers for the first, the second, and the final step. The data demonstrate the sensitivity of the photoelectron angular distributions on the m-quantum number of the intermediate state.

It is obvious that the anisotropy of the photoelectron angular distributions depends on the intermediate state. The maximum order of Legendre polynomials necessary to fit the data is four and six for the $3^2P_{1/2}$ and $3^2P_{3/2}$ intermediate state respectively. As discussed above this is easily explained by the fact that in the case of the $3^2P_{1/2}$-intermediate state no anisotropy is transfered to the atom, i.e. both m-sublevels of the $3^2P_{1/2}$ state are equally populated.

These angular distributions can also be used to show that for Na n^2D states the d $\rightarrow$ p partial wave has a negligible amplitude as compared to

Fig.7: Polar diagrams of photoelectron angular distributions in the ionization of sodium via the 3^2P and the 20^2D states. The polarizations of the three lasers used is indicated.

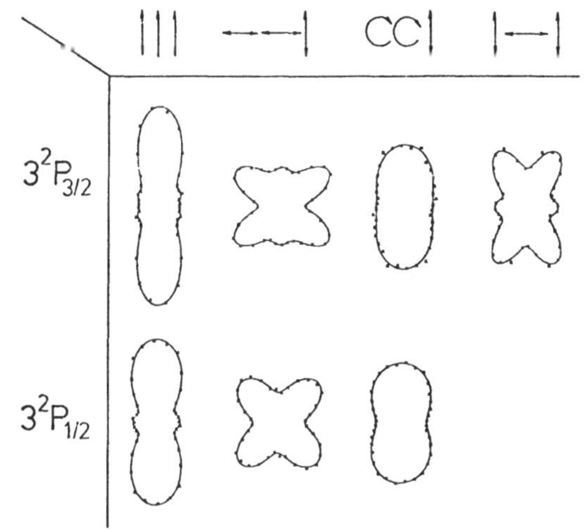

the d → f partial wave [49]. This is a result also obtained in a measurement of the total cross section where the intermediate state was fully polarized [17].

In potassium e.g. the d → f partial wave has an amplitude about four times as large as the one for the d → p partial wave [50]. Consequently the partial cross sections differ by a factor of ten or more. This shows a potential application of angular distribution experiments since weak partial waves appear much stronger than in total cross section experiments. In potassium this could be used to look for the minimum in the wavelength dependance of the d → p partial wave predicted by Aymar [50].

Resonant ionization of barium

As described in the paragraph about relative cross section measurements Ba-atoms have been ionized in a resonant three step process via the $6s6p\,^1P_1$ state and high lying D states [17]. The measurement of electron angular distributions in the photoionization of two electron atoms can be expected to yield especially rich results whenever a doubly excited state perturbs a Rydberg series. In barium this is the case for the $6snd\,^{1,3}D_2$ series which is perturbed by the $5d7d\ ^1D_2$ state lying between n = 26 and n = 27 of the Rydberg series. Fig. 5 has already shown the effect of the perturbing state on the total ionization cross section of the 6s nd-states The possibility of leaving the barium ion either in the ground state or in an excited $5d^2D$ state resulting in fast or in slow electrons, respectively, has been discussed. Photoelectron angular distributions have been recorded again in a plane perpendicular to the direction of propagation of the lasers by rotating half wave plates using the same experimental arrangement as for the experiment on sodium. For the measurements on barium, however, the linear polarizations of the three lasers were always parallel to each other.

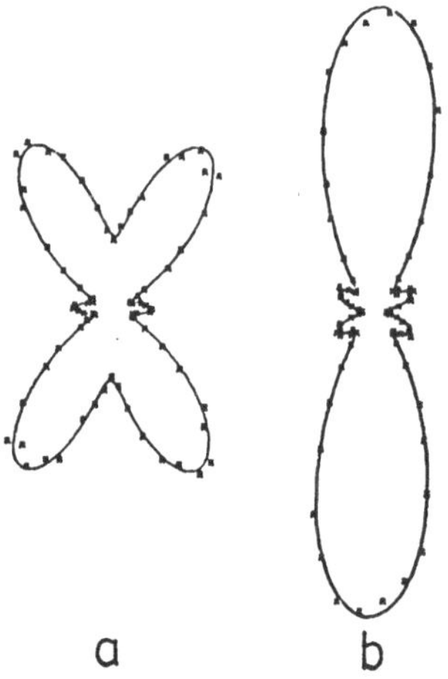

Fig.8: Angular distributions of fast electrons in photoionization of the $6s19d^3D_2$ (a) and $^1D_2$ state (b).

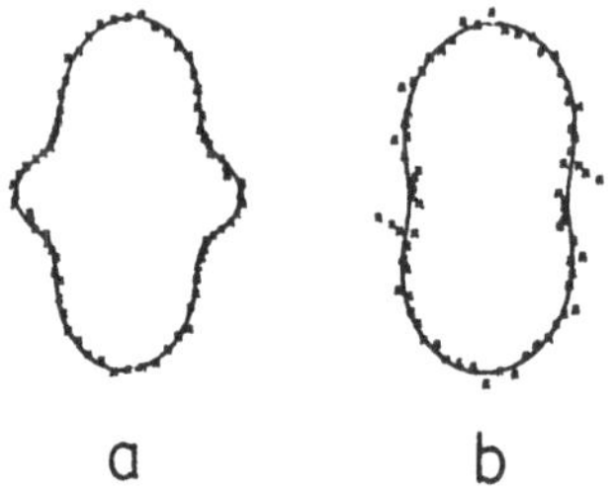

Fig.9: Angular distribution of fast (a) and slow (b) electrons in photoionization of the Ba $5d7d^1D_2$ state [18].

Fig. 8 shows polar diagrams of the angular distributions of fast electrons resulting from the ionization of the 6s 19d $^1D_2$ and $^3D_2$ states. The high anisotropy and the characteristic difference between $^1D_2$ and $^3D_2$ intermediate states is obvious.

Fig. 9 shows angular distributions of fast and slow electrons in the case where the laser for the second step is tuned to the perturbing $5d7d^1D$ state. For the least squares fit of the angular distribution of the fast electrons the fit function (equ. 9) had to include even Legendre polynomials up to the sixth order as expected for three photon ionization. In the case of the slow electrons, however, where the barium ion is left in the excited state the least squares fit yields a β_6 which is zero within the error bars. In general terms this can be understood since the excited ion can also carry some anisotropy. But it also ensures the result of a multichannel quantum defect analysis that the 5d7d state is best described in j-j-coupling and to a great extend has the nature of the $5d_{5/2}\ 7d_{3/2}$ state [51]. In the photoionization of this state the $7d_{3/2}$ electron will be emitted. A $^2D_{3/2}$-state excited with linearly polarized lasers carries only dipole anisotropy and β_6 in the angular distribution is zero. This has already been seen in the sodium experiment, when using the $^2P_{1/2}$ intermediate state. Though the $^2D_{3/2}$ and $^2D_{5/2}$ resonances in sodium had not been resolved spectraly the ΔJ selection rule ensured that only the $^2D_{3/2}$-state was populated. A least squares fit to the corresponding angular distribution also gave $\beta_6 = 0$.

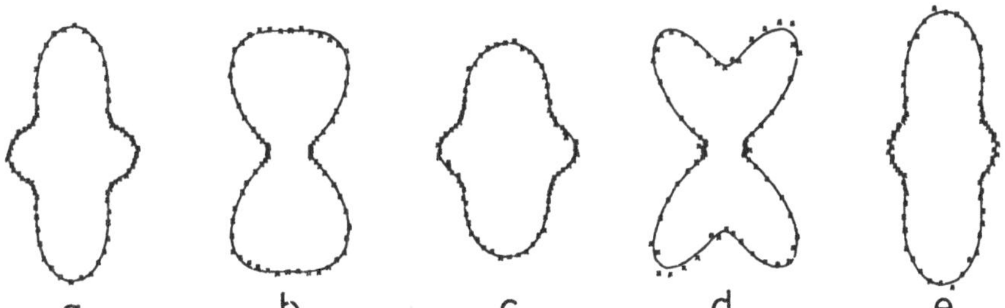

Fig.10: Angular distributions of fast electrons in photoionization of the $6s26d\,^1D_2$, $^3D_2$, $5d7d\,^1D_2$, $6s27d\,^3D_2$ and $^1D_2$ state (a-e) [18].

The angular distribution of the fast photoelectrons has been measured for a series of different intermediate states in the region of the perturbing 5d7d $^1D_2$-state (Fig. 10). Here the perturber is designated in the LS coupling scheme in agreement with the literature. It has to be kept in mind, however, that for this perturber the spin-orbit coupling is very large, so that in the region of the perturber the $^1D_2$ and $^3D_2$-series are strongly mixed. Talking about a $^1D_2$ or $^3D_2$ state only means that the state has either mainly $^1D_2$ or mainly $^3D_2$ character. There has indeed been some discussion in the literature about the assignment of the $^1D_2$ and $^3D_2$ states at n = 26 [19-24]. Far away from the perturber the 6snd $^3D_2$ state lies always at a slightly lower energy than the 6snd $^1D_2$ state. Close to the perturber the resonances can be assigned since the characteristic shapes of angular distributions for pure triplet and singlet intermediate states are known from Fig. 8. As a result for n=26 the $^3D_2$ state has a higher energy than the $^1D_2$ state, as deduced also from lifetime measurements [21]. The assignment in Fig. 5 and 10 that appears to be obvious when the angular distributions are measured, is more arbitrary if only line positions are measured [19,20]. It is also observed that the singlet-triplet splitting is especially large for n = 26 and n = 27. Both observations indicate that the configuration mixing with the perturbing state and the resulting level shifts are larger for the $^1D_2$ than for the $^3D_2$ states. The above comments are of course only qualitative, a detailed analysis of these angular distributions is presently in progress [52].

IV Conclusions

So far only a few laboratory experiments on photoionization out of excited states of atoms have been performed. Reliable cross sections, however, are very important e.g. for the interpretation of astrophysical data. A severe problem in some experiments is the lack of knowledge about the degree of alignment of the atoms in the excited state. This can be solved either by optically pumping the atom into only one magnetic sublevel [17] or by measuring the photoelectron angular distribution in addition to the absolute cross section.

As was pointed out, the characteristic shapes of the angular distributions of photoelectrons may be used to assign excited atomic states. The admixture of different configurations and the coupling scheme can be deduced. As a result one can obtain information needed to calculate atomic wave functions. This information is not available if only the energies of excited states are measured [24].

In the experiments discussed above any complication due to a high laser intensity like light shifts was avoided whenever possible. It is, however, interesting also to exploit photoelectron angular distributions to experimentally study high intensity effects in multiphoton ionization [4]. Another challenge, theoretically and experimentally as well, is the study of laser bandwidth effects [8,53,54]. In this context the recent development of well controlled laser power spectra is very promising [55].

In the case of nonresonant multiphoton ionization photoelectron angular distributions may be useful for the interpretation of free-free transitions, i.e. absorption of additional photons when the electron is already in a continuum state [39,41].

Acknowledgment

It is a pleasure to thank my collaborators D.S. Elliott, E. Matthias and S.J. Smith for making the experiments possible. Furthermore I would like to acknowledge helpful discussions and correspondence with P. Zoller, M. Aymar, S.N. Dixit, P. Lambropoulos and H. Walther. The typing of this manuscript would not have been finished in time without the kind help of M. Schreiber.

The financial support of the Deutsche Forschungsgemeinschaft is gratefully acknowledged.

References

*) Survey of recent experiments at the Joint Institute for Laboratory Astrophysics of the University of Colorado and the National Bureau of Standards in Boulder, U.S.A.

1. G.S. Hurst, M.G. Payne, S.D. Kramer, J.P. Young: Rev. Mod. Phys. 51, 767 (1979)
2. L.A. Lompre, G. Mainfray, C. Manus, J.P. Marinier: J. Phys. B 14, 4307 (1981)
3. P. Lambropoulos: Adv. Atom. Mol. Phys. 12, 87 (1976)
4. S.N. Dixit, P. Lambropoulos: Phys. Rev. Lett. 46, 1278 (1981)
5. S.N. Dixit, P. Lambropoulos: Phys.Rev.A 21, 168 (1980)
6. M. Aymar, M. Crance: J. Phys. B 12, L 667 (1979)
7. L.-A. Lompre, G. Mainfray, B. Mathieu, G. Watel, M. Aymar, M. Crance: J. Phys. B. 13, 1799 (1980)
8. D.E. Nitz, A.V. Smith, M.D. Levenson, S.J. Smith: Phys. Rev. A 24, 288 (1981)
9. G. Mainfray: Comments Atom. Mol. Phys. 9, 87 (1980)
10. T.U. Arslanbekov, V.A. Grinchuk, G.A. Delone, K.B. Petrosyan: Kratkie Soobshchaniya po Fizike No. 10, 33 (1975)
11. J. S. Bakos: Adv. Electron. Electron Phys. 36, 57 (1974)
12. P. Zoller: in Laser Physics, ed. by D.F. Walls, J.D. Harvey, Academic Press, Sydney, 1980.
13. L.-A. Lompre, G. Mainfray, C. Manus, J.P. Marinier: Phys. Lett. 86 A, 17 (1981)
14. J. Lahiri, S.T. Manson: Phys. Rev. Lett. 48, 614 (1982)
15. H.A. Bethe, E.E. Salpeter: Quantum Mechanics of One and Two Electron Atoms" (Plenum Press, New York 1977) p. 296
16. G. Leuchs, S.J. Smith, unpublished material
17. A.V. Smith, J.E.M. Goldsmith, D.E. Nitz, S.J. Smith: Phys. Rev. A 22, 577 (1980)
18. G. Leuchs, E. Matthias, D.S. Elliott, S.J. Smith, P. Zoller, to be published
19. J.R. Rubbmark, S.A. Borgström, K. Bockasten: J. Phys. B 10, 421 (1977)
20. M. Aymar, P. Camus, M. Dieulin, C. Morillon: Phys. Rev. A 18, 2173 (1978)

21. K. Bathia, P. Grafström, C. Levinson, H. Lundberg, L. Nilsson, S. Svanberg: Z. Physik A 303, 1 (1981)
22. T.F. Gallagher, W. Sandner, K.A. Sanfinya: Phys. Rev. A 23, 2969 (1981)
23. M. Aymar, R.-J. Champeau, C. Delsart, J.-C. Keller: J. Phys. B 14, 4489 (1981)
24. P. Grafström, C. Levinson, H. Lundberg, S. Svanberg, P. Grundevik, L. Nilsson, M. Aymar: Z. Physik A 308, 95 (1982)
25. H. Rinneberg, J. Neukammer: Phys. Rev. Lett. 49, 124 (1982)
26. H. Rinneberg, J. Neukammer, E. Matthias: Z. Physik A 306, 11 (1982)
27. W. Bothe: Z. Physik 26, 59 (1924)
28. J. Berkowitz, H. Ehrhardt: Physics Lett. 21, 531 (1966)
29. G. Wentzel: Z. Physik 41, 828 (1927)
30. S. Edelstein, M. Lambropoulos, J. Duncanson, R.S. Berry: Phys. Rev. A 9, 2459 (1974)
31. M. Strand, J. Hansen, R.-L. Chien, R.S. Berry: Chem. Phys. Lett. 59, 205 (1978)
32. G. Leuchs, S.J. Smith, H. Walther, in Laser Spectroscopy IV, ed. by H. Walther, K.W. Rothe (Springer, Berlin, Heidelberg, New York 1979)
33. G. Leuchs, S.J. Smith, E.E. Khawaja, H. Walther: Opt. Commun. 31, 313 (1979)
34. J.C. Hansen, J.A. Duncanson, R.-L. Chien, R.S. Berry: Phys. Rev. A 21, 222 (1980)
35. H. Kaminski, J. Kessler, K.J. Kollath: Phys. Rev. Lett. 45, 1161 (1980)
36. T. Hellmuth, G. Leuchs, S.J. Smith, H. Walther, in Lasers and Applications, ed. by W.O.N. Guimaraes, C.-T. Lin, A. Mooradian (Springer, Berlin, Heidelberg 1981)
37. D. Feldmann, K.-H. Welge: J. Phys. B 15, 1651 (1982)
38. A. Siegel, J. Ganz, W. Bußert, H. Hotop, B. Lewandowski, M.-W. Ruf, M. Waibel, presented at the 8th International Conference on Atomic Physics, Göteborg (1982)
39. F. Fabre, P. Agostini, G. Petite, M. Clement: J. Phys. B 14, L 677 (1981)
40. G. Leuchs, S.J. Smith: J. Phys. B 15, 1051 (1982)
41. P. Kruit, ph. d. thesis, FOM-Institute for Atomic and Molecular Physics, Amsterdam, 1982
42. V.L. Jacobs: J. Phys. B 5, 2257 (1972)
43. K.J. Kollath: J. Phys. B 13, 2901 (1980)
44. H. Klar, H. Kleinpoppen: J. Phys. B 15, 933 (1982)
45. J. Cooper, R.N. Zare, in Atomic Collision Processes, Lectures in Theoretical Physics XI-C, ed. by S. Geltman, K.T. Mahanthappa, W.E. Brittin (Gordon and Breach, New York 1969)
46. C. Manus, G. Mainfray, Proceedings of the 2nd International Conference on Multiphoton Processes, ed. by M. Janossy, S. Varro (Hungarian Academy of Sciences, Budapest 1981)
47. P. Agostini, M. Clement, F. Fabre, G. Petite: J. Phys. B. 14, 1491 (1981)
48. G. Leuchs, D.S. Elliott, S.J. Smith, to be published
49. G. Leuchs, S.J. Smith, Proceedings of the International School on Laser Applications, Vilnius 1981, ed. by V.S. Letokhov, USSR Academy of Sciences, to be published
50. M. Aymar, private communication, see also M. Aymar, E. Luc-Koenig, F. Combet Farnoux: J. Phys. B 9, 1279 (1976)
51. M. Aymar, O. Robaux: J. Phys. B 12, 531 (1979)
52. P. Zoller, private communication
53. A.T. Georges, P. Lambropoulos, P. Zoller, in Laser Spectroscopy IV, ed. by H. Walther, K.W. Rothe (Springer, Berlin, Heidelberg, New York 1979)
54. J.J. Yeh, J.H. Eberly: Phys. Rev. A 24, 888 (1981)
55. D.S. Elliott, Rajarshi Roy, S.J. Smith: Phys. Rev. A 26, 12 (1982)

PUMP DYNAMICAL EFFECTS IN SUPERFLUORESCENT QUANTUM INITIATION AND PULSE EVOLUTION

Charles M. Bowden
Research Directorate, US Army Missile Laboratory
US Army Missile Command
Redstone Arsenal, Alabama 35898, USA

Abstract

Recent work is reviewed where it is shown that even in the regime where the temporal width of the pumping pulse, τ_p , and the characteristic superfluorescence (SF) time, τ_R , are such that $\tau_p/\tau_R < 1$, the effect of coherent pumping on a three-level system can cause a significant contribution to the quantum mechanical SF initiation and corresponding amplified temporal fluctuations. Other recent work shows, furthermore, that for $\tau_p/\tau_R >> 1$, but $\tau_p/\tau_D < 1$, where τ_D is the time delay between the pump pulse peak and the SF peak intensities, initial characteristics of the injected coherent pumping pulse can have distinct deterministic effects on SF pulse longitudinal, transverse and temporal evolution.

I. Introduction

Superfluorescence [1], (SF), is the phenomenon whereby a collection of atoms or molecules is prepared initially in a state of complete inversion, and then allowed to undergo relaxation by collective, spontaneous decay. Since Dicke's initial work [2], there has been a large amount of theoretical and experimental work dealing with this process [3].

With the exception of the more recent work of Bowden and Sung [4], Bowden and Mattar [5], and the even more recent work by Bowden and Sung [6], all theoretical treatments have dealt exclusively with the relaxation process from a prepared state of complete inversion in a two-level manifold of atomic energy levels, and thus do not consider the dynamical effects of the pumping process. Yet, all reported experimental work [3,7-9] has utilized optical pumping on a minimum manifold of three atomic or molecular energy levels by laser pulse injection into the nonlinear medium, which subsequently superfluoresces.

It was pointed out by Bowden and Sung [4] that for a system otherwise satisfying the conditions for superfluorescent emission, unless the characteristic SF time [1], τ_R, is much greater than the pump pulse temporal duration τ_p, i.e., $\tau_R >> \tau_p$, the process of coherent optical pumping on a three-level system can have dramatic effects on the SF. This is a condition which has not been realized over the full range of experimental data [3,7-9].

Using calculational simulation techniques based upon a semiclassical model, Mattar and Bowden [5] have studied in detail the effects of coherent dynamical pumping on an extended cylindrical volume of three-level atoms in the regime where $\tau_p/\tau_R >> 1$, $\tau_p/\tau_D < 1$, where τ_D is the measured delay time between the pump pulse and the SF pulse peak intensities. Their calculation includes propagation as well as transverse and diffraction effects; yet the results demonstrate that specified initial characteristics of the injected pumping pulse, together with the boundary conditions, can have profound deterministic effects upon the SF pulse which evolves, in terms of its on-axis pulse area, temporal and radial shape, and time delay, τ_D. Their results are in qualitative agreement with the predictions made earlier by Bowden and Sung [4] based upon a mean-field approximation.

Not only has it been shown that the dynamics of the pumping can have dramatic effects upon the SF pulse evolution [4,5], but also it is well established that the statistics of the quantum mechanical initiation of the SF has profound effects in terms of macroscopic temporal fluctuations [10,12]. Until quite recently [6], amplified quantum initiation statistics in SF emission have been discussed only with regard to two-level systems with the initial condition of complete inversion [10,12]. In their very recent work, Bowden and Sung [6] have presented a more comprehensive treatment of SF in the linearized regime of SF initiation by combining coherent pump dynamics on the three-level system and simultaneous, as well as subsequent, quantum mechanical initiation of SF emission.

The purpose of this chapter is to present a comprehensive review of the recent work addressed to the issue of the effects of coherent dynamical pumping on SF pulse initiation and evolution in three-level systems. The material for this discussion is drawn largely from the work of Bowden, Sung and Mattar and is contained mainly in the references 4, 5, and 6. Although the degree of spacial, temporal and spectral coherence of the pumping pulse is open to question in most of the reported SF experiments [9], we have assumed full coherence, i.e., a coherent state [13] for the injected pump pulse in order to discuss the effects of coherent dynamical pumping, unencumbered. It is felt that the effects derived from the coherence aspect of the pumping process, and to be discussed here, have been operative to at least some degree in all reported experimental results in SF [3,7-9]. It is hoped that the results of the work reviewed here will stimulate further experimental investigation of dynamic pump effects on SF, and in particular, the very interesting aspect of deterministic pulse shaping [5] which may lead to further developments in the important area of light control by light.

The next section will be used to present the quantum mechanical model from which the quantum initiation results are derived in the linearized regime of SF initiation. Linearization and SF initiation during and subsequent to the pump pulse time frame will be presented in Section III. In that section we will confine our attention to results satisfying the condition $\tau_p/\tau_R < 1$, which is the condition for which SF quantum initiation is expected to be most important in terms of subsequent amplified temporal fluctuations. Section IV will be used to discuss the results of SF evolution in the nonlinear regime which includes propagation, transverse and diffraction effects. The results are calculational and based upon a semiclassical model obtained from the fully quantum mechanical model of Section II. Since the results are semiclassical, we restrict attention in that section to results under the condition that $\tau_p/\tau_R >> 1$, $\tau_p/\tau_D < 1$, where the effects of SF quantum initiation are expected to be minimized. In that section we emphasize the deterministic effects of coherent dynamical pumping on SF pulse evolution. The last section is used for discussion of the main results and implications for further theoretical and experimental investigation.

III. Three-level Model for Superfluorescence

The first model for the study of dynamical effects of coherent pumping on SF evolution was the three-level model proposed by Bowden and Sung [4]. The model is comprised of a collection of identical three-level atoms, each having the energy level scheme shown in Figure 1 such that the $1 \leftrightarrow 3$ transition is induced by a coherent electromagnetic field pulse of frequency ω_0 and wave vector k_0. The transition $3 \leftrightarrow 2$ evolves by spontaneous emission at a much lower frequency ω. It is assumed that the energy level spacing is such that $\varepsilon_3 > \varepsilon_2 >> \varepsilon_1$, and we also retain spontaneous relaxation in the pump transition $1 \leftrightarrow 3$ for generality. The energy levels ε_2 and ε_1 are not coupled radiatively due to parity considerations. The injected pump field is treated as a coherent state [13].

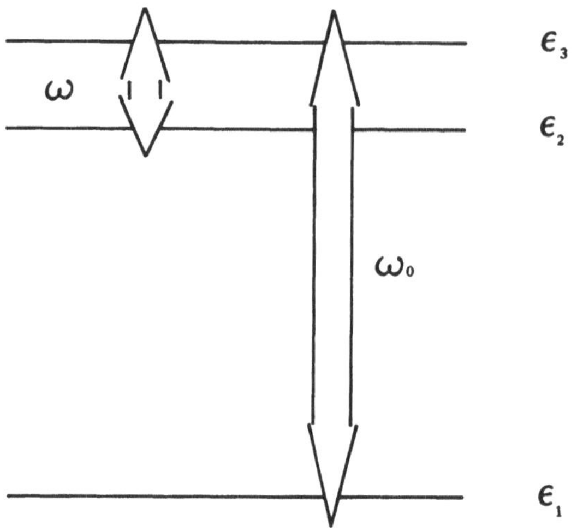

Figure 1. Model three-level atomic system and electric field tunings under consideration. For the results reported here, the injected pulse is tuned to the 1 ↔ 3 transition.

In the electric dipole and rotating wave approximations, the Hamiltonian which describes this system of N dynamically-pumped three-level atoms is:

$$H = \hbar \sum_{r=1}^{3} \sum_{j=1}^{N} \varepsilon_{rj} R_{rr}^{(j)} + \hbar \sum_{\ell} \omega_\ell a_\ell^+ a_\ell + \hbar \sum_{\ell} \hat{\omega}_\ell b_\ell^+ b_\ell$$

$$-i \sum_{j=1}^{N} \sum_{\ell} \left[g_\ell^{(j)} a_\ell R_{32}^{(j)} e^{i\underline{k}_\ell \cdot \underline{r}_j} - g_\ell^{*(j)} a_\ell^+ R_{23}^{(j)} e^{-i\underline{k}_\ell \cdot \underline{r}_j} \right]$$

$$-i \sum_{j=1}^{N} \sum_{\ell} \left[p_\ell^{(j)} b_\ell R_{31}^{(j)} e^{i\hat{\underline{k}}_\ell \cdot \underline{r}_j} - p_\ell^* b_\ell^+ R_{13}^{(j)} e^{-i\hat{\underline{k}}_\ell \cdot \underline{r}_j} \right]$$

$$-i\frac{\hbar}{2} \sum_{j=1}^{N} \left[\omega_R^{(j)} R_{31}^{(j)} e^{-i(\omega_o t - \underline{k}_o \cdot \underline{r}_j)} - \omega_R^{*(j)} R_{13}^{(j)} e^{i(\omega_o t - \underline{k}_o \cdot \underline{r}_j)} \right], \quad (2\text{-}1)$$

where the canonical atomic operators $R_{k\ell}^{(j)}$ obey the Lie algebra defined by the commutation rules [4]

$$\left[R_{ij}^{(m)}, R_{\ell k}^{(n)} \right] = R_{ik}^{(m)} \delta_{j\ell} \delta_{mn} - R_{\ell j}^{(m)} \delta_{ik} \delta_{mn} \quad (2\text{-}2)$$

and the field operators a_ℓ, a_ℓ^+ and b_ℓ, b_ℓ^+ are the usual creation and annihilation operators which obey Bose commutation rules. The sums over the index ℓ are taken as sums over polarizations as well as modes of the electromagnetic field. The quantities $\omega_R^{(j)}$ are Rabi frequencies [14] of the slowly-varying pumping field envelope at atomic positions $\underline{r}_j$. The atom-field coupling factors $p_\ell^{(j)}$ and $g_\ell^{(j)}$ are explicitly stated as

$$g_\ell^{(j)} = \underline{u}^{(j)} \cdot \underline{e}_\ell \quad , \qquad p_\ell^{(j)} = \underline{u}_o^{(j)} \cdot \underline{e}_\ell^{(o)} \tag{2-3}$$

where $\underline{u}^{(j)}$ and $\underline{u}_o^{(j)}$ are the matrix elements of the transition dipole moment for the SF and pump transitions, respectively, and the $\underline{e}_\ell$ and $\underline{e}_\ell^{(o)}$ are given as

$$\underline{e}_\ell = \sqrt{\frac{2\pi\hbar c k_\ell}{v}}\, \hat{\underline{e}}_\ell \quad , \qquad \underline{e}_\ell^{(o)} = \sqrt{\frac{2\pi\hbar c \hat{k}_\ell}{v}}\, \hat{\underline{e}}_\ell \quad , \tag{2-4}$$

where v is the volume of quantization and $\hat{\underline{e}}_\ell$ are unit vectors. We shall present only an outline of the procedures leading to the desired equations of motion. The mathematical details will be presented elsewhere [15].

The Heisenberg equations of motion for the SF fluorescence field obtained from (2-1) are formally integrated, and then separated into the contribution due to the self-field of the atom, the vacuum contribution and the contribution due to the presence of all the other atoms (i.e., the extended dipole contribution). The first mentioned separated field leads to natural atomic relaxation γ^{-1} for the $3 \leftrightarrow 2$ transition in the normally-ordered Heisenberg equations and the vacuum contribution leads to Langevin force terms $f(\tau)$ which satisfy the ensemble averages over the vacuum fluctuations,

$$\langle f(\tau)\, f^+(\tau') \rangle = \frac{1}{N\tau_R}\, \delta(\tau - \tau') \qquad ; \qquad \langle f^+(\tau)\, f(\tau') \rangle = 0 \quad , \tag{2-5a,b}$$

where N is the total number of atoms and τ_R is the characteristic SF time (i.e., the time for which, on the average, one cooperative photon is emitted), and is given by

$$\tau_R = \frac{L}{c}\left[\frac{2\pi|g|^2}{\hbar}\rho\right]^{-1} \quad . \tag{2-6}$$

Here, ρ is the atomic density, L the longitudinal length of the medium, and g is the atom-field coupling in the neighborhood of resonance for the SF transition. The remaining slowly-varying, rightward propagating SF field $A_R^{(-)}$ is given in retarded time $\tau = t - z/c$, by

$$A_R^{(-)}(z,\tau) = \frac{n_s}{h^2}\sum_\ell |g_\ell|^2 \sum_{z_m} \int_{-z_{m/c}}^{\tau+(z-z_m)/c} d\tau'\, e^{ic(k_\ell - k)(\tau-\tau')}\, \bar{R}_{32}(z_m,\tau') \quad . \tag{2-7}$$

In the above expression $\bar{R}_{32}(z_m,\tau')$ is the slice averaged [16], rightward propagating slowly-varying atomic variable, and n_s is a normalization due to the introduction of slice averaged operators [16]. A similar expression is obtained in the same way for the rightward propagating, slowly-varying fluorescence field $A_{RO}^{(-)}$ for the pump transition. The Maxwell equations in retarded time coordinates,

$$\frac{dA_R^{(-)}}{dz} = \frac{1}{\tau_R L}\bar{R}_{32} \quad ; \quad \frac{dA_{RO}^{(-)}}{dz} = \frac{1}{\tau_{RO}L}\bar{R}_{31} \tag{2-8a,b}$$

are derived from (2-7) and its counterpart for $A_{RO}^{(-)}$, respectively, in a manner similar to that leading to Eq. (36) of reference 10. The details of the derivation will be presented elsewhere [15]. The characteristic time τ_{RO} in (2-8b) is defined for the $1 \leftrightarrow 3$ transition in a manner similar to (2-6).

The normally-ordered Heisenberg equations of motion for the rightward propagating atomic variables in the slowly-varying operator representation are,

$$\frac{dR_{33}}{d\tau} = -\Gamma R_{33} - \bar{R}_{32}A_R^{(+)} - A_R^{(-)}\bar{R}_{23} - \bar{R}_{31}A_{OR}^{(+)} - \frac{1}{2}\omega_R\bar{R}_{31} - \frac{1}{2}\omega_R^{*}\bar{R}_{13}$$
$$- \bar{R}_{32}f^{(+)} - f\,\bar{R}_{23} - \bar{R}_{31}f_o^{(+)} - f_o\bar{R}_{13} \tag{2-9a}$$

$$\frac{dR_{22}}{d\tau} = \gamma R_{33} + \bar{R}_{32}A_R^{(+)} + A_R^{(-)}\bar{R}_{23} + \bar{R}_{32}f^{(+)} + f\,\bar{R}_{23} \tag{2-9b}$$

$$\frac{dR_{11}}{d\tau} = \gamma_o R_{33} + \bar{R}_{31}A_{OR}^{(+)} + A_{OR}^{(-)}\bar{R}_{13} + \frac{1}{2}\omega_R\bar{R}_{31} + \frac{1}{2}\omega_R^{*}\bar{R}_{13} + \bar{R}_{31}f_o^{(+)} + f_o\bar{R}_{13} \tag{2-9c}$$

$$\frac{d\bar{R}_{31}}{d\tau} = \left[i(\omega_{31} - \omega_o) - \frac{1}{2}\Gamma\right]\bar{R}_{31} - A_R^{(-)}\bar{R}_{21} + A_{OR}^{(-)}(R_{33}-R_{11})$$
$$+ \frac{1}{2}\omega_R^{*}(R_{33}-R_{11}) - f\,\bar{R}_{21} + f(R_{33}-R_{11}) \tag{2-9d}$$

$$\frac{d\bar{R}_{32}}{d\tau} = \left[i(\omega_{32}-\omega) - \frac{1}{2}\Gamma\right]\bar{R}_{32} + A_R^{(-)}(R_{33}-R_{22}) - A_{OR}^{(-)}\bar{R}_{12}$$
$$- \frac{1}{2}\omega_R^{*}\bar{R}_{12} + f(R_{33}-R_{22}) - f_o\bar{R}_{12} \tag{2-9e}$$

$$\frac{d\bar{R}_{12}}{d\tau} = -i(\omega_{32}-\delta)\bar{R}_{12} + A_R^{(-)}\bar{R}_{13} + \bar{R}_{32}A_{OR}^{(+)} + \frac{1}{2}\omega_R\bar{R}_{32}$$
$$+ f_o\bar{R}_{13} + \bar{R}_{32}f^{(+)} \;. \tag{2-9f}$$

The field variables $A_R^{(-)}$ and $A_{OR}^{(-)}$ and the atomic variables $\bar{R}_{k\ell}$ and R_{kk} are to be understood as functions of z and τ, and the fluctuating force terms f and f_o can be shown to be functions of retarded time τ only [6,15].

The Langevin terms f_o corresponding to the 1 $\leftrightarrow$ 3 transition obey relations identical to (2-5) but with τ_R replaced by τ_{RO}. The Langevin force terms in (2-9) give rise to Gaussian random quantum initiation statistics in both allowed transitions [10-12]. The factors δ and Γ appearing in (2-9) are given by $\delta = \omega_o - \omega$; $\Gamma = \gamma + \gamma_o$, where γ_o^{-1} is the natural lifetime for the 1 $\leftrightarrow$ 3 transition.

The equations (2-8) and (2-9) form the working equations for the calculations presented in the following sections. The pumping field envelope ω_R is taken as a rightward propagating pulse which is injected into the medium with specified initial and boundary conditions and in general is described by a classical Maxwell equation.

III. Pump Dynamic Contributions to Quantum Initiation of SF

This section is used to examine the effect of dynamical coherent excitation of the 3 $\leftrightarrow$ 1 transition on the quantum mechanical initiation of SF in the 3 $\leftrightarrow$ 2 transition. Plane wave propagation is assumed and the excitation (pumping process) is taken as a rightward propagating square pulse of Rabi frequency ω_R and temporal duration τ_p. The pump field is treated classically, i.e, as a coherent state [13], and longitudinally-uniform in the atomic medium, which has been shown [4a] to be justified provided the pump pulse time duration τ_p is much larger than the longitudinal transit time in the medium, τ_E, i.e., $\tau_p >> \tau_E$. The initial condition at retarded time $\tau = 0$ is taken as all the atoms in the ground state ε_1.

In order to direct our attention to the quantum mechanical initiation process during SF fluorescence buildup, we confine our attention to the linearized regime of small SF signal (negligible population in level 2 compared to level 3),

$$\langle R_{22} \rangle \approx \frac{d}{d\tau} \langle R_{22} \rangle \simeq 0 \quad , \tag{3-1a}$$

and strong pump [4],

$$\omega_R \tau_R >> 1 \quad . \tag{3-1b}$$

Since we are concerned about dynamical effects which occur temporally in the time frame $\tau \lesssim \tau_R$ we neglect terms in the equations of motion (2-9) which make temporal contributions on the order of γ^{-1} and γ_o^{-1}. Also, we neglect SF competition in the pump transition 1 $\leftrightarrow$ 3 with the assumption,

$$\omega_R \tau_{RO} >> \omega_R \tau_R \quad . \tag{3-2}$$

If the linearization conditions (3-1) and (3-2) are used in conjunction with (2-8) in the equations of motion (2-9), the following set of coupled equations are generated in the retarded time $\tau = t - z/c$,

$$\frac{dR_{33}}{d\tau} = -\omega_R\bar{R}_{31} \quad ; \quad \frac{d\bar{R}_{31}}{d\tau} = \omega_R\left(R_{33} - \frac{1}{2}\right) \tag{3-3a,b}$$

$$\frac{d\bar{R}_{32}}{d\tau} = -\frac{1}{2}\omega_R\bar{R}_{12} + A_R^{(-)}R_{33} + f_R^{(-)} \tag{3-3c}$$

$$\frac{d\bar{R}_{12}}{d\tau} = \frac{1}{2}\omega_R\bar{R}_{32} + A_R^{(-)}\bar{R}_{13} + h_R^{(-)} \quad ; \quad \frac{d}{dz}A_R^{(-)} = \frac{1}{\tau_R L}\bar{R}_{32} \; . \tag{3-3d,e}$$

The last terms in (3-3c) and (3-3d) are Langevin force terms which give rise to Gaussian random quantum initiation statistics in the SF evolution [10-12] and are written explicitly

$$f_R^{(-)}(z,\tau) = f(\tau)\,R_{33}(z,\tau) \quad ; \quad h_R^{(-)}(z,\tau) = f(\tau)\,\bar{R}_{31}(z,\tau) \quad , \tag{3-4a,b}$$

where $f(\tau)$ is an operator which gives rise to vacuum fluctuation contributions and satisfies the conditions (2-5).

It is noted from (3-3a,b) that the linearized Heisenberg equations of motion for $\bar{R}_{33}$ and $\bar{R}_{31}$ exhibit no operator character and form a closed set in themselves. We thus take them as expectation values, and ω_R and $\bar{R}_{31}$ have been taken as real without loss of generality. Because of the strong pump approximation (3-1b), the collective pump transition variables $\bar{R}_{33}$ and $\bar{R}_{31}$ are dynamically determined by the pumping process entirely and can therefore be replaced by their factorized expectation values in (3-3c,d). Then the only operator character in these equations are the Langevin force terms $f_R^{(-)}$ and $h_R^{(-)}$. Thus, the equations (3-3) can be regarded in terms of expectation values with respect to the atomic system, but not the field reservoir, i.e., the ordering of products of the Langevin force terms in expectation values is essential, as exhibited in (2-5). From this point on, Eqs. (3-3) are regarded in terms of expectation values with respect to the atomic system.

It is re-emphasized that the linearization conditions (3-1) have resulted in a decoupling of the pump term equations (3-3a,b) from the remainder. Thus, for all the atoms initially in the ground state at $\tau = 0$, we have

$$R_{33} = \sin^2\frac{\omega_R}{2}\tau \quad ; \quad \bar{R}_{31} = -\frac{1}{2}\sin\omega_R\tau \quad . \tag{3-5}$$

The remainder of the equations of motion, (2-8a), (3-3c)-(3-3e) can be solved by Laplace transform. These combine in the Laplace regime, in the limit (3-1b) and to first order in τ_p/τ_R , to give the result in terms of $\bar{R}_{32}^{(o)}(s,\tau)$,

$$\frac{d^2\tilde{\bar{R}}_{32}^{(o)}(s,\tau)}{d\tau^2} + \frac{1}{4}\,\omega_R^2\,\tilde{\bar{R}}_{32}^{(o)}(s,\tau) = \frac{u(\tau)}{s} \tag{3-6}$$

where

$$u(\tau) = \frac{d\Gamma_R^{(-)}}{d\tau} - \frac{1}{2}\omega_R h_R^{(-)} \tag{3-7}$$

and

$$\tilde{\bar{R}}_{32}^{(o)}(s,\tau) = \frac{2}{\omega_R}\int_0^{\tau} d\tau' \; \frac{u(\tau')}{s} \sin\frac{\omega_R}{2}(\tau-\tau') \quad . \tag{3-8}$$

The next higher order contribution in $(\omega_R\tau_R)^{-1}$ is $\tilde{\bar{R}}_{32}^{(1)}(s,\tau)$, where

$$\tilde{\bar{R}}_{32}(s,\tau) = \tilde{\bar{R}}_{32}^{(o)}(s,\tau) + \tilde{\bar{R}}_{32}^{(1)}(s,\tau) + \ldots \tag{3-9}$$

and is determined by

$$\left(\frac{d^2}{d\tau^2} + \frac{1}{4}\omega_R^2\right)\tilde{\bar{R}}_{32}^{(1)}(s,\tau) = \frac{1}{\tau_R Ls}\sin^2\frac{\omega_R}{2}\tau\;\frac{d\tilde{\bar{R}}_{32}^{(o)}(s,\tau)}{d\tau} + \frac{\omega_{\bar{R}}}{\tau_R Ls}\tilde{\bar{R}}_{32}^{(o)}(s,\tau) \quad , \tag{3-10}$$

where

$$\tilde{\bar{R}}_{32}^{(1)}(s,\tau) = \frac{2}{\omega_R}\int_0^{\tau} d\tau' \; \sin\frac{\omega_R}{2}(\tau-\tau')\left\{\frac{1}{\tau_R Ls}\sin^2\frac{\omega_R}{2}\tau\;\frac{d\tilde{\bar{R}}_{32}^{(o)}(s,\tau)}{d\tau} + \frac{\omega_R}{\tau_R Ls}\sin\omega_R\tau\;\tilde{\bar{R}}_{32}^{(o)}(s,\tau)\right\} . \tag{3-11}$$

It should be noted that $R_{32}^{(n)}(z,\tau) \sim (\frac{z}{L})^{n-1}$ and that succeeding terms in the expansion are in the ratio $(\omega_R\tau_R)^{-1}$ (i.e., $\omega_R\tau_R$ is the expansion parameter [15]).

The "tipping angle", $\theta(z,\tau)$ is defined as [10,12]

$$\theta^2(z,\tau) = \langle \bar{R}_{23}(z,\tau)\,\bar{R}_{32}(z,\tau) \rangle \tag{3-12}$$

and we are interested in its value in the linearized SF regime $\tau > \tau_p$, $\omega_R = 0$. In the time frame after the pump is turned off, i.e., for $\tau > \tau_p$, the Langevin terms (3-4) become

$$f_R^{(-)}(\tau) = f(\tau)\,R_{33}(\tau_p) \quad ; \quad h_R^{(-)}(\tau) = f(\tau)\,\bar{R}_{31}(\tau_p) \tag{3-13a,b}$$

and equations (3-3) are solved for $\omega_R = 0$ with the initial conditions being the value of each variable at $\tau = \tau_p$. The procedure already outlined leads in a straightforward way to the result in lowest order [6,15], for $\tau' = \tau - \tau_p$, $\tau' > 0$,

$$\theta^2(\tau') = \left\{ I_o \left[2\sqrt{\frac{1}{L\tau_R} R_{33}(\tau_p) z\tau'} \right] \right\}^2 \quad \langle \bar{R}_{23}^{(o)}(\tau_p)\, \bar{R}_{32}^{(o)}(\tau_p) \rangle$$

$$+ \frac{4 R_{33}^2(\tau_p)}{N\tau_R} \int_0^{\tau'} d\tau'' \left\{ I_o \left[2\sqrt{\frac{1}{L\tau_R} R_{33}(\tau_p)\, z(\tau'-\tau'')} \right] \right\}^2 \qquad (3\text{-}14)$$

which is the main result of this section.

It is interesting to compare the result (3-14) with the corresponding result from the two-level model with an initial state of complete inversion (i.e., impulse excitation) of reference 10, Eq. (64). We see that the two cases differ by the first term in (3-14) as well as the factor $R_{33}^2(\tau_p)$ of the square of the pump induced inversion onto the second term and the appearance of $R_{33}(\tau_p)$ in the argument of the Bessel's functions I_o. In the linearized regime of references 10-12, $R_{33} = 1$; in our case this need not be so. But, if $\omega_R\tau_p = \pi$, i.e., a π-pump pulse, then the second term in (3-14) is exactly equivalent to Eq. (64) of reference 10. However, the first term in (3-14) still remains, which arises from spontaneous relaxation of the $3 \leftrightarrow 2$ transition during the dynamics of the pumping process. The first term in (3-14) therefore, uniquely characterizes the effects of dynamical pumping in a three-level system on SF quantum initiation.

The evolution of the expectation value in the first term of (3-14) gives

$$\langle \bar{R}_{23}^{(o)}(\tau_p) \bar{R}_{32}^{(o)}(\tau_p) \rangle \approx \left(\frac{4}{N} \sin^2 \frac{\omega_R\tau_p}{2} \right) \frac{\tau_p}{\tau_R} \qquad (3\text{-}15)$$

to first order in τ_p/τ_R. Thus, for a π-pulse, impulse excitation, i.e., $\omega_R\tau_p = \pi$, $\tau_p \to 0$, all effects of the pump dynamics vanish, and (3-14) reduces to the results of reference 10, Eq. (64).

However, it is apparent that $\tau_p/\tau_R << 1$ is a condition not satisfied in SF experiments over the full range of atomic densities [3,7-9], and therefore, the effects of the dynamical pumping process on amplified quantum initiation could be quite important, depending of course upon the degree of coherence of the pumping pulse.

The nonlinear regime for SF evolution for the conditions $\tau_p/\tau_R >> 1$, $\tau_p/\tau_R < 1$, where amplified quantum initiation effects are expected to be less important, is discussed in the next section in terms of a semiclassical numerical calculation of Mattar and Bowden [5] in which propagation and transverse effects are taken into account.

IV. Deterministic Effects of Pump Dynamics in the Nonlinear Regime of SF

Some aspects of deterministic effects of coherent pump dynamics on SF pulse evolution in the nonlinear regime of SF will be presented in this section. We consider conditions for which $\tau_p/\tau_R >> 1$; $\tau_p/\tau_D < 1$. Quantum initiation is expected to play a less important role in this situation than for the opposite condition, $\tau_p/\tau_R \lesssim 1$ discussed in the previous section for the linear regime of SF. The coherence induced by the pumping process, in this case, is expected to overwhelm the effects of quantum initiation.

The calculation of SF pulse evolution in the nonlinear regime is necessarily a calculational problem if propagation is explicitly included. We use an algorithm presented elsewhere [17] and the model defined by Eqs. (2-9) to analyze the effects of coherent pump dynamics, propagation, transverse and diffraction effects on SF emission. To facilitate numerical calculation, Eqs. (2-9) are taken in their factorized, semiclassical form [5] with the field $A_R^{(-)}$ replaced by its classical representation which is described by Maxwell's equation. The pump field ω_R and fluorescence field $< A_R^{(-)} >$ are determined dynamically and spacially in retarded time, by initial and boundary conditions and the equations

$$\mathcal{F}_p^{-1}\nabla_\rho^2 \begin{Bmatrix} -X_o \\ Y_o \end{Bmatrix} + \frac{\partial}{\partial \eta_p} \begin{Bmatrix} Y_o \\ X_o \end{Bmatrix} = d \begin{Bmatrix} -U_{31} \\ V_{31} \end{Bmatrix} \qquad (4\text{-}1a)$$

$$\mathcal{F}_s^{-1}\nabla_\rho^2 \begin{Bmatrix} -X \\ Y \end{Bmatrix} + \frac{\partial}{\partial \eta_s} \begin{Bmatrix} Y \\ X \end{Bmatrix} = d \begin{Bmatrix} -U_{32} \\ V_{32} \end{Bmatrix} \qquad (4\text{-}1b)$$

In the above equations, X, Y, and X_o, Y_o are the real and imaginary components, respectively, of the SF field $< A_R^{(+)} >$ and ω_R, respectively, i.e.,

$$\frac{< A_R^{(-)} >}{\gamma_\perp} = X + iY \qquad ; \qquad \frac{\omega_R}{\gamma_\perp} = X_o + iY_o \qquad (4\text{-}2a,b)$$

in units of the dephasing rate $\gamma_\perp$ taken now as phenomenological (i.e., homogeneous broadening) and is assumed to be the same in the equations of motion for all off-diagonal matrix elements. The atomic variables appearing on the right-hand side of (4-1) are real quantities and are defined according to

$$R_{k\ell} = \frac{1}{2}(U_{k\ell} + iV_{k\ell}) \qquad k > \ell \; . \qquad (4\text{-}3)$$

The first terms on the left-hand side of Eqs. (4-1) are the transverse parts of Maxwell's equations in cylindrical symmetry, where

$$\nabla_\rho^2 = \frac{1}{\rho}\frac{\partial}{\partial\rho}\left(\rho\frac{\partial}{\partial\rho}\right)$$

and $\rho = r/r_p$ where r is the radial distance and r_p is a characteristic radial width. The longitudinal spacial coordinate $\eta_{p_s} = z\, g_{eff_{p_s}}$, where g_{eff} is the on-axis effective gain,

$$g_{eff_{p_s}} = \frac{\begin{Bmatrix}\omega_o\\ \omega\end{Bmatrix} \begin{Bmatrix}U_{32}\\ U_{31}\end{Bmatrix}^2 N}{n\,\hbar\, c}\, T_2 \tag{4-4}$$

where N is the atomic number density (assumed longitudinally homogeneous) and n is the index of refraction assumed here to be identical for each transition wavelength. The subscripts p and s represent the pump and SF transitions, respectively. The quantity

$$d = \frac{N(r)}{N_o} \tag{4-5}$$

governs the relative radial population density distribution for active atoms, which could have a variation, say, for an atomic beam. Finally, the first factors on the first terms of (4-1) are the reciprocals of the "gain length" Fresnel numbers [5] defined by

$$\mathcal{F}_{p_s} = \frac{2\pi r_p^2}{\lambda_{p_s}\, g_{eff_{p_s}}^{-1}} \quad . \tag{4-6}$$

It is seen from (4-1) that for sufficiently large Fresnel number $\mathcal{F}$ the corrections due to transverse effects become negligible. The "gain length" Fresnel numbers are related to the usual Fresnel numbers $F = 2\pi r_p^2/\lambda L$, where L is the length of the medium, by

$$\mathcal{F}/F = g_{eff} L \tag{4-7}$$

i.e., the total gains of the medium. In the computations, diffraction is explicitly taken into account by the boundary condition that $\rho = \rho_{max}$ corresponds to completely absorbing walls.

The Langevin force fluctuation terms f and f_o appearing in Eqs. (2-9) are taken as complex valued c-numbers,

$$f \rightarrow |f|\, e^{i\phi} \quad ; \quad f_o \rightarrow |f_o|\, e^{i\phi_o} \quad . \tag{4-8a,b}$$

The amplitudes $|f|$ and $|f_o|$ obey Gaussian random probability distributions $P(f)$ and $P_o(f_o)$, where, in accordance with (2-5a),

$$P(|f|^2) = \frac{1}{\pi\langle\sigma\rangle} \exp\left(-|f|^2/\langle\sigma\rangle\right) \quad , \tag{4-9a}$$

$$P_0(|f_0|^2) = \frac{1}{\pi\langle \sigma_0 \rangle} \exp(-|f_0|^2/\langle \sigma_0 \rangle) \quad , \tag{4-9b}$$

where $\langle \sigma \rangle = (N\tau_R L/c)^{-1}$, $\langle \sigma_0 \rangle = (N\tau_{R0} L/c)^{-1}$. Since the phases ϕ and ϕ_0 are initially completely undetermined, their statistical distributions are defined as uniform on a field $0 \leq \begin{Bmatrix} \phi \\ \phi_0 \end{Bmatrix} \leq 2\pi$. The Langevin force contributions to the semiclassical equations of motion give rise to initiation of fluorescence in the $3 \leftrightarrow 2$ transition when the $3 \leftrightarrow 1$ transition is coupled by the pumping field ω_R. Normally, one can ignore fluorescence in the $3 \leftrightarrow 1$ transition, i.e., ignore contributions from f_0.

Thus, by utilizing the relations (4-9), a complete ensemble simulation can be constructed, and in this way the manifestations of amplified quantum initiation can be calculationally analyzed over the full range of dynamical SF evolution. This amounts to generating the calculational results, using the semiclassical representation of Eqs. (2-9) for specified initial and boundary conditions, for each of the values selected for $|f|$, $|f_0|$, ϕ and ϕ_0 according to the statistical distributions (4-9). One then must take the ensemble averages and associated variances. This is necessarily a very expensive calculation. For the results [5] presented here, we have ignored fluorescence in the pump transition and taken what amounts to the average value of f according to (4-9a) and an arbitrary value for ϕ, (see Appendix of reference 5), so these results must be interpreted as ensemble averages. It turns out that for the results to be discussed [5], this is equivalent to an average "tipping angle" on the order [18] of 10^{-3}. The full calculational statistical treatment will be presented elsewhere [19]. The material parameters chosen for these calculations are arbitrary, but correspond roughly to those for optically-pumped metal vapors.

The initial and boundary conditions are such that all the atomic population is in the ground state ε_1 at $\tau = 0$. The pumping pulse which pumps the $1 \leftrightarrow 3$ transition is injected at $z = 0$ and is rightward propagating, and its initial characteristics are specified at $z = 0$. The SF pulse subsequently evolves in z, ρ, and τ due to the initiation of fluorescence instigated by $|f|$ and ϕ as indicated in the equations of motion (2-9) in their semiclassical form discussed above. The pump pulse, whose initial characteristics are specified at injection and the SF pulse co-propagate and interact via the nonlinear medium. We shall show that certain initial characteristics of the injected pumping pulse have deterministic effects upon the SF pulse evolution, thus demonstrating a new manifestation of the phenomenon of light control by light (5].

Figure 2 shows results of the numerical calculations for the transverse integrated intensity profiles for the co-propagating SF and injected pulses at the specified

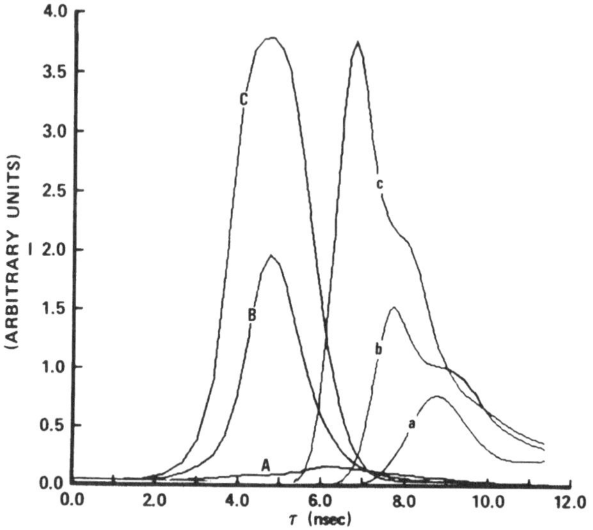

Figure 2. Radially integrated normalized intensity profiles for the SF and injected pulse at z = 5.3 cm penetration depth for three different values for the initial on-axis injection pulse area θ_p. The SF pulses are indicated by a, b, and c, whereas the corresponding injected pump pulses are labeled by A, B, and C. The injected pulses are initially Gaussian in r and τ with widths (FWHM) r_0 = 0.24 cm and τ_p = 4 nsec, respectively. The level spacings are such that $(\varepsilon_3-\varepsilon_1)/(\varepsilon_3-\varepsilon_2)$ = 126.6. The effective gain for the pump transition g_p = 17 cm^{-1} and that for the SF transition g_s = 291.7 cm^{-1}. The gain-length Fresnel numbers for the two transitions are $\mathcal{F}_p$ = 16800 and $\mathcal{F}_s$ = 2278. The relaxation and dephasing times are taken as identical for all transitions and are given as T_1 = 80 nsec and T_2 = 70 nsec, respectively. The injected pulse initial on-axis areas are: (A) θ_p = π , (B) θ_p = 2π and (C) θ_p = 3π. Here, τ_R = 90.5 psec.

penetration depth in the nonlinear medium. These profiles correspond to what would be observed with a wide aperture, fast, energy detector. The pumping pulses are labeled by capital letters, and the corresponding SF pulses are labeled by the corresponding lower case letters. Each set of curves represents a different initial on-axis area for the injected pump pulse, i.e., curve A is the reshaped pump pulse at z = 5.3 cm which had its initial on-axis area specified as θ_p = π , and curve a is the resulting SF pulse which has evolved. All other parameters are identical for each set of pulses. The initial conditions are those already discussed earlier.

These results clearly indicate the coherence effect of the initial pump pulse area on the SF signal which evolves. Notice that the peak intensity of the SF pulses increases monatonically with initial on-axis area for the pump pulse. This is caused by self-focusing due to transverse mode coupling and propagation. For instance, a 2-π injection pulse would generate very small SF response compared to an initial π-injection pulse for these conditions at relatively small penetration z, or for the corresponding case in one spacial dimension. Even so, the peak SF intensity is approximately proportional to the square of the pump-pulse initial on-axis area, whereas the delay time τ_D between the pump-pulse peak and the corresponding SF peak is very nearly inversely proportional to the input pulse area. The temporal SF pulse width at full-width half-maximum (FWHM) τ_s is approximately invariant with respect to the injection pulse area.

Since the average values of τ_D and the peak SF intensity are important quantities for interpreting experimental results with theories of SF, the manner in which the pump pulse coherence and initial on-axis area affect these quantities is seen to be of extreme importance in any analysis.

Figure 3 shows the effect upon the SF pulse of variation in the initial temporal width at half maximum intensity for the pumping pulse. As the initial temporal width of the injected pulse τ_p becomes smaller, the SF delay time τ_D increases, whereas the peak SF intensity decreases, and the SF temporal width τ_s remains very nearly fixed.

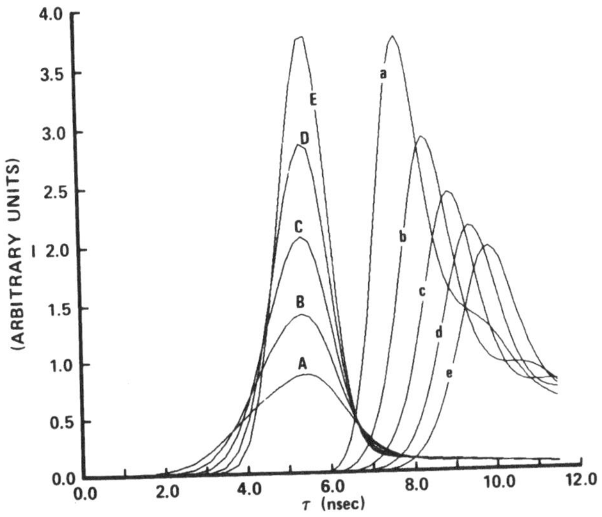

Figure 3. Radially integrated normalized intensity profiles for the SF and injected pulses at z = 5.3 cm penetration depth for five different values for the initial temporal width of the injected pulse. The initial on-axis area of the injected pulse is $\theta_p = \pi$, and the pump transition and SF effective gains are $g_p = 17.5\ cm^{-1}$ and $g_s = 641.7\ cm^{-1}$, respectively. All other parameters except for the Fresnel numbers are the same as those for Fig. 2. The injected pulse initial temporal widths at half maximum are: (A) τ_p = 4 nsec; (B) τ_p = 3.3 nsec; (C) τ_p= 2.9 nsec; (D) τ_p= 2.5 nsec; and (E) τ_p = 2.2 nsec.

It is clear from these results that there exists an approximate linear relationship between the time delay τ_D , between the peak SF intensity and the corresponding pump-pulse intensity, and the initial temporal width τ_p of the pump pulse. These results generate the following empirical formula for τ_D as a function of τ_p:

$$\tau_D = 0.375\ \tau_R \left[\ell n \left(\frac{4\pi}{\phi_0}\right)\right]^2 - 4\ \tau_R \gamma_\perp\ (\gamma_R/4\gamma_\perp - 1)\ \tau_p\ , \tag{4-10}$$

where [20] $\tau_R = 2\ T_2/g_{eff_s}\ z$, and $\gamma_R = \tau_R^{-1}$.

The relation (4-10) is at least in qualitative agreement with the analytical prediction made in reference 4(b), Eq. (5.1), based upon mean-field theory. The first term in (4-10) was chosen to conform with the quantum-mechanical SF initiation result [6,10-12]. The quantity ϕ_0 can be interpreted as the "effective tipping angle" for an equivalent π-initial impulse excitation, i.e., for $\tau_p \to 0$, which initiates subsequent SF. It is to be noted that the value for ϕ_0 is dependent upon our value for $|f|$, Eq. (2-5a); however, τ_D varies less than 25% for order-of-magnitude changes in $|f|^2$.

These results emphasize the importance of the initiating pulse characteristics in SF pulse evolution, and the effect of SF pulse narrowing with approximate pulse shape invariance by increasing the initial temporal width of the injected pulse. It is emphasized that all other parameters, including the initial value for the injected pulse on-axis area, are identical among these sets of curves.

V. Conclusions

The effects on SF evolution of coherent pumping on a three-level system have been discussed in this Chapter. The pump dynamical contribution to SF quantum mechanical initiation for the condition $\tau_p/\tau_R < 1$ was discussed in Section III. The main result is the expression for the "tipping angle", Eq. (3-14) in the linear regime of quantum initiation. The first term in (3-14) is due entirely to spontaneous relaxation in the $3 \leftrightarrow 2$ transition during the time frame of dynamical pumping on the $3 \leftrightarrow 1$ transition. To first order in τ_p/τ_R its evolution is given by (3-15). For a π-impulse excitation, these results reduce to the two-level model SF results [10]. Even though the degree of coherence in most of the reported SF experiments is uncertain and $\tau_p/\tau_R < 1$ is a condition not satisfied in most of the reported SF experiments over the full range of atomic densities, the results reported here indicate that the pump dynamics can have a significant contribution in the effects of observed amplified quantum initiation in SF temporal fluctuations.

Results of numerical simulation in the full SF pulse evolution for several cases where $\tau_p/\tau_R >> 1$, $\tau_p/\tau_D < 1$ were discussed in Section IV. Even though propagation, transverse and diffraction effects were explicitly taken into account in the calculation, it was shown that certain initial characteristics of the injected pump pulse cause deterministic manifestations in the SF which evolves. The results are interpreted as ensemble averages, although the effects of quantum initiation are expected to be relatively less significant in the nonlinear regime in these cases. It is felt that deterministic effects of the type reported and analyzed here may have been operable in some of the data reported in SF experiments. The phenomenon is, however, quite interesting in itself from the standpoint of the physics of co-propagating different wavelength pulses coupled via a nonlinear medium.

References

1. R. Bonifacio and L. A. Lugiato, Phys. Rev. A11, 1507 (1975); A12, 587 (1975).

2. R. H. Dicke, Phys. Rev. 93, 99 (1954).

3. See papers and references in Cooperative Effects in Matter and Radiation, edited by C. M. Bowden, D. W. Howgate and H. R. Robl, Plenum, New York, 1977.

4. (a) C. M. Bowden and C. C. Sung, Phys. Rev. A18, 1558 (1978).
(b) Phys. Rev. A20, 2033 (1979).

5. F. P. Mattar and C. M. Bowden, "Coherent Pump Dynamics, Propagation, Transverse and Diffraction Effects in Three-level Superfluorescence and Control of Light by Light", Phys. Rev. A, January 1983, (to be published).

6. C. M. Bowden and C. C. Sung, "Initiation of Superfluorescence in Coherently-pumped Three-level Systems", Phys. Rev. Lett., January 1983, (to be published).

7. H. M. Gibbs, NATO Advanced Study Institute on "Superfluorescence Experiments", in Coherence in Spectroscopy and Modern Physics, edited by F. T. Arecchi, R. Bonifacio and M. O. Scully, Plenum, New York, 1978, p. 121.

8. M. S. Feld and J. C. MacGillivray, "Superradiance", in Coherent Nonlinear Optics, edited by M. S. Feld and V. S. Letokhov; Topics in Current Physics, Springer-Verlag, Volume 21, 1980, p. 7.

9. Q. H. F. Vrehen and H. M. Gibbs, "Superfluorescence Experiments", in Dissipative Systems in Quantum Optics, edited by R. Bonifacio; Topics in Current Physics, Springer-Verlag, Volume 27, 1982, p. 111.

10. D. Polder, M. F. H. Schuurmans and Q. H. F. Vrehen, Phys. Rev. A19, 1192 (1979).

11. M. F. H. Schuurmans, Q. H. F. Vrehen, D. Polder, and H. M. Gibbs, "Superfluorescence", in Advances in Atomic and Molecular Physics, edited by D. Bates and B. Bederson, Volume 17, Academic Press, 1981, p. 167.

12. R. J. Glauber and F. Haake, Phys. Lett. A68, 29 (1978); Phys. Rev. Lett. 45, 558 (1980); F. Haake, H. King, G. Schröder and J. Haus, Phys. Rev. A20, 2047 (1979).

13. R. J. Glauber, Phys. Rev. 130, 2529 (1963); 131, 2766 (1963).

14. L. Allen and J. H. Eberly, Optical Resonance and Two-level Atoms, John Wiley and Sons, New York, 1975, p. 54.

15. C. C. Sung and C. M. Bowden, to be published.

16. The slice averaged atomic operators $R_{k\ell}$ are obtained by averaging operators $R_{k\ell}^{(j)}$, appearing in (2-1), over a slice of the medium containing many atoms but yet thin in the z direction compared to an atomic transition wavelength. Specifically, $R_{k\ell}(z,t) = \frac{1}{\eta_s} \sum_{j\varepsilon\{j\}_z} R_{k\ell}^{(j)}(t)$, where η_s is the mean number of atoms in a slice and $\{j\}_z$ denotes the collection of atoms in $z - d/2 < z_j < z + d/2$, $d/2\pi\lambda << 1$. The procedure is presented explicitly in reference 10. $\bar{R}_{k\ell}$ is then the rightward propagating collective atomic variable.

17. F. P. Mattar and M. C. Newstein, Reference 3, p. 139; F. P. Mattar, in Optical Bistability, edited by C. M. Bowden, M. Ciftan and H. R. Robl, Plenum, New York, 1981, p. 503; in Proceedings, Tenth Simulation and Modeling Conference, Pittsburgh, 1978, edited by W. Vogt and M. Mickle, (Pub. Inst. Soc. Am., Pittsburgh, PA, 1979).

18. The "tipping angle" determined in the appendix of reference 5 was chosen in a different manner than that outlined here, but essentially amounts to the same procedure as far as the computational results are concerned, although it is not as straightforward.

19. F. P. Mattar, C. M. Bowden and C. C. Sung, to be published.

20. R. Friedberg and S. R. Hartmann, Phys. Rev. A13, 495 (1976).

COHERENT POPULATION TRAPPING AND THE EFFECT OF LASER PHASE FLUCTUATIONS

B.J. Dalton†* and P.L. Knight†

Optics Section, Blackett Laboratory, Imperial College, London SW7 2BZ, England†
Physics Department, University of Queensland, St. Lucia, Queensland 4067, Australia*

1. INTRODUCTION

Coherent population trapping occurs in a wide variety of situations, some of which are listed in Table 1. A more complete survey of the relevent work is given in Reference [5] and within other references listed in Table 1. In each case a rate equation approach based only on populations would lead to the expectation that the populations of certain states would decay to zero, due to the presence of irreversible loss processes. However, the presence of coherence between the states, specified via off diagonal density matrix elements which appear in a master equation analysis, can in certain circumstances lead to a steady state solution where the populations of such potentially decaying states are non zero. Thus coherent population trapping has occured, the final quantum state then being immune from further decay. As the long time spectral features of a system reflect the time constants of its coupled loss processes, then in coherent trapping situations we would expect features with narrower widths than the normal decay widths to appear.

Coherent trapping has ramifications in spectroscopy, in the search for narrow features on which to base improved time standards [7], and in state selective photo-excitation processes where population trapping would restrict the overall yield of the desired product [5].

2. COHERENT TRAPPING CONDITIONS

Some of the essential conditions for coherent trapping phenomena can be understood from prototypes of Type 1 and Type 2 systems, which are analysed using master equation methods [8],[9]. The system (for example an atom plus pump laser mode(s) or an atom in an external static field), containing states $|1>$, $|2>$ for Type 1, and $|1>$, $|2>$ and $|3>$ for Type 2 (all with similar energies) also contains a much lower energy state $|g>$ ($\omega_0 \sim \omega_{ig}$, $i = 1,2,3$). The system also has an interaction V with a large quantum system or reservoir, initially in its lowest state $|\alpha>$ (for example, the unoccupied spontaneous emission modes of the quantum EM field) and into which irreversible system energy losses occur. V is assumed to be of the form $(SR + S^\dagger R^\dagger)$, where S, R are system, reservoir operators. Type 1 loss processes are $|1>|\alpha> \to |g>|\beta>$, $|2>|\alpha> \to |g>|\beta>$, whilst those for Type 2 are $|3>|\alpha> \to |g>|\beta>$ ($|\beta>$ is an excited reservoir state). An internal system interaction V_S (for example, atom-laser mode interactions or atom-external static field interactions) leads to coupling between the system states $|1> \leftrightarrow |2>$ for Type 1, $|1> \leftrightarrow |3>$ and $|2> \leftrightarrow |3>$ for Type 2. We assume all system

Table 1. Examples of Coherent trapping situations

Situation (type 1)	$\|1\rangle$	$\|2\rangle$	$\|L\rangle$	Loss processes
Spontaneous emission from two identical two level ($\|0\rangle$, $\|1\rangle$) atoms A, B. [1]	$\|0\rangle_A\|1\rangle_B$	$\|1\rangle_A\|0\rangle_B$	$\|0\rangle_A\|0\rangle_B\|1_\lambda\rangle$	$\|1\rangle\|0_\lambda\rangle\to\|0\rangle\|1_\lambda\rangle$
Weak interaction decay of $K_0,\bar{K}_0$ mesons [2]	$\|1,0\rangle$ $K_0\ \bar{K}_0$	$\|0,1\rangle$ $K_0\ \bar{K}_0$	$\|0,0\rangle\|1,1\rangle$ $K_0\ \bar{K}_0\ \pi^+\pi^-$	K_0, $\bar{K}_0$ $\to\pi^+ + \pi^-$
Pseudoautoionization. $\|1\rangle$ excited to continuum states $\|f\rangle$ via laser a, $\|0\rangle$ via laser b. [3]	$\|0,n_a,n_b\rangle$	$\|1,n_a+1,n_b-1\rangle$	$\|f,n_a,n_b-1\rangle$	$\|0,n_b\rangle\to\|f,n_b-1\rangle$ $\|1,n_a+1\rangle\to\|f,n_a\rangle$

Situation (type 2)	$\|1\rangle$	$\|2\rangle$	$\|3\rangle$	$\|L\rangle$	Loss processes
Lambda system coupled to lasers a,b. $\|0\rangle$ excited to $\|1\rangle$ via laser a, $\|2\rangle$ excited to $\|1\rangle$ via laser b. [4] [5]	$\|0,n_a,n_b\rangle$	$\|2,n_a-1,n_b+1\rangle$	$\|1,n_a-1,n_b\rangle$	(i) $\|m,n_a-1,n_b\rangle\|1_\lambda\rangle$ (ii) $\|f,n_a-2,n_b\rangle$ $\|f,n_a-1,n_b-1\rangle$	(i) $\|1\rangle\|0_\lambda\rangle$ $\to\|m\rangle\|1_\lambda\rangle$ Spont. emission (ii) $\|1,n_a-1\rangle$ $\to\|f,n_a-2\rangle$ $\|1,n_b\rangle$ $\to\|f,n_b-1\rangle$ Photoionization
Ladder system coupled to lasers a,b. $\|0\rangle$ excited to $\|1\rangle$ via laser a. $\|1\rangle$ excited to $\|2\rangle$ via laser b. [4] [5]	$\|0,n_a,n_b\rangle$	$\|2,n_a-1,n_b-1\rangle$	$\|1,n_a-1,n_b\rangle$	(i) $\|m,n_a-1,n_b\rangle\|1_\lambda\rangle$	(i) As above.

Situation (type 3)	$\|1\rangle$	$\|2\rangle$	$\|3\rangle$	$\|L\rangle$	Loss processes
Laser excitation of two nearby excited states $\|2\rangle$, $\|3\rangle$ from lower state $\|1\rangle$ [6]	$\|2,n-1\rangle$	$\|3,n-1\rangle$	$\|1,n\rangle$	$\|1,n-1\rangle\|1_\lambda\rangle$	$\|2\rangle\|0_\lambda\rangle\to\|1\rangle\|1_\lambda\rangle$ $\|3\rangle\|0_\lambda\rangle\to\|1\rangle\|1_\lambda\rangle$

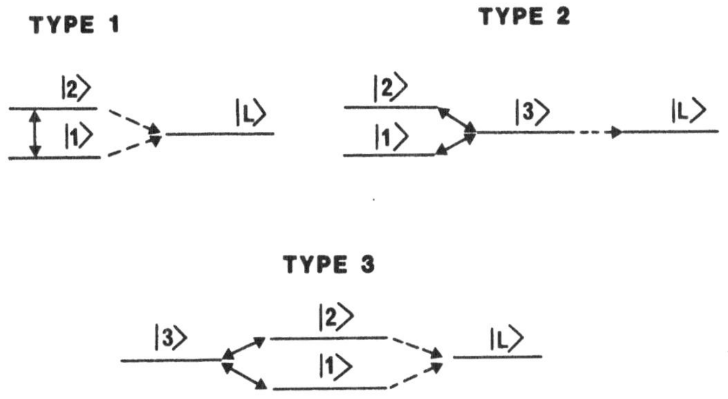

Figure 1. Three types of coherent trapping for states |1>, |2>. Processes |i>--→|L> indicate irreversible losses. Processes |i>↔|j> indicate reversible couplings.

interaction matrix elements are real and that $\langle\alpha|R|\beta\rangle \neq 0$ only if $\omega_{\beta\alpha} \gtrless 0$. Master equations governing the reduced density matrix elements which describe the system behaviour are obtained using the Markoff, weak coupling approximations (for treating relaxation processes) and the secular approximation (rotating wave approximation of the second kind) [8] [9].

For the Type 1 prototype the important master equations are:

$$\begin{bmatrix} i\dot{\sigma}_{11} \\ i\dot{\sigma}_{12} \\ i\dot{\sigma}_{21} \\ i\dot{\sigma}_{22} \end{bmatrix} = \begin{bmatrix} -i\Gamma_1 & -\gamma_{12}^* & \gamma_{12} & 0 \\ -\gamma_{12}^* & \delta-\tfrac{1}{2}i(\Gamma_1+\Gamma_2) & 0 & \gamma_{12} \\ \gamma_{12} & 0 & -\delta-\tfrac{1}{2}i(\Gamma_1+\Gamma_2) & -\gamma_{12}^* \\ 0 & \gamma_{12} & -\gamma_{12}^* & -i\Gamma_2 \end{bmatrix} \begin{bmatrix} \sigma_{11} \\ \sigma_{12} \\ \sigma_{21} \\ \sigma_{22} \end{bmatrix} \tag{1a}$$

$$i\dot{\sigma}_{gg} = i\Gamma_1\,\sigma_{11} + (\gamma_{12}^* - \gamma_{12})\,\sigma_{12} + (\gamma_{12}^* - \gamma_{12})\,\sigma_{21} + i\Gamma_2\,\sigma_{22} \tag{1b}$$

The Γ_i (i=1,2) are the usual decay rates of the states, δ is the transition frequency between |1>, |2>, shifted due the system-reservoir interaction. (See [8] for formulae). The quantity γ_{12} is given in terms of (real) v_{12}, α_{12}, β_{12} as:

$$\gamma_{12} = v_{12} + \alpha_{12} + i\beta_{12} \tag{2a}$$

with $v_{12} = \langle 1|V_S|2\rangle$ (2b)

$$\alpha_{12} = \frac{1}{\hbar^2} \langle 1|S|g\rangle\langle g|S^\dagger|2\rangle \sum_\beta |\langle\alpha|R|\beta\rangle|^2 \left(\frac{P}{\omega_0-\omega_{\beta\alpha}}\right) \tag{2c}$$

$$\beta_{12} = \frac{-\pi}{\hbar^2} \langle 1|S|g\rangle\langle g|S^\dagger|2\rangle \sum_\beta |\langle\alpha|R|\beta\rangle|^2 \delta(\omega_{\beta\alpha}-\omega_0) \tag{2d}$$

It can also be shown that: $\beta_{12}^2 = \tfrac{1}{4}\Gamma_1\Gamma_2$ (3)

The quantities $-i(\alpha_{12} + i\beta_{12})$, $-i(-\alpha_{12} + i\beta_{12})$ are off diagonal Markovian relaxation matrix elements, which couple populations to coherences and arise from the indirect interaction process $|1\rangle|\alpha\rangle \rightarrow |g\rangle|\beta\rangle \rightarrow |2\rangle|\alpha\rangle$. Although known in several contexts ([6],[8],[9],[10]) such off diagonal relaxation terms are usually ignored. However they are vital to the understanding of coherent trapping in the Type 1 (and Type 3) situation.

We search for a steady state solution for which the left hand sides of (1) are zero and in which the populations σ_{11}, σ_{22} are not both zero. By adding the first and fourth of (1a) we see that if (1a) is satisfied then (1b) is also. Secondly, the coherences must be non zero in the steady state solution, thus establishing that population trapping is *coherent* population trapping (if $\sigma_{12}=\sigma_{21}=0$ then $\Gamma_1\sigma_{11}=0$ and $\Gamma_2\sigma_{22}=0$ thereby implying no trapping).

The existence of a steady state solution requires that the determinent of the symmetric 4 x 4 matrix A in (1a) is zero, leading to the trapping condition for this Type 1 prototype:

$$\delta = \frac{-2(\Gamma_1-\Gamma_2)}{\Gamma_1\Gamma_2}\,\beta_{12}(v_{12}+\alpha_{12}) \tag{4}$$

The internal system coupling required is related to the shifted transition frequency via the various relaxation parameters. The vital role of the off diagonal relaxation parameters can be seen by repeating the analysis with $\alpha_{12}=\beta_{12}=0$. The condition for a non zero steady state solution $\delta = \pm(\Gamma_1+\Gamma_2)\sqrt{4v_{12}^4 - (2v_{12}^2 + \Gamma_1\Gamma_2)^2}/2\Gamma_1\Gamma_2$ is impossible. This shows for example that coherent trapping does not occur for the nearby 2s, 2p states in hydrogen coupled by a static electric field. Condition (4) shows that trapping can occur for the case of two identical two level atoms ($\delta=0$, $\Gamma_1=\Gamma_2$).

The full time dependent solution of (1a) is given in terms of the eigenvalues and left and right eigenvectors of A [5],[11] and the initial conditions on σ_{11} etc. For the case where the trapping condition (4) holds, only the single zero eigenvalue contributes at long times. We find that:

$$\begin{bmatrix}\sigma_{11}(\infty)\\ \sigma_{12}(\infty)\\ \sigma_{21}(\infty)\\ \sigma_{22}(\infty)\end{bmatrix} = \frac{\left\{\Gamma_2\sigma_{11}(0) + \frac{\Gamma_1\Gamma_2}{2\beta_{12}}(\sigma_{12}(0) + \sigma_{21}(0))+\Gamma_1\sigma_{22}(0)\right\}}{(\Gamma_1 + \Gamma_2)^2}\begin{bmatrix}\Gamma_2\\ \frac{\Gamma_1\Gamma_2}{2\beta_{12}}\\ \frac{\Gamma_1\Gamma_2}{2\beta_{12}}\\ \Gamma_1\end{bmatrix} \tag{5}$$

Thus for almost all initial conditions there is trapped population at long times if (4) holds, and the non zero σ_{12}, σ_{21} confirm that coherence is present. For the case where $\sigma_{11}(0)=1$, $\sigma_{12}(0)=\sigma_{21}(0)=\sigma_{22}(0)=0$, the total trapped population is $\Gamma_2/(\Gamma_1+\Gamma_2)$.

The Type 2 prototype can be analysed similarly. The important master equations are:

$$i\dot{\sigma}_{11} = - v_{13}\sigma_{13} + v_{13}\sigma_{31} \tag{6a}$$

$$i\dot{\sigma}_{12} = (\delta_{13} + \delta_{32})\sigma_{12} - v_{23}\sigma_{13} + v_{13}\sigma_{32} \tag{6b}$$

$$i\dot{\sigma}_{13} = - v_{13}\sigma_{11} - v_{23}\sigma_{12} + (\delta_{13} - \tfrac{1}{2}i\Gamma_3)\sigma_{13} + v_{13}\sigma_{33} \quad (6c)$$
$$i\dot{\sigma}_{22} = - v_{23}\sigma_{23} + v_{23}\sigma_{32} \quad (6d)$$
$$i\dot{\sigma}_{32} = v_{13}\sigma_{12} + v_{23}\sigma_{22} + (\delta_{32} - \tfrac{1}{2}i\Gamma_3)\sigma_{32} - v_{23}\sigma_{33} \quad (6e)$$
$$i\dot{\sigma}_{33} = v_{13}\sigma_{13} + v_{23}\sigma_{23} - v_{13}\sigma_{31} - v_{23}\sigma_{32} - i\Gamma_3\sigma_{33} \quad (6f)$$
$$i\dot{\sigma}_{gg} = i\Gamma_3\sigma_{33} \quad (6g)$$

where δ_{ij} are the transition frequencies (shifted by system reservoir interactions), v_{13}, v_{23} are real matrix elements of the internal system interaction V_S and Γ_3 is the decay rate of $|3>$. Only the intermediate state $|3>$ undergoes decay, so no off diagonal relaxation elements are involved.

Again a steady state solution is sought for which the left hand sides of (6) are zero. Adding the right hand sides of (6a), (6d), (6f) we see that trapping implies that $\sigma_{33}=0$, the population of the decaying intermediate state is thus zero. (6g) then is satisfied. (6a), (6d) together with $\sigma_{32}=\sigma_{23}{}^*$, $\sigma_{13}=\sigma_{31}{}^*$ then shows that σ_{13} and σ_{32} must be real and (6f) is automatically satisfied. Equations (6b), (6c), (6e) give on being broken up into real and imaginary parts:

$$(\delta_{13} + \delta_{32})\sigma^r_{12} - v_{23}\sigma_{13} - v_{13}\sigma_{32} = 0 \quad (7a)$$
$$(\delta_{13} + \delta_{32})\sigma^i_{12} = 0 \quad (7b)$$
$$- v_{13}\sigma_{11} - v_{23}\sigma^r_{12} + \delta_{13}\sigma_{13} = 0 \quad (7c)$$
$$- v_{23}\sigma^i_{12} - \tfrac{1}{2}\Gamma_3\sigma_{13} = 0 \quad (7d)$$
$$v_{13}\sigma^r_{12} + v_{23}\sigma_{22} + \delta_{32}\sigma_{32} = 0 \quad (7e)$$
$$v_{13}\sigma^i_{12} - \tfrac{1}{2}\Gamma_3\sigma_{32} = 0 \quad (7f)$$

By eliminating σ_{32}, σ_{13} via (7d), (7f) we then obtain a simple determinental condition leading to the following trapping condition for this Type 2 prototype:

$$\delta_{12} = \delta_{13} + \delta_{32} = 0 \quad (8)$$

Thus the shifted transition frequency between the coupled states must be zero for trapping to occur.

Solving for the steady state solution similarly to before we find that:

$$\begin{bmatrix} \sigma_{11}(\infty) \\ \sigma_{12}(\infty) \\ \sigma_{21}(\infty) \\ \sigma_{22}(\infty) \end{bmatrix} = \frac{\left(v_{23}{}^2\sigma_{11}(0) - v_{13}v_{23}(\sigma_{12}(0)+\sigma_{21}(0)) + v_{13}{}^2\sigma_{22}(0)\right)}{(v^2_{13} + v^2_{23})^2} \begin{bmatrix} v^2_{23} \\ -v_{13}v_{23} \\ -v_{13}v_{23} \\ v^2_{13} \end{bmatrix} \quad (9)$$

$$\sigma_{13}(\infty) = \sigma_{31}(\infty) = \sigma_{23}(\infty) = \sigma_{32}(\infty) = \sigma_{33}(\infty) = 0$$

Thus coherent population trapping occurs at long times for almost all initial conditions, provided (8) holds. For the case where $\sigma_{11}(0)=1$, $\sigma_{12}(0)=\sigma_{21}(0)=\sigma_{22}(0)=0$ the total trapped population is $v^2_{23}/(v^2_{13} + v^2_{23})$.

3. COHERENT TRAPPING IN LAMBDA, LADDER SYSTEMS

We now concentrate on a specific trapping problem, the lambda ladder systems interacting with two lasers a, b, whose frequencies, polarizations and electric field operators are ω_c, $\hat{\varepsilon}_c$, $E_c = i(\hbar\omega_c/2\varepsilon_0 V)^{\frac{1}{2}}\hat{\varepsilon}_c(a_c - a_c^\dagger)$ (c=a, b) (see Table 1 and Fig. 2).

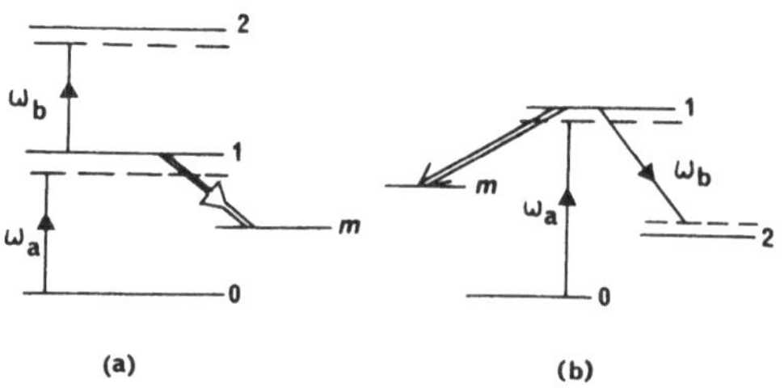

Figure 2. Atomic models considered: (a) ladder, (b) lambda systems showing near-resonant interaction of three-state system $|0\rangle$, $|1\rangle$, $|2\rangle$ with laser modes of frequencies ω_a, ω_b. Spontaneous decay from $|1\rangle$ to a further state $|m\rangle$ is also shown.

From (8) trapping should occur under conditions of two photon resonance:

$$\delta_a + \delta_b = 0 \tag{10}$$

where $\delta_a = \tilde{\omega}_{10} - \omega_a$, $\delta_b = \tilde{\omega}_{21} - \omega_b$ (ladder), $\delta_b = -\tilde{\omega}_{12} + \omega_b$ (lambda) are the laser detunings, given in terms of Lamb shifted atomic transition frequencies. Thus coherent trapping can be achieved here by simply altering laser frequencies. In many lambda cases $\tilde{\omega}_{10} \simeq \tilde{\omega}_{12}$, so that trapping is virtually unaffected by the Doppler effect, hence virtually all atoms in a gas can exhibit trapping effects simultaneously.

Various other effects can also cause coherent trapping and related spectral features to not be fully exhibited. Anything that tends to preferentially destroy coherence between the coupled states, such as elastic collisions would suffice. Also, short experimental time scales (such as transit time effects in the lambda, ladder cases if the atoms and laser fields are confined to narrow intersecting beams) can prevent coherent trapping features from becoming fully developed. However, in lambda, ladder systems, laser band widths are expected to have a major effect [5],[12]: Two photon resonance is crucial for the existence of trapping and finite laser bandwidths imply non two photon resonant Fourier components capable of destroying the trapping.

3.1 Derivation of Optical Bloch equations

Laser fluctuations are treated using a delta-correlated Wiener-Levy phase-diffusion model in which the laser spectrum is lorentzian. The treatment is limited to small detunings from atomic resonance since it severely over-estimates the amount of radiation in the far wings. Our lineshapes are those of near-resonant excitation for which this will not be a limitation. We describe the phase fluctuations by a gaussian Markov stochastic addition to the field hamiltonian,

$$H_S = \hbar\mu_a(t)a_a^{\dagger}a_a + \hbar\mu_b(t)a_b^{\dagger}a_b \tag{11}$$

where $a_i(a_i^{\dagger})$ is the usual annihilation (creation) operator for mode i and the phase velocities $\mu_a = \dot{\Phi}_a$ (frequency fluctuations) are zero-mean gaussian Markov processes, delta-correlated according to $\langle\mu_m(t)\mu_n(t')\rangle = 2\Delta_{mn}\delta(t-t')$, (m,n = a,b). Δ_{mm} is the bandwidth of field m and Δ_{mn}, $(m \neq n)$ is the cross-correlated bandwidth, ignored in most multiphoton theories. If the two lasers are statistically uncorrelated, then we may ignore Δ_{ab}, but if particular care is taken to correlate laser jitter [13], Δ_{ab} will play a crucial role. Following Agarwal [14], we obtain equations of motion for stochastically-averaged density-matrix elements of the atom-laser modes plus spontaneous emission modes system. This leads to an additional term of the form $-\sum_{ab} \Delta_{ab}(n_a - m_a)(n_b - m_b)\, \rho_{in_an_b\{n_\lambda\};jm_am_b\{m_\lambda\}}$ in the Liouville equation for the full density matrix elements $\rho_{in_an_b\{n_\lambda\};jm_am_b\{m_\lambda\}}$ $(i,j = 0,1,2)$.

Master equations for the reduced density matrix elements $\sigma_{in_an_b;jm_am_b}$ for the atom-laser mode system are then obtained using the usual methods [8], [9] based on the Markov, weak coupling, secular and electric dipole interaction approximations. We restrict ourselves to the regime in which the difference between $\tilde{\omega}_{10}$ and $\tilde{\omega}_{12}$ or $\tilde{\omega}_{21}$ is large compared to the decay rates and Rabi frequencies so that laser a excites the $|0\rangle \leftrightarrow |1\rangle$, b the $|1\rangle \leftrightarrow |2\rangle$ transitions only.

The atomic system is initially assumed to be in state $|0\rangle$ and the incident laser light is described by coherent states $|\alpha_c\rangle$ $(c = a,b)$ with $\alpha_c = \bar{n}_c^{\frac{1}{2}} \exp(-i\pi/2)$. The quantity $\bar{n}_c/V$, which determines the laser electric field amplitudes $E_{oc} = 2(\hbar\omega_c \bar{n}_c/2\varepsilon_0 V)^{\frac{1}{2}}$ is finite as $\bar{n}_c \to \infty$ along with the quantization volume V. In this situation the variation of Rabi frequencies ξ_c (Eq.(15)) with n_c can be ignored. In the ladder case terms $\sigma_{1n_an_b;1m_am_b}$ and $\sigma_{2n_a-1n_b;2m_a-1m_b}$, associated with the populating of $\sigma_{0n_an_b;0m_am_b}$ and $\sigma_{1n_a-1n_b;1m_a-1m_b}$ via spontaneous emission, can be approximated (justifiable a posteriori) as $\sigma_{1n_a-1n_b;1m_a-1m_b}$ and $\sigma_{2n_a-1n_b-1;2m_a-1m_b-1}$ respectively. A similar approximation is made in the lambda case.

If the reduced density matrix elements are written as $\sigma_{in_an_b;jn_a+p_a;n_b+p_b}$ and the aforementioned simplifications made, it can easily be shown that the sets of equations involve coefficients which do not depend on n_a and n_b. They break up into coupled sets: in the ladder case, the sets of elements $\sigma_{\alpha\beta}$ in the $(n_an_bp_ap_b)$ set involves $\alpha=0,n_a,n_b;1n_a-1n_b;2n_a-1,n_b-1$, and $\beta=0,n_a+p_a,n_b+p_b;1,n_a+p_a-1,n_b+p_b;\ 2,n_a+p_a-1,n_b+p_b-1$, corresponding to the coupling of sets of the three basis states of the form $|0,n_a,n_b\rangle$, $|1,n_a-1,n_b\rangle$ and $|2,n_a-1,n_b-1\rangle$. In the lambda case, the $(n_an_bp_ap_b)$ set involves $\alpha=0n_an_b;\ 1,n_a-1,n_b;\ 2,n_a-1,n_b+1$ and $\beta=0,n_a+p_a,n_b+p_b;1,n_a+p_a-1,n_b+p_b;2,n_a+p_a-1,n_b+p_b+1$ corresponding to the coupling of sets of the basis states of the form $|0;n_a;n_b\rangle$, $|1,n_a-1,n_b\rangle$ and $|2,n_a-1,n_b+1\rangle$.

These coupled density matrix elements can be written in the form

$$\sigma_{ij}(p_ap_b) \times C(n_an_b) \qquad (12)$$

For example, $\sigma_{0n_an_b;2n_a+p_a-1,n_b+p_b\pm1}=\sigma_{02}(p_ap_b) \times C(n_an_b)$ where the plus (minus) refers to the lambda (ladder) case. The initial conditions along with the choice of $\sigma_{00}(p_ap_b;t=0)=\exp i(p_a \frac{\pi}{2} + p_b \frac{\pi}{2})$ leads to

$$\sigma_{ij}(p_ap_b;t=0)=0 \qquad (i,j \neq 0,0) \qquad (13a)$$

$$C(n_a n_b) = |\langle n_a|\alpha_a\rangle|^2 \times |\langle n_b|\alpha_b\rangle|^2. \tag{13b}$$

The elements $\sigma_{ij}(p_a p_b)$ then satisfy the three-level Bloch equations

$$i\dot{\sigma}_{00} = -[\Omega(p_a p_b) + i\Delta(p_a p_b)]\sigma_{00} + \tfrac{1}{2}i\xi_a\sigma_{01} + \tfrac{1}{2}i\xi_a\sigma_{10} + i\gamma_{10}\sigma_{11} \tag{14a}$$

$$i\dot{\sigma}_{01} = -\tfrac{1}{2}i\xi_a\sigma_{00} - [\Omega(p_a p_b) + i\Delta(p_a-1,p_b) + \delta_a + \tfrac{1}{2}i\gamma_1]\sigma_{01} \pm \tfrac{1}{2}i\xi_b\sigma_{02} + \tfrac{1}{2}i\xi_a\sigma_{11} \tag{14b}$$

$$i\dot{\sigma}_{02} = \mp\tfrac{1}{2}i\xi_b\sigma_{01} - [\Omega(p_a p_b) + i\Delta(p_a-1,p_b\mp 1) + (\delta_a+\delta_b) + \tfrac{1}{2}i\gamma_2]\sigma_{02} + \tfrac{1}{2}i\xi_a\sigma_{12} \tag{14c}$$

$$i\dot{\sigma}_{11} = -\tfrac{1}{2}i\xi_a\sigma_{01} - \tfrac{1}{2}i\xi_a\sigma_{10} - [\Omega(p_a p_b) + i\Delta(p_a p_b) + i\gamma_1]\sigma_{11} \pm \tfrac{1}{2}i\xi_b\sigma_{12} \pm \tfrac{1}{2}i\xi_b\sigma_{21} + i\gamma_{21}\sigma_{22} \tag{14d}$$

$$i\dot{\sigma}_{12} = -\tfrac{1}{2}i\xi_a\sigma_{02} \mp \tfrac{1}{2}i\xi_b\sigma_{11} - [\Omega(p_a p_b) + i\Delta(p_a,p_b\mp 1) + \delta_b + \tfrac{1}{2}i(\gamma_1+\gamma_2)]\sigma_{12} \pm \tfrac{1}{2}i\xi_b\sigma_{22} \tag{14e}$$

$$i\dot{\sigma}_{22} = i\gamma_{12}\sigma_{11} \mp \tfrac{1}{2}i\xi_b\sigma_{12} \mp \tfrac{1}{2}i\xi_b\sigma_{21} - [\Omega(p_a p_b) + i\Delta(p_a p_b) + i\gamma_2]\sigma_{22} \tag{14f}$$

where

$$\Omega(p_a p_b) \equiv p_a\omega_a + p_b\omega_b \tag{15a}$$

$$\Delta(p_a,p_b) \equiv \Delta_{aa}p_a^2 + \Delta_{bb}p_b^2 + 2\Delta_{ab}p_a p_b \tag{15b}$$

$$\xi_a = \langle 1|\underset{\sim}{\mu}\cdot\hat{\underset{\sim}{\varepsilon}}_a|0\rangle\, E_{oa}/\hbar \tag{15c}$$

$$\xi_b = \langle 2|\underset{\sim}{\mu}\cdot\hat{\underset{\sim}{\varepsilon}}_b|1\rangle\, E_{ob}/\hbar \tag{15d}$$

The upper, lowers signs apply in the ladder, lambda cases respectively. The total spontaneous emission decay rates are γ_i $i = 2,1$, with $\gamma_2 = 0$ in the lambda case. Einstein A coefficients for $i \to j$ transitions are denoted γ_{ij}. γ_{12} and γ_{21} are zero in the ladder, lambda cases respectively. Laser bandwidths appear in (14) via $\Delta(p_a p_b)$.

3.2 Interpretation of laser bandwidth factors

Ignoring the atomic system, the density matrix elements for the laser modes system in our Wiener Levy model are given by:

$$\rho_{n_a n_b; n_a+p_a; n_a+p_b} = \exp\left\{[i\Omega(p_a p_b) - \Delta(p_a p_b)]t\right\}\langle n_a|\alpha_a\rangle\langle n_b|\alpha_b\rangle\langle\alpha_a|n_a+p_a\rangle\langle\alpha_b|n_b+p_b\rangle \tag{16}$$

For these to remain finite (and hence also the various multitime quantum correlation functions for the electric field operators) we see that $\Delta(p_a p_b)$ must never be negative.

This leads to restrictions on the bandwidth factors

$$\Delta_{aa} \geq 0 \tag{17a}$$

$$\Delta_{bb} \geq 0 \tag{17b}$$

$$\Delta_{ab}^2 \leq \Delta_{aa}\Delta_{bb} \tag{17c}$$

A calculation of the long time spectrum [8],[15] enables Δ_{aa}, Δ_{bb} to be interpreted as laser bandwidths. The cross spectral bandwidth Δ_{ab} is absent from the long time spectrum, but appears in the expression for the two time quantum correlation function $\langle E_{a\alpha}^-(t_1)E_{b\beta}^+(t_2)\rangle$ [5]. The restriction $\Delta_{ab} \leq \tfrac{1}{2}(\Delta_{aa}+\Delta_{bb})$, which follows from (17) ensures the finiteness of this correlation function.

For two dependent laser fields we would expect $\Delta_{ab}=0$. However for two fields derived from the same source, $\Phi_a = \Phi_b$, and hence $\Delta_{aa} = \Delta_{bb} = \Delta_{ab} = \Delta$. This situation is referred to as (positive) critical cross correlation, (17c) just being satisfied, and has been obtained experimentally [13]. A further interesting possibility, though less easily realized experimentally, could involve deriving two fields from the same source, then reversing the phase fluctuations of one field via reflection from a phase conjugate mirror. For this case $\Phi_a = -\Phi_b$ and hence $\Delta_{aa} = \Delta_{bb} = -\Delta_{ab} = \Delta$ so would be refered to as (negative) critical cross correlation.

3.3 Calculations of atomic state populations

The atomic populations are given by:

$$P_i = \sigma_{ii}(0,0) \tag{18}$$

The Bloch equations (14) have been solved numerically [5],[11] in terms of the eigen values, right and left eigenvectors for the non symmetric 9 x 9 matrix of coefficients associated with the right hand side of (14)(augmented by the conjugates of (14b), (14c), (14e)). Analytic expressions for certain steady state populations have been obtained for zero bandwidths [15].

The results are shown in Figs.(3) and (4). Ladder and lambda cases have been treated along with situations where laser fluctuations are either zero, non zero but uncorrelated and non zero but (positively) critically cross correlated. In Figs.(3), (4) spontaneous decay is ignored except from $|1\rangle$ to a level $|m\rangle$ outside the $|0\rangle$, $|1\rangle$, $|2\rangle$ system, corresponding to a possible nett loss of population.

For the zero laser fluctuation case Figs.3(a), 4(a), 3(d), 4(d) show that coherent population trapping only occurs when the two photon resonance condition (10) is satisfied. For the non zero but uncorrelated case Figs.3(b) and 4(b) show that uncorrelated fluctuations in the two laser fields causes the elimination of coherent population trapping in both lambda and ladder cases.

However, Figs.3(c) and 4(c) show that if the laser fluctuations are positively critically cross correlated then coherent population trapping is fully restored in the lambda case but further destroyed in the ladder case. Since Figs.3(c) and 4(c) also apply to *ladder* and *lambda* systems respectively for the case $\Delta_{aa} = \Delta_{bb} = -\Delta_{ab} = 0.1$ we see also that for *negative* critical cross correlation trapping would be restored the *ladder* case and destroyed in the *lambda* case.

These features can be interpreted in terms of a simple semiclassical two-photon description useful when $|1\rangle$ is far from resonance (δ_a large) and with states $|0\rangle$ and $|2\rangle$ close to two-photon resonance. Under these circumstances, the system is described as an effective two-level system driven by a two-photon Rabi frequency $(\xi_a\xi_b/\delta_a)$. The two-photon coherence ρ_{02} is driven by the two-photon inversion $(\rho_{22}-\rho_{00})$ multiplied by field quantities $E_a^{(-)}(t)E_b^{(\mp)}(t) = E_{oa}E_{ob}\exp\{i(\omega_a-\omega_b)t + i(\Phi_a(t)\pm\Phi_b(t))\}$ where the phases $\Phi_a(t)$ and $\Phi_b(t)$ are stochastic variables (the upper, lower sign refers to the ladder, lambda cases respectively). When stochastic averages are taken, the important quantity is $\langle\langle\exp(i\Phi_a(t)\pm i\Phi_b(t))\rangle\rangle$. If $\Phi_a(t) = \Phi_b(t)$ as in the usual positive critically cross correlated situation, then the two photon (or Raman) coherence is entirely unaffected by laser phase fluctuations in the *lambda* case. On the other hand if $\Phi_a(t) = -\Phi_b(t)$ as in the negatively critically cross correlated case, then phase fluctuations have no effect in the *ladder* case.

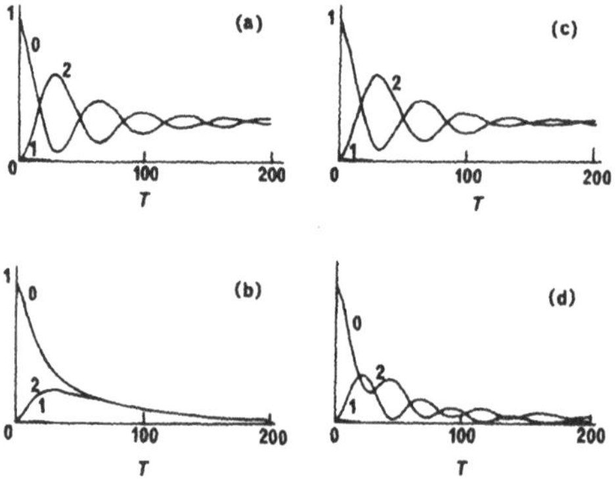

Figure 3. Time-evolution of population in the discrete states $|0\rangle$, $|1\rangle$ and $|2\rangle$ of the lambda system driven by laser fields with Rabi frequencies $\xi_a = \xi_b = 1$, decay rates $\gamma_{10} = \gamma_{12} = \gamma_{21} = 0$, $\gamma_{1m} = 2$. In (a), bandwidths $\Delta_{aa} = \Delta_{bb} = \Delta_{ab} = 0$ and detunings $\delta_a = 5$, $\delta_b = -5$; in (b) $\Delta_{aa} = \Delta_{bb} = 0.1$, $\Delta_{ab} = 0$, $\delta_a = 5 = -\delta_b$; (c) $\Delta_{aa} = \Delta_{bb} = \Delta_{ab} = 0.1$, $\delta_a = 5 = -\delta_b$; (d) $\Delta_{ab} = \Delta_{aa} = \Delta_{bb} = 0$, $\delta_a = 5.1$, $\delta_b = -5$.

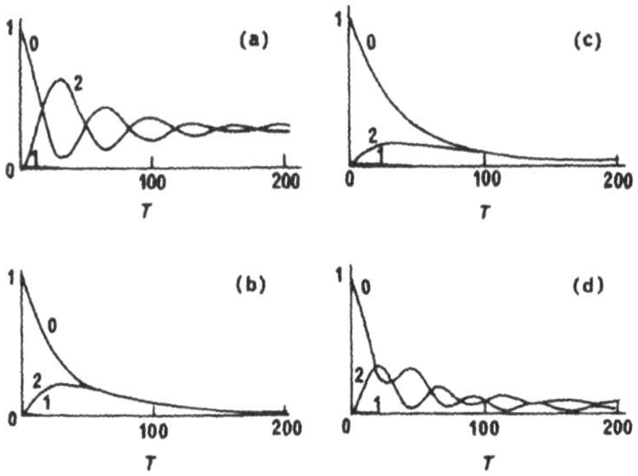

Figure 4. Time evolution of populations for ladder system. Parameters as in Fig. 3.

3.4 Calculations of fluorescence spectra

The total intensity of fluorescent light emitted from $|1\rangle$ is determined by the population $P_1(t)$.

Results of calculations on long time fluorescence spectra as a function of detuning δ_b are shown for the lambda case in Figs.(5) and (6), where the detunings δ_a are

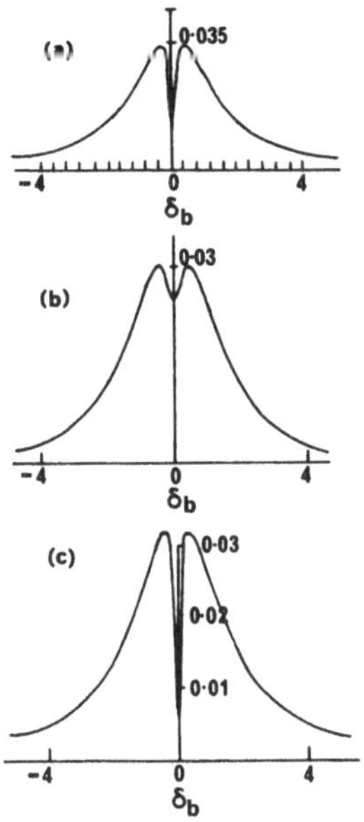

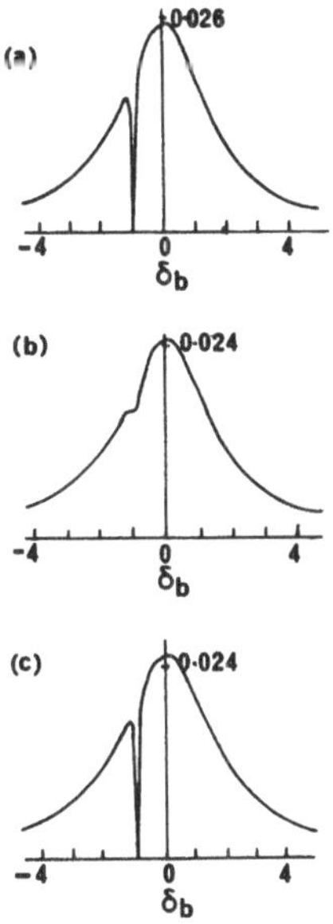

Figure 5. Absorption probability $P_1(t)$ as a function of detuning δ_b for a lambda system with Rabi frequencies $\xi_a = \xi_b = 0.4$, time t = 200, detuning $\delta_a = 0$ for (a) bandwidths $\Delta_{aa} = \Delta_{bb} = \Delta_{ab} = 0$; (b) bandwidths $\Delta_{aa} = \Delta_{bb} = 0.1$, $\Delta_{ab} = 0$; (c) critically cross-correlated fields $\Delta_{aa} = \Delta_{bb} = \Delta_{ab} = 0.1$, with $\gamma_{10} = \gamma_{12} = 1$.

Figure 6. As in figure 5, but with detuning $\delta_a = 1$.

zero, non zero respectively. The only spontaneous emission processes included are $|1\rangle \to |0\rangle$, $|1\rangle \to |2\rangle$ so no nett loss of population is possible. The Rabi frequencies are chosen to be small compared to the decay widths.

Figs.5(a), 6(a) show that a narrow deep minimum occurs (coherence hole) in the fluorescence curve when the condition for two photon resonance is satisfied, corresponding to the formation of coherently trapped population in $|0\rangle$, $|2\rangle$ and where the population of $|1\rangle$ becomes zero.

Figs.5(b), 6(b) show that the narrow minimum is washed out in the case of non zero but uncorrelated laser fluctuations, corresponding to the destruction of coherent population trapping. Figs.5(c), 6(c) show the restoration of the narrow spectral feature for the case of (positive) critically cross correlated fluctuations.

4. EXPERIMENTS ON COHERENT TRAPPING IN LAMBDA SYSTEMS

Two fluorescence spectrum experiments on sodium by Gray et al.[15] and by Ezekiel et al.[13][16][17] are shown in schematic outline in Figs.(7), (8). A more complete survey of experiments is given in [5].

The experiment of Gray et al.(Fig.(7)) is done using two lasers with non zero but uncorrelated fluctuations (Δ_{aa}, Δ_{bb} are $\lesssim$ 1Mhz). Though small compared to the decay

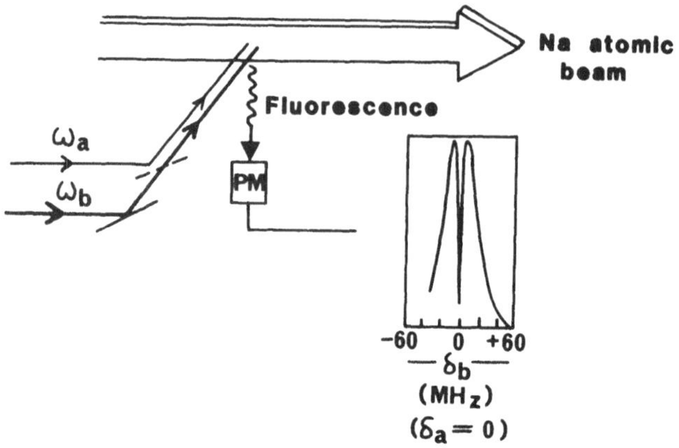

Figure 7. Fluorescence spectrum in sodium lambda system. $|0\rangle \to 3^2S_{\frac{1}{2}}(F=1)$, $|2\rangle \to 3^2S_{\frac{1}{2}}(F=2)$, $|1\rangle \to 3^2P_{\frac{1}{2}}(F=2)$. Uncorrelated laser fields a,b used ($\Delta_{aa} \doteqdot \Delta_{bb} \lesssim 1$Mhz, $\Delta_{ab} = 0$). Schematic outline from Gray et al.[15].

rates the Rabi frequencies are sufficiently large that the theoretical width of the coherence hole ($\sim(\xi_a^2 + \xi_b^2)/2(\gamma_{10} + \gamma_{12})$) is large ($\sim$ 9 Mhz) compared to the laser bandwidths. The experimental results agree fairly well with theoretical predictions [15] ignoring laser fluctuations.

A much narrower coherence hole is seen in the experiments of Ezekiel et al. with less intense fields. The two laser fields satisfy (positive) critical cross correlation conditions, being derived from the same source using accousto-optic modulation. For a single atom-laser beams interaction region, the coherence hole width is reduced [16] to $\sim$ 200 khz, corresponding to the interaction time limit associated with the atomic beam crossing the laser fields, even though the original source has a normal bandwidth. An even narrow spectral feature ($\sim$ 1.3 khz)(shown in Fig.(8)) can be obtained using the Ramsey method [13] involving two atom-laser beams interaction regions.

ACKNOWLEDGEMENTS

The authors wish to thank co-workers P. Radmore and P.E. Coleman. The numerical solution work of C. Penman is also gratefully acknowledged. One of us (P.L.K.) is grateful to the SERC for an Advanced Fellowship and the Nuffield Foundation for their support.

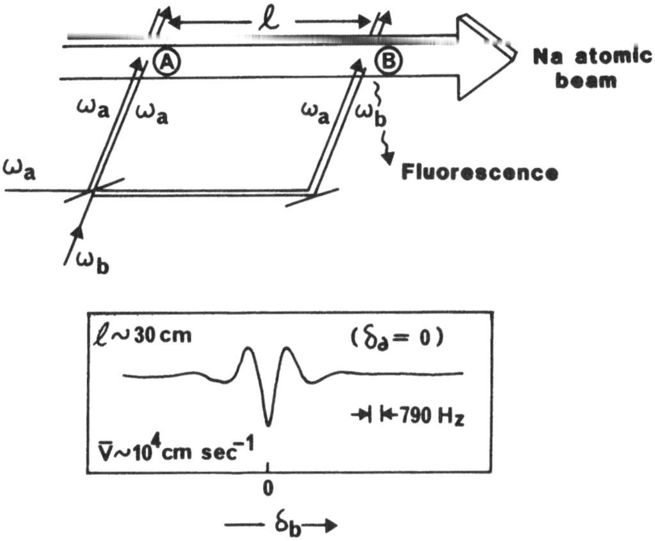

Figure 8. Ramsey fringes in fluorescence spectrum in sodium lambda system. States as in Fig. 7. Laser fields a,b obtained from same source using accousto-optic modulator. Critically cross correlated fields ($\Delta_{aa} \doteqdot \Delta_{bb} \doteqdot \Delta_{ab}$). Schematic outline from Ezekiel et al.[13].

REFERENCES

[1] Milonni P W and Knight P L 1974 Phys Rev A10 1096

[2] Feynman R P, Leighton R B and Sands M 1965 Lectures on Physics (New York: Addison-Wesley) Vol 3, Sect 11.5

[3] Coleman P E and Knight P L 1982 J. Phys.B: At. Mol. Phys 15 L235

[4] Radmore P M and Knight P L 1982 J. Phys.B: At. Mol. Phys 15 561

[5] Dalton B J and Knight P L 1982 J. Phys.B: At. Mol. Phys 15 3997

[6] Cardimona D A, Raymer M G and Stroud C R Jr 1982 J. Phys.B: At. Mol. Phys 15 55

[7] Knight P L 1982 Nature 297 16 6 May

[8] Cohen-Tannoudji C 1977 Frontiers in Laser Spectroscopy,Les Houches Session 27 Ed R Balian, S Haroche and S Liberman (Amsterdam: North-Holland)

[9] Dalton B J 1982 J. Phys.A: Math. Gen. 15 2157, 2177

[10] Knight P L 1979 Opt. Commun. 31 148

[11] Whitley R M and Stroud C R Jr 1976 Phys. Rev. A14 1498

[12] Dalton B J and Knight P L 1982 Opt. Commun. 42 411

[13] Thomas J E, Hemmer P R, Ezekiel S, Leiby C C Jr, Picard R H and Willis C R 1982 Phys. Rev. Lett. 48 867

[14] Agarwal G S 1978 Phys. Rev.A 18 1490

[15] Gray H R, Whitley R M and Stroud C R Jr 1978 Opt. Lett. 3 218

[16] Tench R E, Peuse R W, Hemmer P R, Thomas J E, Ezekiel S, Leiby C C Jr, Picard R H and Willis C R 1981 J. Physique Coll. 42 C8 45

[17] Thomas J E, Ezekiel S, Leiby C C Jr, Picard R H and Willis C R 1981 Opt. Lett. 6 298

AN INTRODUCTION TO THE PHYSICS OF LASER FUSION

by

B. LUTHER-DAVIES
Laser Physics Laboratory
Department of Engineering Physics
Research School of Physical Sciences
The Australian National University
P.O. Box 4
Canberra, A.C.T. 2601
Australia

1. Introduction

The goal of harnessing fusion energy for power production has led to increasingly complex means of generating plasmas whose conditions approach those required for nuclear fusion. In the last decade the concept of compression and heating of inertially-confined plasmas by powerful laser beams has been persued vigorously and with considerable success as an alternative to magnetic-confinement fusion. The basic concept of inertial-confinement fusion (ICF), is straightforward. A spherical pellet containing a Deuterium Tritium fuel mixture is irradiated over its surface by powerful beams from a laser. The pellet surface becomes transformed by absorption of energy into a high temperature plasma which expands rapidly into the surrounding vacuum. A reactive force to this expansion arises which acts to compress the remaining fuel to achieve the densities above one thousand times the initial liquid fuel density and simultaneously, as the compression stagnates, to raise the core temperature to 10^8 ^oK. These are the conditions required to ignite and sustain a thermonuclear burn which proceeds for the short time ($\simeq$ 100psec) during which the pellet is constrained at high density by its own inertia.

The idea and method of compressing the fuel to produce the conditions for fusion, first introduced by Nuckolls [1], led to an upsurge of research in the field since it allowed small pellets ($\simeq$ 1mm diameter) to be used whose energy yield and driver laser requirements fell within the range of conceivable technology. Already lasers capable of delivering powers and energies of 100TW and 100kJ respectively are being constructed for use in the 1980's and these should allow the scientific feasibility of laser-driven ICF to be demonstrated through fuel ignition and possibly significant thermo-nuclear burn.

In this talk I will review some of the physical phenomena which are encountered in laser fusion experiments. In the available time it will be possible to provide only a brief outline and for further information the reader is referred to reviews by Hughes [2] and Max and Ahlstrom [3,4].

In treating the complex physical problems involved in the irradiation, compression and ignition of a laser fusion target, it is convenient to split the pellet into two regions; the outer underdense plasma where the density is less than the critical density for the incoming laser radiation, and the inner overdense core. (The critical density, n_{cr}, corresponds to the density for which the laser frequency equals the local plasma frequency, and represents the classical reflection point in the plasma. As $n \to n_{cr}$, the real part of the plasma refractive index $\eta \to o$, and, hence, the wave can only propagate for $n < n_{cr}$). It is in the underdense region that the incoming laser light is absorbed and direct plasma heating occurs. The processes of ablation and compression are then driven by conduction of the absorbed energy into the high density core. For most of this talk I will concentrate on the physics of the underdense region restricting my comments on processes in the overdense plasma to a very brief description at the end.

2. Characteristics of the Underdense Region

We are interested in the plasma temperature, velocity and density scale length in the underdense plasma. Firstly, by equating the rate of absorption of laser energy near the critical surface with the rate of heat conduction from that region, a rough idea of the plasma temperature can be made. The appropriate expression is:

$$T_e(\text{keV}) \simeq 2.7 \times 10^{-10}\left[\frac{1}{f}\,\{\lambda_L(\mu\text{m})\}^2\, I_{abs}\right]^{2/3} \quad (1)$$

where I_{abs} is the absorbed laser flux density, T_e is the temperature, λ_L the laser wavelength, and the factor f, is the "flux limit" introduced to indicate that in practice the heat flow is reduced below classical values. Using f=0.03, values in agreement with those necessary to explain experimental data, temperatures are in the keV region for absorbed intensities around 10^{14} W/cm^2 and visible wavelengths.

Secondly, the expansion velocity is approximately equal to the local sound speed at the critical density given by the expression:

$$V \simeq C_s = (ZT_e/M_i)^{\frac{1}{2}} = 3 \times 10^7 \left(\frac{Z}{A}\right)^{\frac{1}{2}} T_e(\text{keV})^{\frac{1}{2}} \text{ cm sec}^{-1} \quad (2)$$

which has values around 10^7 - 10^8 cm/sec.

Thirdly, the density scale length of the plasma can be estimated for short laser pulses as approximately equal to the product of the laser pulse duration and the sound speed (a few microns for picosecond laser pulses). For long laser pulses, however, this is no longer appropriate and the scale lengths are limited by the laser and target geometry and the divergence of the plasma flow. In that case the scale lengths are approximately equal to the laser beam diameter for planar geometry or the radius of the pellet in spherical geometry (100μm-1mm for typical laser fusion targets). Since, as will be seen later, the density scale length is a crucial factor in determining the exact interaction processes that will occur, it is worth noting that the trend in fusion experiments is away from the short scale lengths encountered in the so-called exploding pusher regime where short picosecond laser pulses were used, towards the long scale length plasmas obtained using large targets and nanosecond duration pulses.

It is most important to understand the mechanisms of absorption of the laser light by the plasma. Efficient absorption is crucial since the overall efficiency of a power plant is directly proportional to the fraction of the laser light absorbed by the target. In this context, mechanisms that can decouple the incident light by reflection and scattering must also be investigated.

There are thought to be three principal absorption mechanisms in laser fusion plasmas, those being inverse Bremsstrahlung, resonanance absorption and absorption due to ion-acoustic turbulence.

Inverse Bremsstrahlung occurs due to collisions between electrons and ions during which the oscillation energy of the electrons in the electromagnetic wave is converted into random (thermal) motion, causing the wave to be damped. The coefficient of absorption, for weak fields and a Maxwellian electron velocity distribution can be shown to be given by:

$$K' = (2\pi)^{\frac{1}{2}} \left(\frac{16\pi}{3}\right) \frac{Zn^2 e^6 \ln\Lambda}{c(m_e T_e)^{\frac{3}{2}} \omega_L^2 (1-n/n_{cr})^{\frac{1}{2}}} \qquad (3)$$

where $\ln\Lambda$ is the coulomb logarithim, c is the velocity of light, ω_L is the laser frequency, m_e the electron mass, z the ionic charge, and n the electron density. Equation 3 shows that the absorption increases with plasma density, average ionic charge, z, and decreases with increasing plasma temperature. In general, long scale length plasmas enhance the absorption by this mechanism.

As the incident field strength increases the oscillation energy of the electrons, v_{os}, exceeds the electron thermal energy, v_{th}, ($v_{os}/v_{th} = [eE_L/m_e\omega_L]/(kTe/m_e)^{\frac{1}{2}} > 1$) and the effective electron-ion collision frequency decreases. This causes the efficiency of inverse Bremsstrahlung absorption to decrease at high laser irradiances. A further non-linear reduction in the efficiency of inverse Bremsstrahlung absorption has been discussed by Langdon [5] and arises due to the formation of a non-Maxwellian electron distribution at high laser intensities due to the electron-electron thermalisation rate becoming smaller than the absorption rate.

The second absorption mechanism, resonance absorption, occurs when an electromagnetic wave is obliquely incident at p-polarisation on a plasma density gradient with density scale length L. The situation is illustrated in figure 1. The incoming beam is refracted as it passes through the plasma and can only penetrate to its turning point density, $n_t = n_{cr}\cos^2(\Theta)$, where Θ is the angle of incidence of the beam. At the point of reflection there exists a component of the electric field

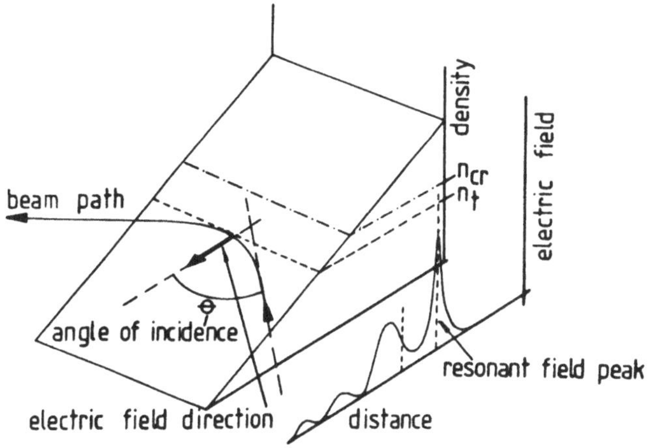

Fig. 1 Schematic showing the configuration in which a p-polarised electromagnetic wave incident upon a density gradient undergoes resonance absorption.

of the laser beam down the plasma density gradient. This electric field can tunnel to the critical surface an excite resonant longitudinal plasma oscillations (Langmuir waves) at that point. Damping of these resonant waves can result in absorption of energy from the laser beam. In the case of normal incidence or s-polarisation there is no longitudinal component of the E-M wave and hence no resonance absorption (in an unmagnetised plasma). The level of resonance absorption is sensitive to the angle of the incoming laser radiation and an optimum angle for resonance absorption exists. For angles less than this optimum the field component that drives the Langmuir waves is reduced whilst for angles greater than the optimum the turning point density moves further away from the critical density and the coupling between the driving field and the resonance region becomes weaker. The absorption can be characterised by the resonance function [6], $\phi(\tau)$, where $\tau = (L\omega_{L}/c)^{1/3}\mathrm{Sin}\Theta$, which has a maximum for $\tau = 0.8$ (for relatively weak plasma density gradients such that $k_0L>1$). The level of absorption may be as high as 50% in short scale length plasmas where the radiation is non-normally incident, and is independent on the strength of the damping of the Langmuir wave.

An important feature of resonance absorption is that it produces high energy electrons with an approximately Maxwellian distribution but with a temperature much higher than the thermal plasma temperature. Friedberg [7] has given an approximate relation for this temperature as:

$$T_h \simeq (m_e\omega_L I_{abs}L/n_{cr})^{\frac{1}{2}} \tag{4}$$

which for typical laser parameters of $I_{abs} = 10^{15}\ \mathrm{W/cm^2}$, $L \simeq \lambda_L$, gives $T_h \simeq 20\mathrm{keV}$.

The number of these hot electrons n_h, produced can be estimated using similar arguments to those for eqn. 1 but in this case the flux limiter, f, is that appropriate to the hot electrons which may be very much larger than that for the thermal distribution. We obtain:

$$n_h/n_{cr} \approx \frac{0.4}{f} I_{abs} \left(\frac{V_{osc}}{V_{th}} \right) \tag{5}$$

where V_{os} is the oscillation velocity of the electrons in the field of the E-M wave. The condition $n_h/n_{cr} \approx 1$ is particularly easy to satisfy for long wavelength lasers since at constant irradiance the electron oscillation energy is proportional to wavelength squared, and hence we can expect the low density region of plasmas generated by long wavelength lasers to be dominated by superthermal electrons. There are two deleterious properties of superthermal electrons in the fusion context. Firstly, their production leads to high velocity, low mass plasma expansion which reduces the momentum available for compression, and secondly, hot electrons can penetrate and pre-heat and target core making subsequent compression more difficult.

The third mechanism arises in the presence of ion-acoustic turbulence which can be shown to effect a number of absorption and scattering mechanisms in the plasmas. As an example, consider inverse Bremsstrahlung absorption. The absorption coefficient of eqn. 3 was calculated on the assumption that there existed a random distribution of background ions with which a Maxwellian electron distribution interacts. In the presence of ion turbulence an element of coherence is introduced into the ion distribution which can lead to increased damping of the laser light. In the presence of such coherent ion turbulence the energy damping rate $(\nu_E)_{EFF}$ becomes [8]:

$$(\nu_E)_{EFF} = \frac{\omega_L}{2} \sum_k \left(\frac{\delta n_i(k)}{n_{cr}} \right)^2 I_m \left| \frac{1}{\varepsilon(k,\omega_L)} \right| (\hat{k}\cdot\hat{E}_L)^2 \tag{6}$$

where $\hat{E}_L$ is the direction of the laser electric field, k is the k-vector of the ion waves, where $\delta n_i(k)/n_i$ is the ion fluctuation spectrum. Values of $(\nu_E)_{EFF}$ are greater than those for an uncorrelated ion background.

Although ion-acoustic turbulence can increase inverse Bremsstrahlung absorption, recent theoretical work from our group has shown that the level of resonance absorption is reduced by the presence of such turbulence isolated in the vicinity of the critical density surface [9]. In that case there arises an anomalous reflectivity for the plasma given by the expression:

$$R_{an} = (1 - R)Q/(1 + Q) \tag{7}$$

where R is the normal plasma reflectivity, and Q is given by the expression:

$$Q = \tfrac{1}{2} \frac{L}{\lambda_L} \left(\frac{\Delta}{\Delta z} \right) \left(\frac{\Delta n}{n_c} \right)^2 \left(\frac{\lambda}{\lambda_D \cos\Theta_0} \right)^2 \tag{8}$$

where Δn is the fluctuation amplitude, with Δ the width of the region containing the fluctuations, and Δz is the peak width of the resonant electric field, and Θ_0 is the angle of incidence of the incoming p-polarised beam and λ_D is the Debye length. The anomalous reflectivity can be quite large (≃ 50%), substantially reducing the efficiency of resonance absorption.

It should be clear from these two points that evaluating the effect of ion turbulence on the overall plasma absorption is a complex matter and can only be made through a self-consistent treatment of all possible mechanisms including the influence of the hydrodynamic motion.

Obviously, for either of these mechanisms to operate there must be some well-established reasons to expect the plasmas to be turbulent. Recently experimental evidence has been presented that demonstrates the presence of such modulations [10]. Theoretically, the origin for the fluctuations and their amplitude is less well defined although several mechanisms have been proposed including heat-flow driven ion acoustic instabilities [11] or ion-ion streaming [12]. Further discussion of this point can be found in the references.

To summarise the results on absorption, we note that resonance absorption produces hot electrons which are undesirable in the fusion context. It is excited efficiently in short scale length plasmas where the range of possible angles over which the mechanism can operate becomes large, and when the incident radiation strikes the plasma density gradient at non-normal incidence due either to the focussing behaviour of the laser beam or the target geometry. Inverse Bremsstrahlung absorption, on the other hand, occurs in long scale length plasmas particularly at low to moderate intensities. It can also be shown that inverse Bremsstrahlung becomes more efficient as the laser wavelength decreases whilst resonance absorption becomes more prevalent for long wavelengths. Both mechanisms can be effected by the presence of ion-acoustic turbulence, although evaluating its effect on the overall absorption is a complex matter.

3. The Ponderomotive Force in Laser-Plasma Interactions

So far we have assumed that the hydro-dynamic motion of the plasma is uneffected by the absorption and propagation of the incoming (and reflected) electromagnetic waves. This approximation is in fact a poor one in many situations since, for example, at high laser intensities the radiation pressure of the incoming laser beam can be large in comparison with the plasma pressure and, hence, we can expect the plasma flow to be modified by that pressure. A basic condition for the plasma dynamics to be modified by the radiation pressure is that the ratio $V_{os}/V_{th}>1$, which is easy to achieve for long wavelength lasers at moderate to high intensities.

The effect on the plasma dynamics is usually evaluated in terms of the ponderomotive force in the plasma and has been extensively treated in the literature [13]. Following a simple analysis as given by Chen [14], the force on an electron in an E-M wave can be written as:

$$f = \frac{-e^2}{m_e \omega_L^2} \frac{1}{2} \left(\underline{E} \cdot \underline{\nabla}\, \underline{E} + \underline{E} \times \underline{\nabla} \times \underline{E} \right) \tag{9}$$

Here, the Ex∇ x E term gives rise to the drift of the electrons in the direction of the wave, whilst the E.∇E term results in acceleration down the electric field gradient. The ponderomotive force, F_{nL}, is the force per unit volume, and can be written:

$$F_{nL} = - \frac{\omega_p^2}{\omega_L^2} \frac{\nabla \langle E^2 \rangle}{8\pi} \qquad (11)$$

after averaging over one optical cycle, with $\omega_p^2 = 4\pi n e^2/m_e$.

The importance of the ponderomotive force in laser-plasma interactions will be illustrated by three examples, firstly, density profile modification, secondly filamentation or self-focussing and thirdly the excitation of parametric instabilities.

When a laser beam is incident upon a plasma profile modification occurs because the spatial distribution of the plasma dielectric constant gives rise to electric field gradients of the incident wave in the direction of the density gradient. With reference to figure 2a, these electric field gradients act to modify the plasma flow, particularly in the vicinity of the critical density surface where $\eta \to o$, and this changes the density profile as shown in figure 2b. The change in the density profile in turn effects the electric field distribution and thus the ponderomotive force. To predict the plasma density profile in steady conditions this coupling must be taken into account.

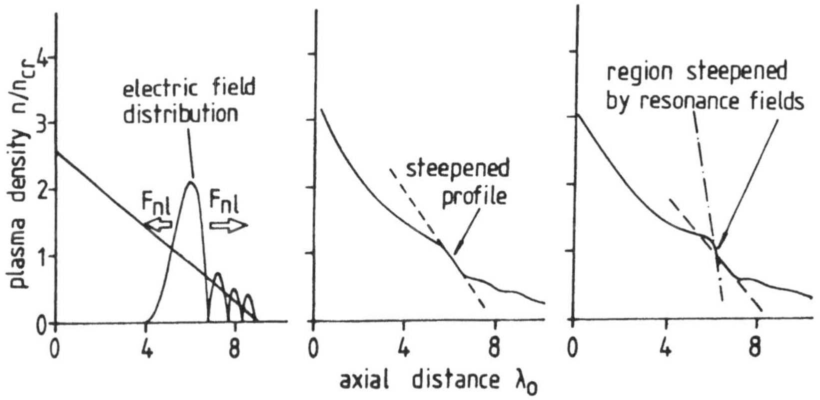

Fig. 2

2(a) Electromagnetic field distribution and induced ponderomotive forces on a linear density ramp, 2(b) the resulting modified density profile for a normally incident radiation, and 2(c) the resulting density profile for non-normal incidence in the presence of resonance absorption.

Direct evidence of such profile modification has been obtained from experiments where interferometry has been used to measure the plasma density distribution in the vicinity of the critical density surface [15]. An important consequence of density profile modification is that the density gradients become very steep near the critical surface, typically of the order of the wavelength of the laser light, and hence the volume of plasma with its density close to critical is very small. Recalling the results above, this means that inverse Bremsstrahlung absorption becomes insignificant, whilst the range of angles over which resonance absorption is possible becomes large.

Several theories have been developed to explain the experimental observations of profile modification involving, in general, a steady balance of the ponderomotive and thermo-kinetic forces at the critical surface. The problem is complicated in the presence of resonance absorption since the local fields associated with the electrostatic wave also give rise to ponderomotive forces. The resultant density profile in this situation is shown in figure 2c and differs from the case of normal incidence (fig. 2b), by the presence of an additional steepened region due to the resonance fields. Even at nominally normal incidence resonance absorption can be important in determining the steepened density profile if the critical surface is rippled due to the presence of ion acoustic turbulence. Combining all these factors has enabled us to recently develop a model of profile modification which is consistent with experimental data obtained at our laboratory [16].

The ponderomotive force can also give rise to self-focussing or filamentation of the laser beam as it propagates up the plasma density gradient. Consider a gaussian beam propagating through an initially uniform plasma. The radial variation in the beam intensity results in a radial component of the ponderomotive force which expels plasma from the beam centre. As the plasma density there drops, the refractive index increases (recall $\eta = (1-n/n_{cr})^{\frac{1}{2}}$), and this results in the formation of a positive lens-like refractive index profile and focussing of the incoming beam. This mechanism can, similarly, result in initial spatial intensity modulations on a laser beam becoming enhanced in a plasma. If, in this manner, the laser intensity in the plasma becomes very non-uniform this could have serious consequences in the fusion context by modifying the absorption process, increasing hot electron generation and giving rise to local small scale magnetic fields. The stability of such self-focussed filaments is open to some question since the interaction becomes highly non-linear in the vicinity of the focus. The process will however, become of increasing importance in long scale length plasmas irradiated at high intensities.

A third area where the ponderomotive force has a marked effect is through the excitation of parametric instabilities. For an introduction to this topic the reader is referred to Chen [14]. The instabilities can be grouped as either of the electrostatic or backscatter type, the former including the two plasmon decay and the parametric decay whilst the latter includes stimulated Brillouin and Raman scattering. Such processes result in the decay of the incident wave into electrostatic (e), ion acoustic (i) or additional electromagnetic (t) waves as indicated in the following relations:

Two Plasmon Decay	$t \rightarrow e + e$	
Parametric Decay	$t \rightarrow e + i$	(12)
Raman Scattering (SRS)	$t \rightarrow t + e$	
Brillouin Scattering (SBS)	$t \rightarrow t + i$	

To illustrate the role of the ponderomotive force we shall consider the example of stimulated Brillouin backscatter. In this case the laser (pump) wave, E_t, enters the plasma where it interacts with plasma density fluctuations (noise) to produce a weak reflected wave, E_{Rt}. The electric fields of the two waves interact giving rise to a component of the ponderomotive force proportional to $E_t \times E_{Rt}$, which has the correct phase to increase the initial plasma density fluctuations, which, in turn, increases the amplitude of the reflected light. This feedback process results in exponential growth of both the perturbation and the reflected light provided the appropriate energy and momentum conservation conditions are satisfied.

Obviously the fluctuation level cannot grow indefinitely, and the study of the saturation behaviour of SBS has been the subject of much theoretical work. Experiments have, however, shown large levels of SBS in plasmas with long density scale lengths (where ample distance exists for the backscattered wave to grow). One possible way of stabilising SBS is through competition between SBS and inverse Bremsstrahlung absorption, since in that case the distance over which growth of the backscatter can occur is limited not by the plasma scale length but by the inverse Bremsstrahlung absorption length. This would again suggest that short wavelength lasers should be used as fusion drivers to avoid the problems of SBS.

A further damaging effect of some parametric instabilities can arise due to the fact that they excite plasma waves which, in a manner similar to that of resonance absorption, can cause laser energy to be absorbed into high energy superthermal electrons. As an example, we consider the Raman instability which occurs for plasma densities less than the quarter critical and produces a backscattered wave with a frequency close to the sub-harmonic of the laser light.

We have recently performed measurements of SRS in plasmas irradiated by short 20psec duration pulses from a Neodymium laser and recorded both the Raman spectra and X-ray emission from the plasmas over a very wide range of laser intensities. It was found that above the Raman threshold an energetic tail could be identified in the X-ray emission spectra equivalent to the presence of electrons with a temperature in excess of 80keV. Such temperatures are greater than those normally obtained via resonance absorption and are, therefore, potentially of greater importance in causing core pre-heat in the fusion context. As for SBS, SRS is expected to be important in long scale length plasmas at moderate laser intensities and recent experiments from the Livermore Laboratories in the USA have measured Raman backscatter of around 10% in such plasmas emphasing that the potential damaging effects of SRS [17].

4. Problems in the Overdense Region

We have concentrated above on some of the physical processes which are being studied to understand the interaction of intense laser light in the underdense region of a laser fusion target. As is evident the subject contains a rich variety of complex interacting physical phenomena. The topics treated above form only the very briefest summary of the physics involved and of necessity some major areas have been overlooked.

In the overdense region, the laser radiation is no longer present, and the physical problems involve the mechanisms of heat transport from the absorption region to the ablation front, the effect of fast electron pre-heat of the pellet core, the effect of fluid instabilities (Rayleigh-Taylor growth) on the symmetry of the implosion, and the implosion efficiency and densities and temperatures that are achieved.

As already indicated above, there are advantages of using short wavelength lasers in simplifying the physics of the underdense plasma, and, broadly speaking, this also applies to the overdense region through an increase in the ablation pressure and mass ablation rate, a reduction in transport inhibition (recall the factor, f, in equation 1), and an overall increase in the hydrodynamic efficiency. The only area where short wavelengths appear to be at a disadvantage is that concerning the implosion symmetry. Here it is usual to assume that the laser uniformity on the target surface need not be better than about 10% since lateral heat flow reduces this non-uniformity to a tolerable level at the ablation surface. As the wavelength is reduced, however, this thermal smoothing is also reduced and, hence, to obtain good implosion symmetry implies a smaller intensity variation is needed at the pellet surface. Although this might be difficult to achieve in a reactor geometry, where the need to shield the reaction region by a neutron absorbing blanket reduces access to the pellet, in small scale experiments it has recently been demonstrated with the OMEGA laser at the University of Rochester that uniformities of a few percent can be achieved using multiple overlapping laser beams.

Another method of overcoming the problem of radiation uniformity has recently been described in which x-rays are used to drive the implosion. Unfortunately, few details of this work are yet available because of its security classification, but the idea is to suspend the pellet inside a high-Z outer shell which is then irradiated by a short wavelength laser. In these conditions very high conversion efficiency can be obtained from laser light to x-rays and it is, therefore, the x-ray emission which drives the implosion. Since the x-rays are emitted isotropically, geometrical factors can be used to increase the irradiation uniformity at the ablation front. Using such targets core densities of about 200 times liquid D-T have been obtained.

In conclusion, there is still a long way to go before the complex physics of the laser fusion process is fully understood. In recent years understanding has increased to the stage where there is confidence that experiments to demonstrate the scientific feasibility of the scheme can be completed in the next decade. The main thrust of this programme will be made within the major fusion laboratories in the USA, Japan and Europe where funding for this research maintains a high level. Lasers such as the Neodymium glass laser NOVA at the LLNL which will deliver 100kJ pulses at powers in excess of 100TW at infra-red and visible wavelengths should ignition to be demonstrated in the 1980's.

It should be remembered, however, that although such systems are impressive in the technology they employ and their sheer size, they are incapable of operating at the efficiencies required for a reactor. It will take massive advances in the development of new high energy lasers such as the Krypton Fluoride or free-electron laser before a reactor becomes a reality. In that time it may well occur that alternative drivers such as accelerators delivering the energy to the pellet in the form of heavy ion beams may well prove more practical. Nevertheless, it will be the use of lasers through which the basic physics of ICF will continue to be addressed in the next decade.

References.

1. Nuckolls, J., Wood, L., Thiessen, A., Zimmerman, G., Nature 239, 139 (1972).
2. Hughes, T.P., Plasmas and Laser Light: Adam Hilger Press, U.K. (1975).
3. Max, C.E., The Physics of Laser Fusion Vol 1; Lawrence Livermore Laboratories Report UCRL-53107 (1981).
4. Ahlstrom, H.G., The Physics of Laser Fusion Vol 2, Lawrence Livermore Laboratories Report UCRL-53106 (1982).
5. Langdon A.B., Phys. Rev. Letts. 44, 575 (1980).
6. Ginzberg, V.L., Propagation of Electromagnetic Waves in Plasmas, translated by J.B. Sykes and R.J. Taylor, Pergamon Press, New York (1964).
7. Friedberg, J.P., Mitchell, R.W., Morse, R.L., Rudsinski, L.J., Phys. Rev. Letts. 28, 795 (1972).
8. Faehl, R., Kruer, W.L., Phys. Fluids 20, 55 (1977), Manheimer, W., Phys. Fluids 20, 265 (1977), Manheimer, W., Colombant, D., Ripin, B., Phys. Rev. Letts. 38, 1135 (1977), Manheimer, W., Colombant, D., Phys. Fluids 21, 1818 (1978).
9. Dragila, R., to be published, Phys. of Fluids, Laser Physics Laboratory technical report LPL 8207 (1982).
10. Gray, D.R., Kilkenny, J.D., White, M.S., Blyth, P., Hull, D., Phys. Rev. Letts. 39, 1270 (1977).
    Gray, D.R., Kilkenny, J.K., Plasma Physics 22, 81 (1980).
11. Colombant, D.G., Manheimer, W.M., Phys. Fluids 23, 2512 (1980) also ref. 8.
12. Forslund, D.W., Kindel, J.M., Lee, K., Lindman, E.L., Bull. Am. Phys. Soc. 20, 1377 (1975).
13. See for example Hora, H., Nonlinear Plasma Dynamics at Laser Irradiation, Springer-Verlag, Berlin, (1979).
14. Chen, F.F., Laser Interaction and Related Plasma Plenomena, Volume 3A,

Plenum Press, New York, p. 291-313 (1974).

15. Attwood, D.T., Sweeney, D.W., Averback, J.M., Lee, P.H.Y., Phys. Rev. Letts. 40, 184 (1978).
Raven, A., Willi, O., Phys. Rev. Letts. 43, 278 (1979).

16. Burgess, M.D.J., Dragila, R., Luther-Davies, B., Nugent, K.A., Tallents, G.J., Proceedings of the Sixth Workshop on Laser Interaction with Matter, Monteray, U.S.A. October 1982.

17. Phillion, D.W., Campbell, E.M., Estabrook, K.G., Phillips, G.E., Ze, F. Phys. Rev. Letts. 49, 1405 (1982).

DISTRIBUTION FUNCTIONS IN QUANTUM OPTICS

R. F. O'CONNELL

Department of Physics and Astronomy, Louisiana State University
Baton Rouge, LA 70803

Perhaps the simplest way of including quantum mechanics, in various problems in quantum optics, is by use of the classical-quantum-correspondence method, by means of which one replaces quantum-mechanical operators by complex numbers. This is carried out by means of quasi-classical distribution functions and here we address the question of which is the best choice of function from the large selection available. Whereas Glauber's P(α) distribution is useful in many applications, it does not exist as a well-behaved function in many others. In such cases, a more useful function is the generalized P-representation of Drummond and Gardiner. However, based on simplicity and overall applicability, we conclude that Wigner's function also has a claim to be the optimum choice.

1. Introduction

Many problems in quantum optics involve dissipative processes as an essential element. The usual approach to such problems is to consider a system of interest, S say, coupled to a reservoir (R), and interacting with it via a potential V which results in S losing energy to R. From the fluctuation-dissipation theorem [1], we know that the dissipation is related to the fluctuations of the system in equilibrium. Of paramount interest is the time development of the distribution functions for the system. These remarks make it clear that we are in the province of irreversible statistical mechanics and thus--both for the purpose of putting quantum optics problems in perspective and also in the hope of finding the optimum way of solving such problems--we present in Section 2 an overview of the usual approach to traditional problems in irreversible statistical mechanics for comparison with the quantum optics situation.

In particular, we point out in Section 2 that perhaps the simplest way of including quantum mechanics is by means of the so-called classical-quantum correspondence (CQC), a technique originally introduced by Wigner [2], which makes use of quasi-classical distribution functions. In Section 3 we point out that there is literally an infinite choice of such functions and thus we are led to make a choice on the basis of what properties we would like our functions to have. This leads us into a discussion of the more widely used functions. For example, we argue that Wigner's choice of distribution function [2] has particularly desirable properties as far as traditional problems are concerned which might suggest that this should also be the choice in the solution of problems in quantum optics. This leads us to examine, in Section 4, a miscellany of topical problems in quantum optics which have been treated by the use of various distribution functions. We conclude that strong claims can be made for the use of either the so-called generalized P-representation or the Wigner distribution for treating problems in quantum optics--our own feeling being that further investigations are required before definitive conclusions can be reached.

2. Irreversible Statistical Mechanics/Quantum Optics/...:an Overview

The time evolution of a large system of particles is a common theme in a broad area of studies in non-equilibrium statistical mechanics [3], quantum optics [4-7], and in the many phenomena which are grouped under the common umbrella of synergetics [8,9]. The central equation of classical statistical mechanics is the Liouville equation

$$\frac{\partial P(q,p,t)}{\partial t} = \left\{H(q,p),\ P(q,p,r)\right\}, \tag{1}$$

where P is a time-dependent distribution function and (q,p) denotes the phase-space coordinates $(q_1 --- q_N, p_1 --- p_N)$ where N is the number of particles. Also, H is the Hamiltonian and the curly bracket denotes the Poisson bracket. The state of the system at a given time t is completely specified by P(q,p,t), which is the probability density for finding the particle at the point q,p in phase space.

In quantum-statistical mechanics, the basic equation is that of von Neumann:

$$\frac{\partial}{\partial t}\,\hat{\rho}(t) = (ih)^{-1}[\hat{H},\hat{\rho}(t)] \tag{2}$$

where $\hat{\rho}$ (t) is the density matrix, $\hat{H}$ is the Hamiltonian operator and the square bracket denotes the commutator. Since, as we will demonstrate in Section 3, the use of a CQC can be used to convert Eq. (2) into a quantum

Section 3, the use of a CQC can be used to convert Eq. (2) into a quantum Liouville equation i.e. a Liouville equation with quantum corrections taken into account exactly, by the use of c-numbers as distinct from quantum-mechanical operators, the starting point of our discussion will be Eq. (2), as we illustrate in Fig. 1. The rest of this section will be devoted to a discussion of Fig. 1.

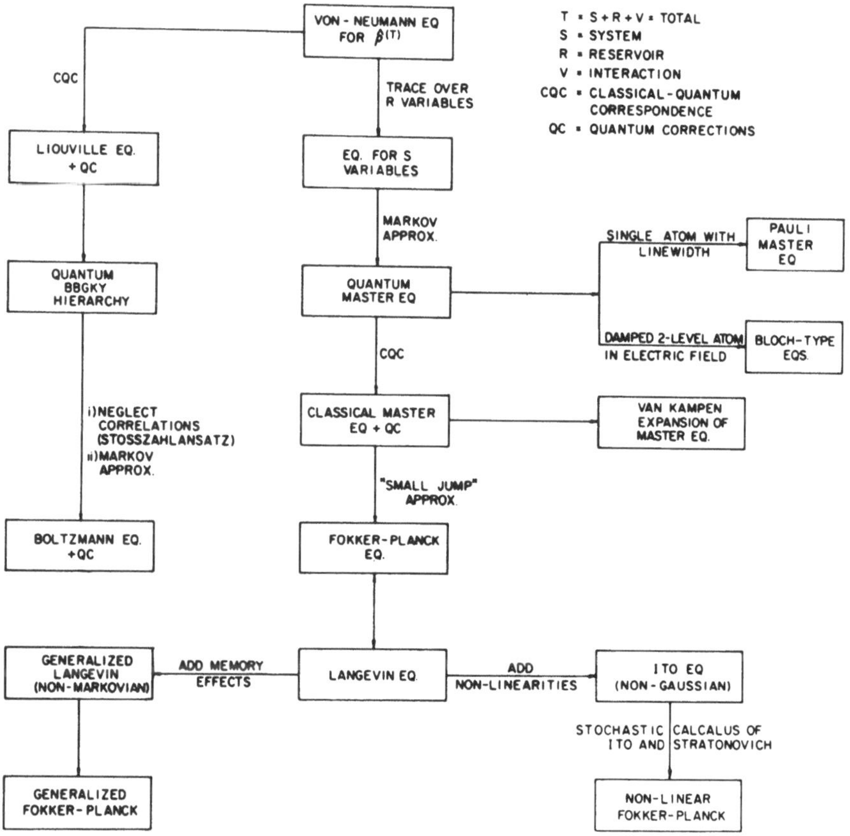

Fig. 1 Irreversible Statistical Mechanics/Quantum Optics/...: an Overview

The left-hand chain of Fig. 1 illustrates a route towards the traditional kinetic equations of statistical mechanics. First of all, the generalized Liouville equation is rewritten exactly in terms of the BBGKY hierarchy [3,10], which is a set of N equations for the reduced distribution functions: the rate of change of the distribution function, P_M say, for M particles ($M \leq N-1$) depends on P_{M+1} as well as P_M. However, because of the complexity of these exact equations, various approximations are usually made to reduce them to kinetic equations i.e. equations (generally non-linear) for the time-evolution of the one-particle reduced distribution function. The most famous of such equations is the celebrated Boltzmann equation [3,11,12,13]

$$\frac{\partial P}{\partial t} + \frac{\vec{p}}{m} \cdot \frac{\partial P}{\partial \vec{q}} + \vec{F} \cdot \frac{\partial P}{\partial \vec{p}} = C(P) \tag{3}$$

where $\vec{q}$ and $\vec{p}$ now refer to the coordinates and momenta of a single particle (in contrast to the case of the N particles considered in Eq. (1)), where $\vec{F}$ is the external force acting on the particle, and C(P) denotes the rate of change of P due to collisions with the other particles.

Inherent in a kinetic description is the use of a Markovian approximation i.e. the rate of change of P at time t depends only on its value at that time and not on its previous history (no memory effects). Another assumption underlying the derivation of the Boltzmann equation is the neglect of correlations caused by the interactions. This is referred to as Boltzmann's Stosszahlansatz and, in essence, it means that the time duration of a collision (i.e. the time during which the particle trajectories differ significantly from a straight line) is much smaller than the time between collisions. As a result, the interaction process is confined to two-body collisions and the Boltzmann equation is only applicable to a dilute gas. Another point of note is the time reversible character of the BBGKY hierarchy vis-à-vis the irreversible character of the Boltzmann equation, a necessary but not sufficient condition for the latter being the presence of interactions [3,13].

There is, of course, a host of other kinetic equations [13], such as the Landau equation and the Vlasov equation but the Boltzmann equation is a prototype for displaying all the general features of kinetic equations relevant to our present purposes. All of these kinetic equations treat a "system" of one particle interacting via collisions with a reservoir of N-1 particles. This observation brings us to the second branch of Fig. 1 where the total system of N particles is considered as consisting of a system S and a reservoir R interacting via a potential V. Since our interest lies in the system and not in the reservoir, a trace is carried out over the R variables, resulting in an equation for the S variables. This is the starting point of many investigators in a variety of disciplines such as, for example, Zwanzig, Nakajima, Mori and Prigogine [14] in the general area of statistical mechanics; Lax [5], Louisell [4], and Haken [7-9] in quantum optics and synergetics; and Barker and Ferry [15] in their investigations of transport properties in small devices.

Next, applying a Markov approximation leads to a quantum master equation, particular examples of same being the original master equation derived by Pauli for a single atom with line-width and the Bloch-type equations (familiar from NMR studies) for the case of a damped two-level atom in an electric field. The quantum master equation is essentially the differential form of the Chapman-Kolmogorov equation for Markov processes [16] and it amounts to having carried out a coarse-grained averaging in time. It is at this stage that many authors use the CQC to get a c-number equation with c-number quantum corrections. Whereas a general expansion of the master equation is possible [16], the next reduction is more commonly that of making the so-called "small-jump" approximation to obtain the Fokker-Planck equation, from which the stochastic Langevin equation may be derived and vice versa. The Fokker-Planck and/or Langevin equations have been applied not only to a large selection of problems in quantum optics--which constitutes our present interest--but also to areas as diverse as condensed matter physics [17] and nuclear physics [18]. It should be emphasized that both methods should lead to the same result, as was found, for example, in investigations of rotational Brownian motion [19]. It is of interest to note that a conclusion arising from the latter work is that the solution of the Langevin stochastic equation presented less mathematical difficulties than the corresponding Fokker-Planck equation but that the latter allowed results to be calculated to a higher order of approximation. Also, Agarwal [20] has concluded that Langevin equations are easier to interpret than the master equation.

Generalizations of the Langevin equation (and hence the Fokker-Planck equation) include the addition of memory (non-Markovian effects) and non-linearities, the latter addition resulting in a complete new stochastic calculus due to Ito and Stratonovich [21].

The Langevin and Fokker-Planck equations and their generalizations have found wide application in quantum optics, some examples of which we will discuss in Section 4 but before doing so, we will first of all discuss (in

Section 3) the method of the CQC which leads to such equations.

In closing this section, we should emphasize that the scenario outlined in the block diagram of Fig. 1 is not the only route to the "Holy Grail". However, our purpose will be fulfilled if it enables the non-expert to obtain some insight into the interconnection of the many different approaches to a myriad of problems which possess, however, certain unifying aspects.

3. Classical-Quantum Correspondence (CQC)

In classical mechanics the average of any function of q and p, A(q,p) say, may be written as

$$\langle A\rangle = \int_{-\infty}^{\infty} dq \int_{-\infty}^{\infty} dp \quad A(q,p)P_{cl}(q,p) \tag{4}$$

where for clarity we have added a subscript "cl" to P(q,p). The latter quantity will now be used to denote the so-called quasi-classical distribution function and, in general, it depends on $\hbar$.

In 1932, Wigner [2] presented an exact reformulation of non-relativistic quantum mechanics in terms of classical concepts. In particular, he showed that the ensemble average of a function of the position and momentum operators, $\hat{A}(\hat{q},\hat{p})$ say, may be written in a form similar to Eq. (4), as follows:

$$\langle\hat{A}\rangle = \int_{-\infty}^{\infty} dq \int_{-\infty}^{\infty} dp \; A(q,p)P(q,p) \tag{5}$$

The next question to be considered relates to the choice of P. As it turns out, there is literally an infinite choice of such functions. However, if one demands some compelling properties then the choice considerably narrows. In particular, Wigner [2] chose the function

$$P_W(q,p) = (\pi\hbar)^{-1}\int_{-\infty}^{\infty} dy\psi^*(q+y)\psi(q-y)e^{2ipy/\hbar} \tag{6}$$

where we have assumed for simplicity that the system is in a pure state

$\psi(q)$. Since the properties of P_W have been discussed in detail elsewhere [22,23] we focus our attention on the property most relevant to our present considerations viz. that in the force-free case the equation of motion is the classical one i.e.

$$\frac{\partial P_W}{\partial t} = - \frac{p}{m} \frac{\partial P_W}{\partial q} , \tag{7}$$

for a one-dimensional configurational problem (to which we will confine ourselves from henceforth since generalization is straightforward).

No other distribution function that we are aware of enjoys this property. For example, the non-negative distribution function of Husimi [24] contains an additional $\hbar^2$ term which is not of quantum origin [23]. In addition, when an external potential is included [23], the quantum corrections which result are much simpler if one uses P_W in contrast to Husimi's function.

Let us turn now to some distribution functions which, in addition to P_W, are those most commonly used in quantum optics. A convenient way of introducing these functions is by the use of characteristic functions [25,4], which are simply the Fourier transforms of the respective distribution functions. Thus, for $P_W(q,p)$, the corresponding characteristic function is

$$C_W(\sigma,\tau) = \langle\psi| \exp [\tfrac{i}{\hbar} (\sigma\hat{q}+\tau\hat{p})] |\psi\rangle , \tag{8}$$

or, equivalently, in terms of creation and annihilation operators $\hat{a}$ and $\hat{a}^+$ (defined in terms of $\hat{p}$ and $\hat{q}$ in the usual way):

$$C_W(\eta,\eta^*) = \langle\psi| \exp [\tfrac{i}{\hbar} (\eta\hat{a}+\eta^*\hat{a}^+)] |\psi\rangle . \tag{9}$$

In quantum optics, other common choices are the normal and anti-normal functions, denoted by C_n and C_a respectively, where

$$C_n(\eta,\eta^*) = \langle\psi| \exp [\tfrac{i}{\hbar} \eta^*\hat{a}^+] \exp [\tfrac{i}{\hbar} \eta\hat{a}] |\psi\rangle , \tag{10}$$

and C_a is the same as C_n with the order of the exponential factors interchanged. We remark that the distribution function corresponding to C_a, P_a say, is the same as Glauber's widely-used $P(\alpha)$ distribution [4] if, in the evaluation of the expectation value of an arbitrary operator, we first write it in normal ordering sequence (creation operators precede annihilation operators) prior to carrying out the CQC.

If we now use the Baker-Hausdorff theorem and then convert back from the $\hat{a}$, $\hat{a}^+$ language to the $\hat{q}$, $\hat{p}$ language, we find that the corresponding distribution functions may be written in terms of P_W as follows:

$$P_{a,n}(q,p) = \exp\left[\pm \frac{1}{2} q_0^2 \frac{\partial^2}{\partial q^2} \pm \frac{1}{2} p_0^2 \frac{\partial^2}{\partial p^2}\right] P_W , \tag{11}$$

where $q_0^2 = \hbar/2m\omega$ and $P_0^2 = m\hbar\omega/2$. Next, using Eq. (7), it immediately follows that the time derivative of $P_{n,a}$ is not as simple as that of P_W since it contains also second derivative terms. Thus, quantum corrections to Eq. (3), the Boltzmann equation, are simpler in form if one chooses P_W. In other words, if we solve a problem via the left-hand route of Fig. 1 it is clearly best to choose P_W. Thus, we might expect the same to hold if we choose the right-hand route--the route of quantum optical investigations. In the next section, we examine briefly some specific applications to enable us to judge what happens in practice.

4. Examples from Quantum Optics

Glauber's $P(\alpha)$ function (P_a) has been the most widely used distribution function in quantum optics, primarily because of its convenience in averaging the normally ordered operator products that often arise in problems in this area. In particular, it was applied to the laser by Weidlich et al. [26], to dispersive optical bistability by Drummond and Walls [27], to the damped harmonic oscillator [28], and to a variety of other problems. However, it was long recognized [4,5,7] that the corresponding Fokker-Planck equations often have non-positive-definite diffusion coefficients. Nevetheless, it was only recently that this problem was taken seriously, principally in the papers of Drummond, Gardiner, and Walls [29,30]. This problem basically arises in dealing with intrinsically non-classical effects, such as photon anti-bunching which occurs in atomic flourescence experiments [31]. For such nonclassical photon fields, P_a does not exist as a well-behaved function. In order to avoid such problems, Drummond and Gardiner [29], introduced a class of generalized P-representations, which include the complex P representation, in which (α,α^*) are replaced by the independent complex variables (α,β). Such generalized representations were applied successfully to non-linear problems in quantum optics (two-photon absorption; dispersive bistability; degenerate parametric amplifier) and chemical reaction theory [29,30,32].

However, calculations using the generalized P representations are relatively complicated. Since they were designed in effect to handle situations for which the Glauber distribution does not exist the following question naturally arises: why not use the Wigner distribution since it always exists and its equation of motion is simple in that it possesses the desirable property discussed in the last section? While we were pondering such a question in the course of completing the present paper, a paper by Lugiato, Casagrande, and Pizzuto [33] just appeared in which the Wigner distribution was used in the consideration of a system of N two-level atoms interacting with a resonant mode radiation field and coupled to suitable reservoirs. The presence of an external CW coherent field injected into the cavity is also included, which allows for the possibility of treating optical bistability as well as a laser with injected signal. They then carry out very detailed calculations to obtain Fokker-Planck equations, which they compare with the corresponding ones obtained using the Glauber function, and conclude that the use of the Wigner function is preferable to the Glauber function.

In the problems considered by Lugiato et al., the existence of a smallness parameter $N_S^{-\frac{1}{2}}$ (where N_S is the saturation photon number) enabled them to truncate at the second order, which is the essence of the Fokker-Planck approximation. On the other hand, Walls and Milburn [32] conclude that, in the case of dispersive bistability and two-photon absorption, the use of the generalized P representation is preferable to the use of the Wigner function because the latter gives rise to equations containing third order derivatives. It is thus clear that further investigations are required before definitive conclusions can be reached. In particular, does the flexibility which occurs by the use of a complex phase space give the generalized P representation some unique and desirable property which accounts for its success in applications?

Acknowledgements

The author is pleased to acknowledge enlightening comments received from Professor Dan Walls on the initial draft of this manuscript. This research was partially supported by the Department of Energy, Division of Materials Science, under contract no. DE-AS05-79ER10459.

References

1. H. B. Callen and T. A. Welton, Phys. Rev. 83, 34 (1951).
2. E. Wigner, Phys. Rev. 40, 749 (1932).
3. R. Balescu, Equilibrium and Nonequilibrium Statistical Mechanics (Wiley-Interscience, New York, 1975).
4. W. H. Louisell, Quantum Statistical Properties of Radiation (Wiley, New York, 1973).
5. M. Lax in Statistical Physics, Phase Transitions, and Superfluidity, ed. by M. Chretien, E. P. Gross, and S. Deser (Gordon and Breach, New York, 1968).
6. M. Sargent, M. O. Scully, and W. E. Lamb, Laser Physics (Addison-Wesley, Reading, Mass. 1974).
7. H. Haken, Laser Physics, Handbuch der Physik 15, 2c (Springer, Berlin, 1970).
8. H. Haken, Rev. Mod. Phys. 47, 67 (1975).
9. H. Haken, Synergetics, 2nd ed. (Springer, New York, 1978).
10. K. Imre, E. Özizmir, M. Rosenbaum, and P. F. Zweifel, J. Math. Phys. 8, 1097 (1967).
11. E. M. Lifshitz and L. P. Pitaevskii, Physical Kinetics (Pergamon Press, 1981).
12. E. G. D. Cohen and W. Thirring, eds., The Boltzmann Equation Theory and Applications (Springer, New York, 1973).
13. H. Spohn, Rev. Mod. Phys. 53, 569 (1980).
14. R. Zwanzig, Phys. Rev. 124, 983 (1961); ibid., Physica 30, 1109 (1964); S. Nakajima, Prog. Theor. Phys. 20, 948 (1958); H. Mori, Prog. Theor. Phys. 33, 423 (1965); I. Prigogine and P. Resibois, Physica 27, 629 (1961).
15. J. R. Barker and D. K. Ferry, Solid St. Electron 23, 531 (1980).
16. N. G. van Kampen, Stochastic Processes in Physics and Chemistry (North-Holland, New York, 1981).
17. D. Forster, Hydrodynamic Fluctuations, Broken Symmetry, and Correlation Functions (Benjamin, Reading, Mass. 1975).
18. E. P. Wigner, Ann. Math. 62, 548 (1955); T. A. Brody, J. Flores, J. B. French, P. A. Mello, A. Pandey, and S. S. M. Wong, Rev. Mod. Phys. 53, 385 (1981); R. U. Haq, A. Pandey, and O. Bohigas, Phys. Rev. Lett. 48, 1086 (1982).
19. G. W. Ford, J. T. Lewis, and J. McConnell, Phys. Rev. A19, 907 (1979); J. McConnell, Rotational Brownian Motion and Dielectric Theory (Academic, New York, 1980).
20. G. S. Agarwal, Quantum Statistical Theories of Spontaneous Emission and their Relation to Other Approaches (Springer, New York, 1974).
21. K. Itô, Mem. Amer. Mathem. Soc. 4, 51 (1951); R. L. Stratonovich, SIAM J. Control 4, 362 (1966); S. Chaturvedi and C. W. Gardiner, J. Phys. B 14, 1119 (1981), see Appendix.
22. E. P. Wigner in Perspectives in Quantum Theory, ed, W. Yourgrau and A. van der Merwe (Dover, New York, 1979); R. F. O'Connell and E. P. Wigner, Phys. Lett. 83A, 145 (1981); R. F. O'Connell, Found. Phys. 13, 83 (1983).
23 R. F. O'Connell and E. P. Wigner, Phys. Lett. 85A, 121 (1981).
24. K. Husimi, Proc. Phys. Math. Soc. Japan 22, 264 (1940).
25. J. E. Moyal, Proc. Cambridge Phil. Soc. 45, 99 (1949).
26. W. Weidlich, H. Hisken, and H. Haken, Z. Phys. 186, 85 (1965).
27. P. D. Drummond and D. F. Walls, J. Phys. A 13, 725 (1980); ibid. Phys. Rev. A 23, 2563 (1981).
28. W. H. Louisell and J. H. Marburger, IEEE J. Quantum Electron. 3, 348 (1967).
29. P. D. Drummond and C. W. Gardiner, J. Phys. A13, 2353 (1980).
30. P. D. Drummond, C. W. Gardiner, and D. F. Walls, Phys. Rev. A24, 914 (1981).
31. D. F. Walls, Nature 280, 451 (1979) reviews this topic.

32. D. F. Walls and G. J. Milburn, in Proceedings of the NATO ASI in Bad Windsheim, West Germany, 1981, ed. by P. Meystre (Plenum, New York, 1982).
33. L. A. Lugiato, F. Casagrande, and L. Pizzuto, Phys. Rev. A26, 3438 (1982).

QUANTUM NON DEMOLITION MEASUREMENTS

D.F. Walls and G.J. Milburn
Physics Department
University of Waikato
Hamilton, New Zealand

§1 INTRODUCTION

Current attempts to detect gravitational radiation have to take into account the quantum uncertainties in the measurement process. Considering that the detectors are macroscopic objects as large in some cases as a 10 ton bar the fact that the quantum fluctuations in the detector must be taken into account seems surprising. However the strength of the gravity waves is so weak that a displacement of the order of 10^{-19}cm is expected. To illustrate how the measurement process may introduce uncertainties which obscure the signal we consider the following simple example. Let us consider as our detector a free mass. A measurement of the position of the free mass with a precision $\Delta x_i \simeq 10^{-19}$cm will perturb the momentum by an amount given by Heisenberg's uncertainty relation $\Delta p \geqslant \frac{\hbar}{2\Delta x_i}$. The period of the gravity waves is expected to be $\sim 10^{-3}$sec hence a second measurement of position should be made in a time interval $\tau = 10^{-3}$sec. During this time the mass will move from its initial position by an amount $\Delta x = \frac{\Delta p}{m}\tau \geqslant \frac{\hbar\tau}{2m\Delta x_i}$.

Taking the detector mass equal to 10 tons, we find $\Delta x \geqslant 5 \times 10^{-19}$cm. That is, the uncertainty introduced by the first measurement has made it impossible for a second accurate measurement to determine with certainty whether a gravity wave has acted or not.

It is instructive to consider measurements of momentum instead of position of the free mass. The first measurement of momentum causes an uncertainty in position, this however does not feed back to disturb the momentum of a free mass. Hence subsequent determination of the momentum may be made with complete predictability. The momentum of a free mass is an example of a quantum non demolition (Q.N.D.) variable. The concept of quantum non demolition measurements has been introduced over the past few years[1-6] to allow the detection in principle of very weak forces below the level of the quantum noise of the detector. Such Q.N.D. measurement procedures have been the subject of a number of review articles.[5-6] We shall give a brief review of the concepts of Q.N.D. measurements in the next two sections.

§2 MEASUREMENTS OF A CLASSICAL FORCE

We wish to detect a gravitational wave which we may represent by a classical force $F(t)$. As our detector we may use a bar detector which we may represent by a harmonic oscillator. (The harmonic oscillator may also represent a cavity mode of the electromagnetic field in interferometric detection schemes)

The Hamiltonian for the coupled force-detector system is

$$H = \hbar\omega a^{\dagger}a + F(t)\hat{x} \tag{1}$$

where $[a,a^{\dagger}] = 1$

and

$$\hat{x} = \left(\frac{\hbar}{2m\omega}\right)^{\frac{1}{2}} (a + a^{\dagger})$$

ω is the fundamental frequency of the oscillator and m is its mass.

The equation of motion for the oscillator's amplitude in the interaction picture is

$$\frac{da}{dt} = -i \frac{1}{(2\hbar m\omega)^{\frac{1}{2}}} F(t)e^{i\omega t} \tag{2}$$

This has solution

$$a(\tau) = a(0) + \alpha(\tau) \tag{3}$$

where

$$\alpha(\tau) = -i \frac{1}{\sqrt{2\hbar m\omega}} \int_0^{\tau} F(t)e^{i\omega t}dt$$

A sinusoidal force on resonance $F(t) = F_0\sin\omega t$ acting for a time $t > \frac{2\pi}{\omega}$ induces a displacement of the oscillators position $\delta x = \frac{F_0 t}{2m\omega}$. Let us assume that the oscillator is in its ground state or in fact any coherent state. The uncertainty in the momentum and position are then given by

$$\Delta x = \frac{\Delta p}{m\omega} = \left(\frac{\hbar}{2m\omega}\right)^{\frac{1}{2}} \tag{4}$$

For typical parameters of a gravitational wave bar detector $(m \simeq 10^6 g, \frac{\omega}{2\pi} \simeq 1kHz)$ $\Delta x \simeq 3 \times 10^{-19}$cm. This is of the same order as the expected displacement of the gravitational wave. In order that the force be unambiguously detected we require that the the displacement produced δx, be at least twice the standard error Δx. This leads to the condition

$$F_0 > \frac{2}{\tau} (2\hbar m\omega)^{\frac{1}{2}} \tag{5}$$

In general we require that $|\alpha(\tau)| \gtrsim 1$.

This represents the standard quantum limit (S.Q.L.)[5] for detecting a classical force against a background of the detector's zero point noise.

§3 Q.N.D. MEASUREMENTS

The basic requirement of a Q.N.D. measurement is the availability of a variable which may be measured repeatedly, giving completely predictable results in the absence of a gravitational wave. Clearly this requires that the act of each measurement itself does not degrade the predictability of subsequent measurements. This requirement is satisfied if for an observable $\hat{A}^I(t)$ (in the interaction picture)

$$[\hat{A}^I(t), \hat{A}^I(t')] = 0 \tag{6}$$

This condition ensures that if the system is in an eigenstate of $A^I(t_0)$ it remains in this eigenstate for all subsequent times t even though the eigenvalues may change. Such observables are called Q.N.D. observables.

Thus for a free particle, energy and momentum are Q.N.D. observables, position is not since

$$\hat{x}(t + \tau) = \hat{x}(t) + \hat{p}\frac{\tau}{m} \tag{7}$$

$$[\hat{x}(t), \hat{x}(t + \tau)] = \frac{i\hbar\tau}{m} \tag{8}$$

For a harmonic oscillator

$$\begin{aligned} [\hat{x}(t), \hat{x}(t + \tau)] &= \frac{i\hbar}{m\omega}\sin\omega\tau \\ [\hat{p}(t), \hat{p}(t + \tau)] &= i\hbar m\omega\sin\omega\tau \end{aligned} \tag{9}$$

thus $\hat{x}$ and $\hat{p}$ are not Q.N.D. observables for the harmonic oscillator.

We may introduce the quadrature phase amplitudes of the harmonic oscillator $\hat{X}_1$ and $\hat{X}_2$ defined by

$$\left(\frac{2\hbar}{m\omega}\right)^{\frac{1}{2}} a = (\hat{X}_1 + i\hat{X}_2)e^{-i\omega t} \tag{10}$$

$\hat{X}_1$ and $\hat{X}_2$ satisfy the requirement (6) for Q.N.D. variables.

The $\hat{X}_1$ and $\hat{X}_2$ axes rotate with angular frequency ω with respect to the x and p axes

$$\begin{aligned} \hat{X}_1(t) &= \hat{x}\cos\omega t - \frac{\hat{p}}{m\omega}\sin\omega t \\ \hat{X}_2(t) &= \hat{x}\sin\omega t - \frac{\hat{p}}{m\omega}\cos\omega t \end{aligned} \tag{11}$$

The behaviour of $\hat{X}_1$, $\hat{X}_2$ are most easily discussed with reference to an amplitude and phase diagram. In such a diagram the state of the system is represented by a set of points centred on the mean and contained within an error box determined by the variance of $\hat{X}_1$, $\hat{X}_2$. Alternatively the error box may be regarded as a contour of the Wigner function.

In Figure 1 an error ellipse for the oscillator is shown. The error ellipse is stationary with respect to the $\hat{X}_1$ and $\hat{X}_2$ axes but rotates with respect to the $\hat{x}$ and $\hat{p}$ axes. This clearly illustrates how the uncertainties in $\hat{p}$ feed back into $\hat{x}$ as the ellipse rotates with time.

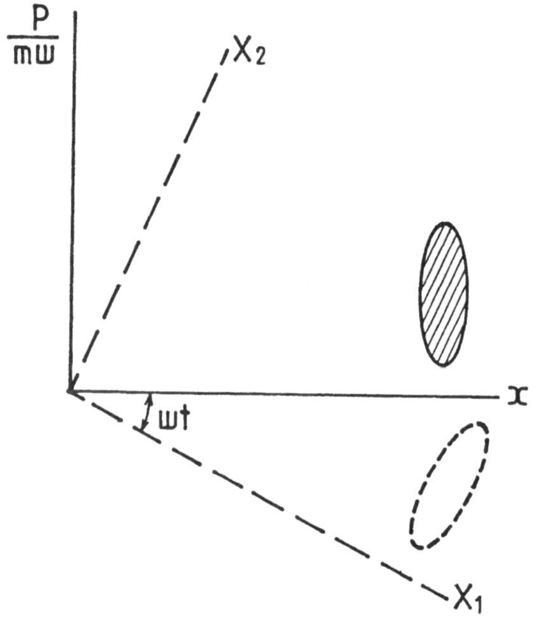

Figure 1 : *Error box in the phase plane for a harmonic oscillator. The error box rotates with respect to the* x, p/mω *axes but is stationary with respect to the* $\hat{X}_1$, $\hat{X}_2$ *axes.*

We shall now consider two possible Q.N.D. measurement schemes on the harmonic oscillator.

(a) Squeezed State Method

The Q.N.D. variables $\hat{X}_1$ and $\hat{X}_2$ obey the commutation relation

$$[\hat{X}_1(t), \hat{X}_2(t)] = \frac{i\hbar}{m\omega} \tag{12}$$

This implies the uncertainty relation

$$\Delta X_1(t)\Delta X_2(t) \geqslant \frac{\hbar}{2m\omega} \tag{13}$$

where

$$\Delta X_i(t) = V(\hat{X}_i(t))^{\frac{1}{2}}$$

and (13)

$$V(X_i(t)) = \langle\hat{X}_i{}^2(t)\rangle - \langle\hat{X}_i(t)\rangle^2$$

i.e. the variance in $\hat{X}_i(t)$.

The standard quantum limit arises when one attempts to measure $\hat{X}_1$ and $\hat{X}_2$ with equal accuracy. The limit to the accuracy of such a measurement (called amplitude and phase measurements by Thorne et.al[4]) is given by the S.Q.L.

$$\Delta X_1 = \Delta X_2 \geqslant \left(\frac{\hbar}{2m\omega}\right)^{\frac{1}{2}} \qquad (14)$$

The method (first proposed by Thorne et.al[4]) to beat the S.Q.L. is clear. One must not attempt to measure both $\hat{X}_1$ and $\hat{X}_2$ but instead measure only one component. One can measure $\hat{X}_1$ with an accuracy $\Delta X_1 < \left(\frac{\hbar}{2m\omega}\right)^{\frac{1}{2}}$ at the expense of increased uncertainty in X_2 $\left(\Delta X_2 > \left(\frac{\hbar}{2m\omega}\right)^{\frac{1}{2}}\right)$. Such a measurement places the oscillator in a state with an elliptic error box $\left(\Delta X_1 < \left(\frac{\hbar}{2m\omega}\right)^{\frac{1}{2}} < \Delta X_2\right)$. Such states are known as squeezed states. Squeezed states of light are presently a subject of intensive research in quantum optics.[7]

This measurement technique which places the oscillator in a squeezed state is an example of a general class of quantum non demolition measurements. In such a Q.N.D. measurement scheme the first measurement places the detector in a squeezed state with $\Delta X_1 \ll \left(\frac{\hbar}{2m\omega}\right)^{\frac{1}{2}}$. Since $\hat{X}_1$ is a constant of the motion the oscillator will remain in the near eigenstate under free evolution. The classical force then displaces the error ellipse without changing its size, shape or orientation. A second measurement of $\hat{X}_1$ can detect the force provided that the displacement $\delta X_1 > 2\Delta X_1$. Such a scheme beats the S.Q.L. provided that $\Delta X_1 < \left(\frac{\hbar}{2m\omega}\right)^{\frac{1}{2}}$. (see Figure 2). A particular example of such a Q.N.D. measurement scheme will be discussed in §4.

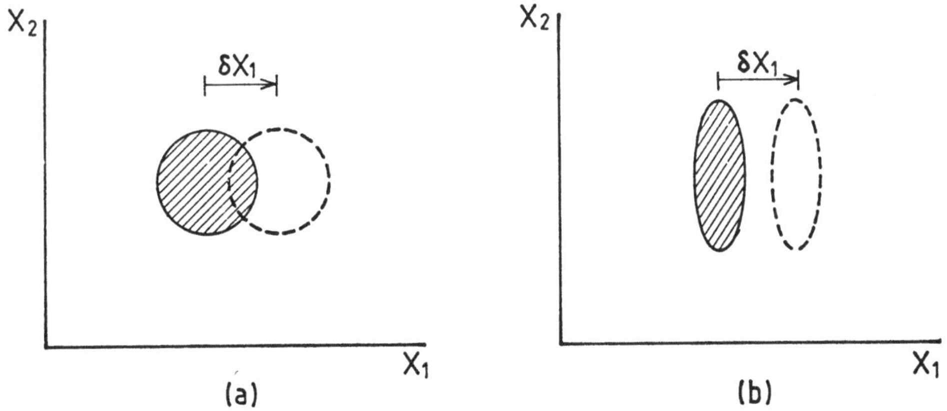

Figure 2 : *Displacement* δX_1 *of error box by a force.*

(a) Oscillator initially in a coherent state $\delta X_1 < 2\Delta X_1 = 2\left(\frac{\hbar}{2m\omega}\right)^{\frac{1}{2}}$

(b) Oscillator initially in a squeezed state $2\Delta X_1 < \delta X_1 < 2\left(\frac{\hbar}{2m\omega}\right)^{\frac{1}{2}}$

A Q.N.D. measurement using the squeezed state technique may monitor the time evolution of the force. The Q.N.D. operator evolves as

$$\hat{X}_1(t) = \hat{X}_1(t_0) - \int_{t_0}^{t} \frac{F(t')}{m\omega} \sin\omega t' dt' \tag{15}$$

For a system initially in an eigenstate $|\varepsilon_0\rangle$ of $X_1(t_0)$ it remains in this eigenstate with eigenvalue given by

$$\varepsilon(t) = \varepsilon_0 + \left(\frac{2\hbar}{m\omega}\right)^{\frac{1}{2}} \mathrm{Re}(\alpha(t)) \tag{16}$$

As $\hat{X}_1$ can be measured repeatedly over smaller and smaller intervals $(\Delta t \rightarrow 0)$ we find

$$\lim_{\Delta t \rightarrow 0} \left\{ \frac{\mathrm{Re}(\alpha(t))}{\Delta t} \right\} = \frac{F(t)}{\sqrt{2\hbar m\omega}} \sin\omega t \tag{17}$$

which enables the precise time development of $F(t)$ to be determined.

(b) Quantum Counting Measurements

The earliest suggested quantum non demolition measurements (12) involved counting the number of quanta without destroying any, hence the origin of the name.

Consider once again the classical force acting on the harmonic oscillator initially in an energy eigenstate $|n_0\rangle$. The change in the mean number of quanta induced by the force

$$\delta n = \langle n\rangle - n_0 \tag{18}$$

$$= |\alpha|^2 = \frac{F_0{}^2 t^2}{8m\hbar\omega}$$

The variance in the number $V(n)$ is

$$\langle n| D^\dagger(\alpha) V(a^\dagger a) D(\alpha) |n\rangle = |\alpha|^2 (n_0 + 1) \tag{19}$$

where

$$D(\alpha) = e^{\alpha a^\dagger - \alpha^\dagger a}$$

In order to detect a force we require $V(n) > 1$, that is,

$$F_0 > \frac{2}{t} \frac{(2m\omega\hbar)^{\frac{1}{2}}}{(n_0 + 1)^{\frac{1}{2}}} \tag{20}$$

or

$$|\alpha(\tau)| \gg \frac{1}{\sqrt{n_0 + 1}}$$

Comparing this with the S.Q.L. given by Eq.(5) we see that the quantum counting method reduces the S.Q.L. by a factor $\frac{1}{(n_0 + 1)^{\frac{1}{2}}}$. Hence for a highly excited oscillator an arbitrarily weak force may be detected. However since there is no unique relationship between the measured energy and F_0 $(\sigma(E) \gg \langle E\rangle - E_0)$ this method cannot tell us the precise magnitude of F_0. This occurs because the energy is not a Q.N.D.

variable in the presence of the force and the system moves out of its original energy eigenstate into a superposition of eigenstates.

If there is no classical force acting on an oscillator in a number state a series of measurements of n must yield a constant sequence of results. If at any time during a sequence of measurements a classical force of sufficient strength was present (i.e. $|\alpha(\tau)| > \frac{1}{\sqrt{n_0+1}}$) then one must eventually obtain a result different from the previous constant values and conclude a classical force had been detected.

A proposed scheme to carry out a quantum counting measurement will be discussed in §5.

(c) Meter Readout

Having first determined the Q.N.D. variable of the detector it is necessary to couple the detector to a readout system or meter. In order that the meter does not introduce additional fluctuations into the Q.N.D. variable of the detector it is sufficient that the only detector operator appearing in the detector meter interaction is the Q.N.D. operator of the detector. The detector readout system is then said to be "back action evading". If the back action evasion criterion is satisfied then the Q.N.D. variable obeys the criterion (6) in the presence of the detector meter interaction. This is a sufficient condition for a Q.N.D. measurement.

An analysis of a Q.N.D. measurement process may be divided into two stages. The first stage involves solving for the time dependent unitary evolution of the coupled detector meter system. During this stage correlations between the state of the detector and meter build up. At some point the free evolution is suspended and a readout of the meter is made, whereupon the meter state is reduced. The second stage of the analysis then involves a determination of the nonunitary effect of meter state reduction upon the detector.

We shall now consider two examples of Q.N.D. measurements which illustrate the squeezed state and quantum counting techniques respectively.

§4 SQUEEZED STATE METHOD IN A PARAMETRIC AMPLIFIER

We consider as our model of the detector meter system two coupled harmonic oscillators. We take as the Hamiltonian for the system

$$\begin{aligned} H &= H_0 + H_I \\ H_0 &= \hbar\omega_a a^\dagger a + \hbar\omega_b b^\dagger b \\ H_I &= -\hbar\kappa(a^\dagger b^\dagger e^{-i(\omega_a+\omega_b)t} + h.c.) \end{aligned} \tag{21}$$

where a, (b) is the boson operator for the detector (meter). This Hamiltonian represents a parametric amplifier coupling in quantum optics.[8] Analyses of a Q.N.D. measurement scheme based on this coupling are given in references (9) and (10). In terms of the quadrature phase amplitudes

$$a = \left(\frac{\omega_a}{2\hbar}\right)^{\frac{1}{2}} (\hat{X}_1 + i\hat{X}_2)e^{-i\omega_a t}$$

$$b = \left(\frac{\omega_b}{2\hbar}\right)^{\frac{1}{2}} (\hat{Y}_1 + iY_2)e^{-i\omega_b t} \tag{22}$$

the interaction Hamiltonian may be written

$$H_I = -K\sqrt{\omega_a\omega_b}\,(\hat{X}_1\hat{Y}_1 - \hat{X}_2\hat{Y}_2) \tag{23}$$

$\hat{X}_1$ is chosen as the Q.N.D. observable of the detector. It is clear that the interaction Hamiltonian does not satisfy the general back action evading criterion that $\hat{X}_1$ be the only detector observable to appear in the interaction Hamiltonian. However solving for $\hat{X}_1(t)$ we find

$$\hat{X}_1(t) = \hat{X}_1(0)\cosh Kt - \sqrt{\frac{\omega_b}{\omega_a}}\,\hat{Y}_2(0)\sinh Kt \tag{24}$$

Clearly $[\hat{X}_1(t), \hat{X}_1(t')] = 0$, thus $\hat{X}_1(t)$ is a Q.N.D. variable including the interaction with the meter. We shall see that although the interaction is not back action evading in free evolution, it satisfies the requirements for a Q.N.D. measurement when the meter state reduction is performed.

We may express $\hat{X}_1(t)$ as

$$\hat{X}_1(t) = \sqrt{\frac{\omega_b}{\omega_a}}\left(\hat{Y}_2(t)\coth Kt - \frac{\hat{Y}_2(0)}{\sinh Kt}\right) \tag{25}$$

A value for $\hat{X}_1(t)$ may be inferred from a measurement of $\hat{Y}_2(t)$ made on the meter. Under free evolution the variance in $\hat{X}_1(t)$ grows as

$$V(\hat{X}_1(t)) = V(\hat{X}_1(0))\mathrm{Cosh}^2 K\tau + \frac{\omega_b}{\omega_a} V(\hat{X}_2(0))\sinh^2 K\tau \tag{26}$$

Hence for an initial measurement with $V(\hat{X}_1(0)) \neq 0$ the variance grows with time demonstrating the failure of back action evasion. In order to study in full the Q.N.D. measurement one must include the non unitary evolution due to the meter state reduction. We shall carry out this procedure below.

The time evolution of the density operator for the coupled detector meter system is given by the master equation (in the interaction picture)

$$\frac{\partial\rho}{\partial t} = \frac{1}{i\hbar}[H_I, \rho] \tag{27}$$

Standard techniques[11] enable one to convert the operator master equation into a Fokker Planck equation using the Glauber-Sudarshan P representation.[12,13] However since a Q.N.D. measurement process involves squeezed states which do not have a non-singular representation in terms of the Glauber-Sudarshan P representation it is necessary to use the complex P representation.[14] The density operator ρ may be expressed as

$$\rho = \int_{C_i} d\alpha_1 d\beta_1 d\alpha_2 d\beta_2 P(z,t) \frac{|\alpha_1,\alpha_2\rangle\langle\beta_1^*,\beta_2^*|}{\langle\beta_1^*\beta_2^*|\alpha_1,\alpha_2\rangle} \tag{28}$$

where $\underset{\sim}{z}^T = (\alpha_1,\beta_1,\alpha_2,\beta_2)$ and we have the following correspondences

$$\begin{aligned}
a\rho &\leftrightarrow \alpha_1 P(\underset{\sim}{z}) \\
a^\dagger\rho &\leftrightarrow (\beta_1 - \frac{\partial}{\partial\alpha_1})\, P(\underset{\sim}{z}) \\
\rho a^\dagger &\leftrightarrow P(\underset{\sim}{z})\beta_1 \\
\rho a &\leftrightarrow (-\frac{\partial}{\partial\beta_1} + \alpha_1) P(\underset{\sim}{z})
\end{aligned}$$

with similar correspondences between $(b,b^\dagger)$ and (α_2,β_2).

There are actually four independent contour integrals $(i = 1,4)$ involved in Eq.(28) in the complex space of each variable. We are free to choose these contours to obtain a normalizable P function, providing partial integration is defined. Substituting Eq. (28) into Eq.(27) gives the following Fokker Planck equation for the complex P function.

$$\frac{\partial P}{\partial t}(\underset{\sim}{z},t) = \left\{\nabla_{\underset{\sim}{z}}^T A\underset{\sim}{z} + \frac{1}{2}\nabla_{\underset{\sim}{z}}^T D\nabla_{\underset{\sim}{z}}\right\} P(\underset{\sim}{z},t)$$

where

$$\nabla_{\underset{\sim}{z}}^T = \left(\frac{\partial}{\partial\alpha_1}, \frac{\partial}{\partial\beta_1}, \frac{\partial}{\partial\alpha_2}, \frac{\partial}{\partial\beta_2}\right) \tag{29}$$

and

$$A = \begin{pmatrix} 0 & 0 & 0 & -i\kappa \\ 0 & 0 & i\kappa & 0 \\ 0 & -i\kappa & 0 & 0 \\ i\kappa & 0 & 0 & 0 \end{pmatrix}$$

and

$$D = \begin{pmatrix} 0 & 0 & i\kappa & 0 \\ 0 & 0 & 0 & -i\kappa \\ i\kappa & 0 & 0 & 0 \\ 0 & -i\kappa & 0 & 0 \end{pmatrix}$$

We shall consider the detector and meter to be initially in coherent states. The solution for $P(\underset{\sim}{z},t)$ assumes a multivariate Gaussian form

$$P(z,t) = \exp\left\{-\frac{1}{2}(\underset{\sim}{z} - \langle\underset{\sim}{z}\rangle)^T\sigma^{-1}(z)(\underset{\sim}{z} - \langle\underset{\sim}{z}\rangle)\right\} \tag{30}$$

where

$$\underset{\sim}{z}^T(t) = \left(\langle a(t)\rangle, \langle a^\dagger(t)\rangle, \langle b(t)\rangle, \langle b^\dagger(t)\rangle\right)$$

and the covariance matrix

$$\sigma(z) = \frac{1}{2}\begin{pmatrix} 0 & \mathrm{Cosh}2kt - 1 & i\mathrm{Sinh}2kt & 0 \\ \mathrm{Cosh}2kt - 1 & 0 & 0 & -i\mathrm{Sinh}2kt \\ i\mathrm{Sinh}2kt & 0 & 0 & \mathrm{Cosh}2kt - 1 \\ 0 & -i\mathrm{Sinh}2kt & \mathrm{Cosh}2kt - 1 & 0 \end{pmatrix}$$

At this stage we are able to include the nonunitary effect of meter state reduction upon readout of a meter variable.

Readout of the meter observable $\hat{Y}_2(t)$ is made with result $y_2(t)$. The meter thus collapses into an eigenstate of $\hat{Y}_2(t)$ with eigenvalue $y_2(t)$. This causes a nonunitary change in the state of the detector, which then becomes the initial state for the next measurement. We analyze the effect of meter state reduction by a variant of von Neumann's projection postulate.[15]

In the Schrodinger picture the total system is represented at time t by the density operator $\rho^S(t)$. After readout at time τ the meter is in an eigenstate of $\hat{Y}_2(\tau)$. The density operator (in the Schrodinger picture) for the total system after readout $\bar{\rho}^S(\tau)$ may be written as

$$\bar{\rho}^S(\tau) = N\hat{P}(y_2(\tau))\rho^S(\tau)\hat{P}(y_2(\tau)) \tag{31}$$

where $\hat{P}(y_2(\tau))$ is a projector onto the subspace spanned by $|y_2(\tau),\tau\rangle$, a $\hat{Y}_2(\tau)$ eigenstate and $N^{-1} = \mathrm{Tr}(\rho^S(\tau)P(y_2(t)))$. The state of the detector after readout is then given by tracing out over the meter variables

$$\bar{\rho}_D^S(\tau) = \mathrm{Tr}_m(\bar{\rho}^S(\tau)) \tag{32}$$

Thus

$$\bar{\rho}_D^S(\tau) = N\langle y_2(\tau),\tau|\rho^S(\tau)|y_2(\tau),\tau\rangle \tag{33}$$

In the interaction picture defined by

$$\rho^I(t) = \exp(\frac{i}{\hbar}H_0t)\rho^S(t)\exp(-\frac{i}{\hbar}H_0t)$$

the state of the detector after readout is

$$\bar{\rho}_D^I(\tau) = N\langle y_2(\tau),0|\rho^I(\tau)|y_2(\tau),0\rangle \tag{35}$$

where we have used the property

$$|y_2(t),t\rangle = \exp(-\frac{i}{\hbar}\omega b^\dagger bt)|y_2(t),0\rangle \tag{36}$$

and $|y_2(t),0\rangle$ is an eigenstate of $\hat{Y}_2(0)$.

Expanding $\rho^I(\tau)$ in terms of the complex P representation we find the complex P representation for the detector after readout is

$$\bar{P}(\alpha_1,\beta_1,\tau) = \oint_{C_2}\oint_{C_2^\dagger} d\alpha_2 d\beta_2 P(\underset{\sim}{z},\tau)\langle y_2(\tau),0|\alpha_2\rangle \frac{\langle\beta_2^*|y_2(\tau),0\rangle}{\langle\beta_2^*|\alpha_2\rangle}$$

where $P(z,t)$ is given by Eq.(30). Details of the integration are given in Ref.(10).

We shall quote here the results for the variances in the detector observables after readout.

$$\overline{V_2(\hat{X}_1(\tau))} = \frac{\hbar}{2\omega_a} \frac{1}{\cosh 2\kappa t}$$
$$\overline{V(\hat{X}_2(\tau))} = \frac{\hbar}{2\omega_a} \cosh 2\kappa t \qquad (38)$$

$$\overline{\langle\hat{X}_1(\tau)\rangle} = \frac{\langle\hat{X}_1(\tau)\rangle}{\cosh 2\kappa t} + \frac{2}{\coth^2\kappa\tau + 1}\, x_1(\tau) \qquad (39)$$

where $x_1(\tau)$ is the inferred value.

After the measurement the detector is in a minimum uncertainty squeezed state.

In the limit of $\kappa\tau \to \infty$ we find

$$\overline{V(\hat{X}_1(\tau))} \to 0$$
$$\overline{\langle\hat{X}_1(\tau)\rangle} \to x_1(\tau) \qquad (40)$$

Thus no matter how small the measurement time τ, the coupling κ may be made sufficiently large to ensure that upon readout the detector is placed in an eigenstate of the measured observable, with eigenvalue equal to the measured result. This is the usual limit for a perfect measurement. The detector meter system will then remain in an $\hat{X}_1$, $\hat{Y}_2$ eigenstate for all time.

As in practice the limit $\kappa\tau \to \infty$ does not apply the detector will not be placed in a perfect eigenstate. Due to the failure of back action evasion it will move out of this eigenstate during the free evolution stage of the next measurement (see Eq.(26)). Despite this, after a second measurement (with the meter reprepared in a coherent state)

$$\overline{V(\hat{X}_1(\tau))} = \frac{\hbar}{2\omega_a} \frac{1}{(2\cosh^4\kappa\tau - 1)} \qquad (41)$$

which tends to zero for $\kappa\tau \to \infty$. The possible error in the inferred value of $\hat{X}_1$ at a third measurement (after the same time τ) is

$$V(\hat{X}_1(\tau)) \simeq \frac{\hbar}{2\omega_a} \frac{\cosh^2\kappa\tau}{(2\cosh^4\kappa\tau - 1)} \qquad (42)$$

This error may be made arbitrarily small, no matter how short the measurement times. Thus the value obtained from the third measurement can be predicted with arbitrary certainty from the result of the second measurement. This is precisely what is required of a Q.N.D. measurement. The failure of back action evasion was not crucial in this system since $\hat{X}_1(t)$ remained a Q.N.D. variable of the detector plus meter system.

Damping has not been included in the above analysis. The effects of damping were discussed in Ref.(10) where it was shown that damping even for a zero temperature heat bath will degrade the squeezing as e^{-t/τ_D} where τ_D is the characteristic damping time. In order to obtain accurate Q.N.D. measurements in the presence of damping it is necessary that the time between measurements be much less than τ_D.

§5 QUANTUM COUNTING MEASUREMENTS IN FOUR WAVE MIXING

A different kind of Q.N.D. measurement is provided by the quantum counting method. Unruh[3] pointed out that this would require a quadratic coupling to the oscillator coordinate.

As an example of a quantum counting measurement we consider the following Hamiltonian.

$$H = \hbar\omega_a a^\dagger a + \hbar\omega_b b^\dagger b + \hbar\chi' a^\dagger a(b\,\varepsilon(t) + b^\dagger \varepsilon^*(t)) \tag{43}$$

where a and b are bose operators for the detector and meter modes respectively. $\varepsilon(t)$ is a classical field. Such a Hamiltonian may arise in quantum optics in a four wave mixing process.[16]

Choosing $a^\dagger a$ as the Q.N.D. variable of the detector the interaction Hamiltonian satisfies the back action evading criterion.

The evolution of the number operator for the meter mode is given by the equation

$$\hat{N}_b(t) = (\chi t)^2 \hat{G}_a + i\chi t \hat{N}_a (b(0) - b^\dagger(0)) + \hat{N}_b(0) \tag{44}$$

where

$$\hat{N}_b(t) = b^\dagger(t) b(t), \qquad \hat{N}_a = a^\dagger a$$

$$\hat{G}_a = (a^\dagger a)^2, \qquad \chi = \chi'\varepsilon$$

and we have assumed the classical field is resonant with the b mode $(\varepsilon(t) = \varepsilon e^{i\omega_b t})$. We shall choose $\hat{G}_a$ as our Q.N.D. variable rather than $\hat{N}_a$. If we assume the meter is initially in a number state $|n_b(0)\rangle$ we have

$$\langle \hat{N}_b(t)\rangle = (\chi t)^2 \langle \hat{G}_a\rangle + n_b(0) \tag{45}$$

Thus from a measurement of $\hat{N}_b(t)$ at time t with result $n_b(t)$ we may infer a value g_a for $\hat{G}_a$ given by

$$g_a = \frac{n_b(t) - n_b(0)}{(\chi t)^2} \tag{46}$$

The possible error in the inferred value is given by Δg_a where

$$\Delta g_a = \frac{\Delta n_b(t)}{(\chi t)^2} \qquad \text{and} \qquad \Delta n_b(t) = V^{1/2}(\hat{N}_b(t))$$

where $V(\hat{N}_b(t))$ is the variance in $\hat{N}_b(t)$. For the initial meter state $|n_b\rangle$ we find

$$(\Delta g_a)^2 = V(\hat{G}_a) + \frac{2\langle \hat{G}_a\rangle}{(\chi t)^2}\,(n_b(0) + \tfrac{1}{2}) \tag{47}$$

The first term is the intrinsic uncertainty in g_a, the second term is the additional uncertainty contributed by the measurement. Thus provided the intrinsic uncertainty is small we may determine a value g_a accurately provided χt is sufficiently large. The above considerations apply to the unitary evolution. We shall now consider the non unitary effect of the readout of the meter.

If $\rho^S(t)$ is the Schrodinger picture density operator of the coupled detector meter system the density operator for the total system after readout $\bar{\rho}^S(t)$ is given by

$$\bar{\rho}^S(t) = N\hat{P}(n_b(t))\rho^S(t)\hat{P}(n_b(t)) \tag{48}$$

where $\hat{P}(n_b(t))$ is a projector onto the one dimensional subspace spanned by $|n_b(t)>$ an eigenstate of $\hat{N}_b(t)$ and $N^{-1} = Tr(\rho^S(t)\hat{P}(n_b(t))$.

A similar relation holds in the interaction picture. Using

$$e^{-i\omega b^\dagger bt}|n_b(t)> = e^{-i\omega n_b(t)}|n_b(t)>$$

we may show that

$$\bar{\rho}^I(t) = N\hat{P}(n_b(t))\rho^I(t)\hat{P}(n_b(t)) \tag{49}$$

The density operator for the detector after readout is

$$\bar{\rho}^I_D(t) = Tr_m(\bar{\rho}^I(t) = <n_b(t)|\rho^I(t)|n_b(t)> \tag{50}$$

We choose as the initial density operator of the system

$$\rho(0) = |\psi>|n_b(0)><n_b(0)|<\psi| \tag{51}$$

where $|\psi>$ is the initial state of the detector. The initial number distribution of the detector is

$$P(n_a) = |<n_a|\psi>|^2 \tag{52}$$

The postreadout number distribution of the detector is

$$\bar{P}(n_a) = <n_a|\bar{\rho}^I_D(t)|n_a> \tag{53}$$

This expression may be evaluated using the unitary evolution operator in the interaction picture

$$U(t,0) = \exp\left\{\hat{N}_a(\varepsilon(t)b^\dagger - \varepsilon^*(t)b)\right\} \tag{54}$$

where

$$\varepsilon(t) = -i\chi t$$

Thus we find

$$\bar{P}(n_a) = N\left|<n_b(t)|\exp n_a(\varepsilon(t)b^\dagger - \varepsilon^*(t)b)\right\}|n_b>|^2 P(n_a) \tag{55}$$

Evaluating this expression for the meter initially in the ground state $(n_b(0) = 0)$ yields

$$\bar{P}(n_a) = N\frac{1}{n_b(t)!} x^k e^{-x} P(n_a) \tag{56}$$

where

$$x = (n_a\chi t)^2$$
$$k = g_a(\chi t)^2$$

The distribution for $\bar{P}(n_a)$ is peaked around $n_a = \sqrt{g_a}$ and becomes more sharply peaked as χt increases. In Fig.(3), $\bar{P}(n_a)$ is plotted for two different values of χt showing the narrowing of the distribution as χt is increased. Thus for χt large it is possible to place the detector in a near eigenstate $|n_a>$ of $\hat{G}_a$ where $n_a = \sqrt{g_a}$. Since the interaction is back action evading the detector once placed in a near eigenstate of $\hat{G}_a$ will remain in this eigenstate unless acted on by an external force. Thus subsequent measurements must yield the constant sequence $\{g_a\}$ of results. Any departure from this sequence may be taken as evidence of the presence of an external force. As stated in §3(b) it is possible only to detect that a force has acted and it is not possible to reconstruct the time dependence of the force since the number operator and $\hat{G}_a$ are not Q.N.D. operators in the presence of the force.

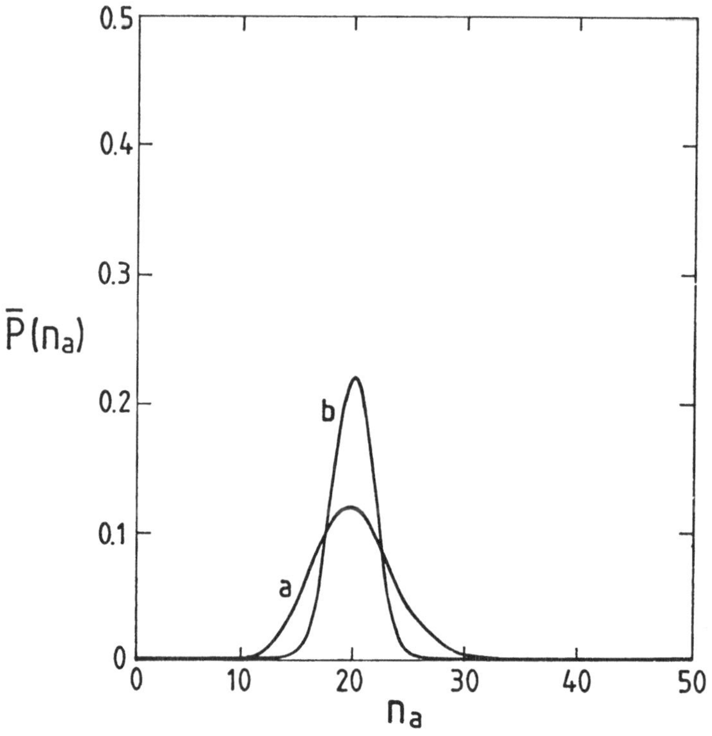

Figure 3 : Detector number distribution $\bar{P}(n_a)$ *after readout.*

Parameters

(a) $n_b(t) = 4$, $n_b^0 = 0$

$\chi t = 0.1$, $\bar{n}_a = 20$

(b) $n_b(t) = 25$, $n_b^0 = 0$

$\chi t = 0.25$, $\bar{n}_a = 20$

CONCLUSION

The effort to detect gravitational radiation has motivated renewed interest in the quantum limitations to measurements. We have shown that in principal quantum mechanics does not preclude the detection of gravitational radiation.

We have also given a complete analysis, including state reduction, of two possible schemes to make Q.N.D. measurements. These are based on a "squeezed state" detection scheme and a quantum counting detection scheme.

It has been demonstrated that despite initial misgivings the parametric amplifier is capable of making "squeezed state" Q.N.D. measurements. This conclusion is reached by taking fully into account the reduction of state which occurs in a measurement sequence.

REFERENCES

1. V.B. Braginsky, ZH.Eksp.Theor.Fiz., 53, 1434, (1968), [Sov.Phys. JETP, 26, 831].

2. W.G. Unruh, Phys.Rev., D18, 1764, (1978).

3. W.G. Unruh, Phys.Rev., D19, 2888, (1979).

4. K.S. Thorne, R.W.P. Drever, C.M. Caves, M. Zimmerman and V.D. Sandberg, Phys.Rev.Lett., 40, 667, (1978).

5. C.M. Caves, K.S. Thorne, R.W.P. Drever, V.D. Sandberg and M. Zimmerman, Rev.Mod.Phys., 52, 341, (1980).

6. C.M. Caves, In "Quantum Optics, Experimental Gravitation and Measurement Theory", Eds. P. Meystre and M.O. Scully, Plenum (in press), (1981).

7. D.F. Walls, Nature, (to be published).

8. W.H. Louisell, A. Yariv and A.E. Siegman, Phys.Rev., 124, (1961).

9. M. Hillery and M.O. Scully, In "Quantum Optics, Experimental Gravitation and Measurement Theory, Eds. P. Meystre and M.O. Scully, Plenum, (in press), (1981).

10. G.J. Milburn, A.S. Lane and D.F. Walls, Phys.Rev.A., (in press).

11. W.H. Louisell, "Quantum Statistical Properties of Radiation, Wiley, (1973).

12. R.J. Glauber, Phys.Rev., 131, 2766, (1963).

13. E.C.G. Sudarshan, Phys.Rev.Lett., 10, 277, (1963).

14. P.D. Drummond and C.W. Gardiner, J.Phys., 13A, 2353, (1980).

15. E. Beltrametti and G. Casinelli, "The Logic of Quantum Mechanics", Encyclopedia of Mathematics and its Applications, V15, Addison Wesley, (1981).

16. N. Bloembergen, In "Proceedings of the International School 'Enrico Fermi'", Course LXIV, Ed. N. Bloembergen, North Holland, (1977).

ACKNOWLEDGEMENT

This work was supported in part by the United States Army through its European Research Office.

R. G. Hunsperger

Integrated Optics: Theory and Technology

1982. 167 figures. XIV, 299 pages.
(Springer Series in Optical Sciences, Volume 33)
ISBN 3-540-11667-2

Contents: Introduction. - Optical Waveguide Modes. - Theory of Optical Waveguides. - Waveguide Fabrication Techniques. - Losses in Optical Waveguides. - Waveguide Input and Output Couplers. - Coupling Between Waveguides. - Electro-Optic Modulators. - Acousto-Optic Modulators. - Basic Principles of Light Emission in Semiconductors. - Semiconductor Lasers. - Heterostructure, Confined-Field Lasers. - Distributed Feedback Lasers. - Direct Modulation of Semiconductor Lasers. - Integrated Optical Detectors. - Applications of Integrated Optics and Current Trends. - References. - Subject Index.

B. R. Frieden

Probability, Statistical Optics, and Data Testing

A Problem Solving Approach

1983. 99 figures. XVII, 404 pages.
(Springer Series in Information Sciences, Volume 10). ISBN 3-540-11769-5

Contents: Introduction. - The Axiomatic Approach. - Continuous Random Variables. - Fourier Methods in Probability. - Functions of Random Variables. - Bernoulli Trials and its Limiting Cases. - The Monte Carlo Calculation. - Stochastic Processes. - Introduction to Statistical Methods: Estimating the Mean, Median, Variance, S/N, and Simple Probability. - Estimating a Probability Law. - The Chi-Square Test of Significance. - The Student *t*-Test on the Mean. - The *F*-Test on Variance. - Least-Squares Curve Fitting - Regression Analysis. - Principal Components Analysis. - The Controversy Between Bayesians and Classicists. - References. - Subject Index.

W. Demtröder

Laser Spectroscopy

Basic Concepts and Intrumentation

2nd corrected printing. 1982. 431 figures.
XIII, 696 pages. (Springer Series in Chemical Physics, Volume 5). ISBN 3-540-10343-0

Contents: Introduction. - Absorption and Emission of Light. - Widths and Profiles of Spectral Lines. - Spectroscopic Instrumentation. - Fundamental Principles of Lasers. - Lasers as Spectroscopic Light Sources. - Tunable Coherent Light Sources. - Doppler-Limited Absorption and Fluorescence Spectroscopy with Lasers. - Laser Raman Spectroscopy. - High-Resolution Sub-Doppler Laser Spectroscopy. - Time-Resolved Laser Spectroscopy. - Laser Spectroscopy of Collision Processes. - The Ultimate Resolution Limit. - Applications of Laser Spectroscopy. - References. - Subject Index.

Picosecond Phenomena III

Proceedings of the Third International Conference on Picosecond Phenomena
Garmisch-Partenkirchen, Federal Republic of Germany, June 16–18, 1982
Editors: **K. B. Eisenthal, R. M. Hochstrasser, W. Kaiser, A. Laubereau**

1982. 288 figures. XIII, 401 pages. (Springer Series in Chemical Physics, Volume 23)
ISBN 3-540-11912-4

Contents: Advances in the Generation of Ultrashort Light Pulses. - Ultrashort Measuring Techniques. - Advances in Optoelectronics. - Relaxation Phenomena in Molecular Physics. - Picosecond Chemical Processes. - Ultrashort Porcesses in Biology. - Applications in Solid-State Physics. - Index of Contributors.

Laser Spectroscopy V

Proceedings of the Fifth International Conference
Jasper Park Lodge, Alberta, Canada,
June 29 - July 3, 1981
Editors: **A. R. W. McKellar, T. Oka, B. P. Stoicheff**

1981. 319 figures. XI, 495 pages. (Springer Series in Optical Sciences, Volume 30). ISBN 3-540-10914-5

Contents: Introduction: Progress and Perspectives in Laser Spectroscopy. - Fundamental Applications of Laser Spectroscopy.- Laser Spectroscopic Applications. - Double Resonance. - Collision-Induced Phenomena. - Nonlinear Processes. - Rydberg States (Panel Discussion). - Methods of Studying Unstable Species. - Cooling, Trapping and Control of Ions, Atoms, and Molecules. - Surface and Solid State. - Vacuum Ultraviolet. - Progress in New Laser Sources. - List of Contributors.

Springer-Verlag
Berlin
Heidelberg
New York
Tokyo

Lecture Notes in Physics

Vol. 144: Topics in Nuclear Physics I. A Comprehensive Review of Recent Developments. Edited by T.T.S. Kuo and S.S.M. Wong. XX, 567 pages. 1981.

Vol. 145: Topics in Nuclear Physics II. A Comprehensive Review of Recent Developments. Proceedings 1980/81. Edited by T. T. S. Kuo and S. S. M. Wong. VIII, 571-1.082 pages. 1981.

Vol. 146: B. J. West, On the Simpler Aspects of Nonlinear Fluctuating. Deep Gravity Waves. VI, 341 pages. 1981.

Vol. 147: J. Messer, Temperature Dependent Thomas-Fermi Theory. IX, 131 pages. 1981.

Vol. 148: Advances in Fluid Mechanics. Proceedings, 1980. Edited by E. Krause. VII, 361 pages. 1981.

Vol. 149: Disordered Systems and Localization. Proceedings, 1981. Edited by C. Castellani, C. Castro, and L. Peliti. XII, 308 pages. 1981.

Vol. 150: N. Straumann, Allgemeine Relativitätstheorie und relativistische Astrophysik. VII, 418 Seiten. 1981.

Vol. 151: Integrable Quantum Field Theory. Proceedings, 1981. Edited by J. Hietarinta and C. Montonen. V, 251 pages. 1982.

Vol. 152: Physics of Narrow Gap Semiconductors. Proceedings, 1981. Edited by E. Gornik, H. Heinrich and L. Palmetshofer. XIII, 485 pages. 1982.

Vol. 153: Mathematical Problems in Theoretical Physics. Proceedings, 1981. Edited by R. Schrader, R. Seiler, and D.A. Uhlenbrock. XII, 429 pages. 1982.

Vol. 154: Macroscopic Properties of Disordered Media. Proceedings, 1981. Edited by R. Burridge, S. Childress, and G. Papanicolaou. VII, 307 pages. 1982.

Vol. 155: Quantum Optics. Proceedings, 1981. Edited by C.A. Engelbrecht. VIII, 329 pages. 1982.

Vol. 156: Resonances in Heavy Ion Reactions. Proceedings, 1981. Edited by K.A. Eberhard. XII, 448 pages. 1982.

Vol. 157: P. Niyogi, Integral Equation Method in Transonic Flow. XI, 189 pages. 1982.

Vol. 158: Dynamics of Nuclear Fission and Related Collective Phenomena. Proceedings, 1981. Edited by P. David, T. Mayer-Kuckuk, and A. van der Woude. X, 462 pages. 1982.

Vol. 159: E. Seiler, Gauge Theories as a Problem of Constructive Quantum Field Theory and Statistical Mechanics. V, 192 pages. 1982.

Vol. 160: Unified Theories of Elementary Particles. Critical Assessment and Prospects. Proceedings, 1981. Edited by P. Breitenlohner and H.P. Dürr. VI, 217 pages. 1982.

Vol. 161: Interacting Bosons in Nuclei. Proceedings, 1981. Edited by J.S. Dehesa, J.M.G. Gomez, and J. Ros. V, 209 pages. 1982.

Vol. 162: Relativistic Action at a Distance: Classical and Quantum Aspects. Proceedings, 1981. Edited by J. Llosa. X, 263 pages. 1982.

Vol. 163: J. S. Darrozes, C. Francois, Mécanique des Fluides Incompressibles. XIX, 459 pages. 1982.

Vol. 164: Stability of Thermodynamic Systems. Proceedings, 1981. Edited by J. Casas-Vázquez and G. Lebon. VII, 321 pages. 1982.

Vol. 165: N. Mukunda, H. van Dam, L.C. Biedenharn, Relativistic Models of Extended Hadrons Obeying a Mass-Spin Trajectory Constraint. Edited by A. Böhm and J.D. Dollard. VI, 163 pages. 1982.

Vol. 166: Computer Simulation of Solids. Edited by C.R.A. Catlow and W.C. Mackrodt. XII, 320 pages. 1982.

Vol. 167: G. Fieck, Symmetry of Polycentric Systems. VI, 137 pages, 1982.

Vol. 168: Heavy-Ion Collisions. Proceedings, 1982. Edited by G. Madurga and M. Lozano. VI, 429 pages. 1982.

Vol. 169: K. Sundermeyer, Constrained Dynamics. IV, 318 pages. 1982.

Vol. 170: Eighth International Conference on Numerical Methods in Fluid Dynamics. Proceedings, 1982. Edited by E. Krause. X, 569 pages. 1982.

Vol. 171: Time-Dependent Hartree-Fock and Beyond. Proceedings, 1982. Edited by K. Goeke and P.-G. Reinhard. VIII, 426 pages. 1982.

Vol. 172: Ionic Liquids, Molten Salts and Polyelectrolytes. Proceedings, 1982. Edited by K.-H. Bennemann, F. Brouers, and D. Quitmann. VII, 253 pages. 1982.

Vol. 173: Stochastic Processes in Quantum Theory and Statistical Physics. Proceedings, 1981. Edited by S. Albeverio, Ph. Combe, and M. Sirugue-Collin. VIII, 337 pages. 1982.

Vol. 174: A. Kadić, D.G.B. Edelen, A Gauge Theory of Dislocations and Disclinations. VII, 290 pages. 1983.

Vol. 175: Defect Complexes in Semiconductor Structures. Proceedings, 1982. Edited by J. Giber, F. Beleznay, J. C. Szép, and J. László. VI, 308 pages. 1983.

Vol. 176: Gauge Theory and Gravitation. Proceedings, 1982. Edited by K. Kikkawa, N. Nakanishi, and H. Nariai. X, 316 pages. 1983.

Vol. 177: Application of High Magnetic Fields in Semiconductor Physics. Proceedings, 1982. Edited by G. Landwehr. XII, 552 pages. 1983.

Vol. 178: Detectors in Heavy-Ion Reactions. Proceedings, 1982. Edited by W. von Oertzen. VIII, 258 pages. 1983.

Vol. 179: Dynamical Systems and Chaos. Proceedings, 1982. Edited by L. Garrido. XIV, 298 pages. 1983.

Vol. 180: Group Theoretical Methods in Physics. Proceedings, 1982. Edited by M. Serdaroğlu and E. İnönü. XI, 569 pages. 1983.

Vol. 181: Gauge Theories of the Eighties. Proceedings, 1982. Edited by R. Raitio and J. Lindfors. V, 644 pages. 1983.

Vol. 182: Laser Physics. Proceedings, 1983. Edited by J. D. Harvey and D. F. Walls. V, 263 pages. 1983.

Selected Issues from

Lecture Notes in Mathematics

Vol. 836: Differential Geometrical Methods in Mathematical Physics. Proceedings, 1979. Edited by P. L. García, A. Pérez-Rendón, and J. M. Souriau. XII, 538 pages. 1980.

Vol. 837: J. Meixner, F. W. Schäfke and G. Wolf, Mathieu Functions and Spheroidal Functions and their Mathematical Foundations Further Studies. VII, 126 pages. 1980.

Vol. 838: Global Differential Geometry and Global Analysis. Proceedings 1979. Edited by D. Ferus et al. XI, 299 pages. 1981.

Vol. 840: D. Henry, Geometric Theory of Semilinear Parabolic Equations. IV, 348 pages. 1981.

Vol. 841: A. Haraux, Nonlinear Evolution Equations- Global Behaviour of Solutions. XII, 313 pages. 1981.

Vol. 842: Séminaire Bourbaki vol. 1979/80. Exposes 543–560. IV, 317 pages. 1981.

Vol. 843: Functional Analysis, Holomorphy, and Approximation Theory. Proceedings. Edited by S. Machado. VI, 636 pages. 1981.

Vol. 845: A. Tannenbaum, Invariance and System Theory: Algebraic and Geometric Aspects. X, 161 pages. 1981.

Vol. 849: P. Major, Multiple Wiener-Itô Integrals. VII, 127 pages. 1981.

Vol. 851: Stochastic Integrals. Proceedings, 1980. Edited by D. Williams. IX, 540 pages. 1981.

Vol. 852: L. Schwartz, Geometry and Probability in Banach Spaces. X, 101 pages. 1981.

Vol. 856: R. Lascar, Propagation des Singularités des Solutions d'Equations Pseudo-Différentielles à Caractéristiques de Multiplicités Variables. VIII, 237 pages. 1981.

Vol. 858: E. A. Coddington, H. S. V. de Snoo: Regular Boundary Value Problems Associated with Pairs of Ordinary Differential Expressions. V, 225 pages. 1981.

Vol. 861: Analytical Methods in Probability Theory. Proceedings 1980. Edited by D. Dugué, E. Lukacs, V. K. Rohatgi. X, 183 pages. 1981.

Vol. 866: J.-M. Bismut, Mécanique Aléatoire. XVI, 563 pages. 1981.

Vol. 878: Numerical Solution of Nonlinear Equations. Proceedings, 1980. Edited by E. L. Allgower, K. Glashoff, and H.-O. Peitgen. XIV, 440 pages. 1981.

Vol. 881: R. Lutz, M. Goze, Nonstandard Analysis. XIV, 261 pages. 1981.

Vol. 888: Padé Approximation and its Applications. Proceedings, 1980. Edited by M. G. de Bruin and H. van Rossum. VI, 383 pages. 1981.

Vol. 898: Dynamical Systems and Turbulence, Warwick, 1980. Proceedings. Edited by D. Rand and L.-S. Young. VI, 390 pages. 1981.

Vol. 901: Séminaire Bourbaki vol. 1980/81 Exposés 561–578. III, 299 pages. 1981.

Vol. 904: K. Donner, Extension of Positive Operators and Korovkin Theorems. XII, 182 pages. 1982.

Vol. 905: Differential Geometric Methods in Mathematical Physics. Proceedings, 1980. Edited by S.J. Andersson, H.-D. Doebner, and H.R. Petry. VI, 309 pages. 1982.

Vol. 909: Numerical Analysis. Proceedings, 1981. Edited by J.P. Hennart. VII, 247 pages. 1982.

Vol. 912: Numerical Analysis. Proceedings, 1981. Edited by G. A. Watson. XIII, 245 pages. 1982.

Vol. 920: Séminaire de Probabilités XVI, 1980/81. Proceedings. Edité par J. Azéma et M. Yor. V, 622 pages. 1982.

Vol. 921: Séminaire de Probabilités XVI, 1980–81 Supplément: Géométrie Différentielle Stochastique. Proceedings. Edité par J. Azéma et M. Yor. III, 285 pages. 1982.

Vol. 922: B. Dacorogna, Weak Continuity and Weak Lower Semicontinuity of Non-Linear Functionals. V, 120 pages. 1982.

Vol. 923: Functional Analysis in Markov Processes. Proceedings, 1981. Edited by M. Fukushima. V, 307 pages. 1982.

Vol. 926: Geometric Techniques in Gauge Theories. Proceedings, 1981. Edited by R. Martini and E.M.de Jager. IX 219 pages. 1982.

Vol. 927: Y. Z. Flicker, The Trace Formula and Base Change for GL (3). XII, 204 pages. 1982.

Vol. 928: Probability Measures on Groups. Proceedings 1981. Edited by H. Heyer. X, 477 pages. 1982.

Vol. 929: Ecole d'Eté de Probabilités de Saint-Flour X – 1980. Proceedings, 1980. Edited by P.L. Hennequin. X, 313 pages. 1982.

Vol. 930: P. Berthelot, L. Breen, et W. Messing, Théorie de Dieudonné Cristalline II. XI, 261 pages. 1982.

Vol. 931: D.M. Arnold, Finite Rank Torsion Free Abelian Groups and Rings. VII, 191 pages. 1982.

Vol. 932: Analytic Theory of Continued Fractions. Proceedings, 1981. Edited by W.B. Jones, W.J. Thron, and H. Waadeland. VI, 240 pages. 1982.

Vol. 934: M. Sakai, Quadrature Domains. IV, 133 pages. 1982.

Vol. 935: R. Sot, Simple Morphisms in Algebraic Geometry. IV, 146 pages. 1982.

Vol. 936: S.M. Khaleelulla, Counterexamples in Topological Vector Spaces. XXI, 179 pages. 1982.

Vol. 937: E. Combet, Intégrales Exponentielles. VIII, 114 pages. 1982.

Vol. 938: Number Theory. Proceedings, 1981. Edited by K. Alladi. IX, 177 pages. 1982.

Vol. 942: Theory and Applications of Singular Perturbations. Proceedings, 1981. Edited by W. Eckhaus and E.M. de Jager. V, 363 pages. 1982.

Vol. 953: Iterative Solution of Nonlinear Systems of Equations. Proceedings, 1982. Edited by R. Ansorge, Th. Meis, and W. Törnig. VII, 202 pages. 1982.

Vol. 956: Group Actions and Vector Fields. Proceedings, 1981. Edited by J.B. Carrell. V, 144 pages. 1982.

Vol. 957: Differential Equations. Proceedings, 1981. Edited by D.G. de Figueiredo. VIII, 301 pages. 1982.

Vol. 963: R. Nottrot, Optimal Processes on Manifolds. VI, 124 pages. 1982.

Vol. 964: Ordinary and Partial Differential Equations. Proceedings, 1982. Edited by W.N. Everitt and B.D. Sleeman. XVIII, 726 pages. 1982.

Vol. 968: Numerical Integration of Differential Equations and Large Linear Systems. Proceedings, 1980. Edited by J. Hinze. VI, 412 pages. 1982.

Vol. 970: Twistor Geometry and Non-Linear Systems. Proceedings, 1980. Edited by H.-D. Doebner and T.D. Palev. V, 216 pages. 1982.

Vol. 972: Nonlinear Filtering and Stochastic Control. Proceedings, 1981. Edited by S.K. Mitter and A. Moro. VIII, 297 pages. 1983.

Vol. 978: J. Ławrynowicz, J. Krzyż, Quasiconformal Mappings in the Plane. VI, 177 pages. 1983.

Vol. 979: Mathematical Theories of Optimization. Proceedings, 1981. Edited by J.P. Cecconi and T. Zolezzi. V, 268 pages. 1983.